Springer Series in Molecular Biology

Series Editor: Alexander Rich

Springer Series in Molecular Biology

Series Editor: ALEXANDER RICH

J. F. T. SPENCER, DOROTHY M. SPENCER, A. R. W. SMITH (Eds.)
Yeast Genetics · Fundamental and Applied Aspects (1983)

EUGENE ROSENBERG (Ed.)
Myxobacteria · Development and Cell Interactions (1984)

AHARON RAZIN, HOWARD CEDAR, ARTHUR D. RIGGS (Eds.)
DNA Methylation · Biochemistry and Biological Significance (1984)

TERRELL L. HILL
Cooperativity Theory in Biochemistry · Steady-State and Equilibrium Systems (1985)

ROGER L. P. ADAMS, ROY H. BURDON
Molecular Biology of DNA Methylation (1985)

BOYD HARDESTY, GISELA KRAMER (Eds.)
Structure, Function and Genetics of Ribosomes (1986)

ARNOLD W. STRAUSS, IRVING BOIME, GÜNTHER KREIL (Eds.)
Protein Compartmentalization (1986)

THEODOR WIELAND
Peptides of Poisonous Amanita Mushrooms (1986)

AKIRA WAKE, HERBERT R. MORGAN
Host-Parasite Relationships and the Yersinia Model (1986)

TERRELL L. HILL
Linear Aggregation Theory in Cell Biology (1987)

GRAHAM W. PETTIGREW, GEOFFREY R. MOORE
Cytochromes c · Biological Aspects (1987)

Graham W. Pettigrew · Geoffrey R. Moore

Cytochromes c

Biological Aspects

With 68 Figures

Springer-Verlag
Berlin Heidelberg New York
London Paris Tokyo

Dr. GRAHAM W. PETTIGREW
Department of Biochemistry
Royal (Dick) School
of Veterinary Studies
Summerhall
Edinburgh EH9 1QH
U.K.

Dr. GEOFFREY R. MOORE
University of East Anglia
School of Chemical Sciences
Norwich NR4 7TJ
U.K.

Series Editor:
ALEXANDER RICH
Department of Biology
Massachusetts Institute of Technology
Cambridge, Massachusetts 02139, USA

ISBN-13: 978-3-642-72700-9 e-ISBN-13: 978-3-642-72698-9
DOI: 10.1007/978-3-642-72698-9

Library of Congress Cataloging-in-Publication Data. Pettigrew, Graham W. (Graham Walter), 1948– . Cytochromes C. (Springer series in molecular biology) Bibliography: p. . Includes index. 1. Cytochrome c. I. Moore, Geoffrey R. (Geoffrey Robert), 1950– . II. Title. III. Series. [DNLM: 1. Cytochrome C–biosynthesis. WH 190 P511c] QP 552.C94P48 1987 574.1'33 87-12921

Softcover reprint of the hardcover 1st edition 1987

Typesetting: K+V Fotosatz GmbH, Beerfelden

2131/3130-543210

Series Preface

During the past few decades we have witnessed an era of remarkable growth in the field of molecular biology. In 1950 very little was known of the chemical constitution of biological systems, the manner in which information was transmitted from one organism to another, or the extent to which the chemical basis of life is unified. The picture today is dramatically different. We have an almost bewildering variety of information detailing many different aspects of life at the molecular level. These great advances have brought with them some breath-taking insights into the molecular mechanisms used by nature for replicating, distributing and modifying biological information. We have learned a great deal about the chemical and physical nature of the macromolecular nucleic acids and proteins, and the manner in which carbohydrates, lipids and smaller molecules work together to provide the molecular setting of living systems. It might be said that these few decades have replaced a near vacuum of information with a very large surplus.

It is in the context of this flood of information that this series of monographs on molecular biology has been organized. The idea is to bring together in one place, between the covers of one book, a concise assessment of the state of the subject in a well-defined field. This will enable the reader to get a sense of historical perspective – what is known about the field today – and a description of the frontiers of research where our knowledge is increasing steadily. These monographs are designed to educate, perhaps to entertain, certainly to provide perspective on the growth and development of a field of science which has now come to occupy a central place in all biological studies.

The information in this series has value in several perspectives. It provides for a growth in our fundamental understanding of nature and the manner in which living processes utilize chemical materials to carry out a variety of activities. This information is also used in more applied areas. It promises to have a significant impact in the biomedical field where an understanding of disease processes at the molecular level may be the capstone which ultimately holds together the arch of clinical research and medical therapy. More recently in the field of biotechnology, there is another type of growth in which this science can be used with many practical consequences and benefit in a variety of fields ranging from agriculture and chemical manufacture to the production of scarce biological compounds for a variety of applications.

This field of science is young in years, but it has already become a mature discipline. These monographs are meant to clarify segments of this field for the readers.

Cambridge, Massachusetts

ALEXANDER RICH
Series Editor

Foreword

It is no easy task to bring order into the inchoate mass of current data relating to structure and function of cytochromes c. A vast literature has accumulated in the last few decades consequent on the application of new methodologies based on improved X-ray diffraction procedures and ever more increasingly sophisticated versions of a great variety of spectroscopies (NMR, EPR, infrared, EXAFS, laser Raman, etc.). Understandably, this bewilders the novice and often proves baffling to the expert who ventures outside an area of specialisation. It is fortunate that there are some willing and competent to accept the challenge to produce a text which can serve as an adequate guide. None are better equipped, by virtue of familiarity with the older knowledge, and personal experience as contributors to the new, than the authors of the present text. They have produced an all-inclusive coverage of relevant data, and – more important – thoughtful summaries of current knowledge which provide a basis for not only a better understanding of cytochrome c function but also for reasonable assessment of results of future research. Thus the text is not merely a compilation of data but an integrated presentation of findings evaluated as to reliability and significance. There will be some effort involved in reading the text but it will be well rewarded as an increasing comprehension of the material is attained.

Investigations on cytochrome c structure and function through the years have reflected accurately the general history of protein research. The quarter century, which began in the 1920s with David Keilin's epochal findings, can be considered a classical era centered mainly on the nature of cytochrome c in eukaryotic mitochondrial systems. Transition to a comparative biochemistry of cytochromes c began in the 1950s with attention paid to non-mitochondrial systems. The explosive development of physical biochemistry and molecular biology produced a flood of new information bringing promise of better perceptions of structural bases for the function of cytochrome c and heightened appreciation of the many mechanisms made possible by these heme proteins in supporting their general function as mediators of coupled electron transfer in both prokarytic and eukaryotic bioenergetic transduction. Thus, the appearance of this text is most timely. It meets the urgent need of researchers and students at all levels of sophistication for an adequate aid in assessing progress toward an eventual understanding of how cytochromes c function.

M. D. KAMEN

Preface

Cytochromes c play a central role in biological electron transport systems and, because of their small size and their stability, have been a popular subject for study in the general areas of protein chemistry and redox reactions. Our knowledge of cytochrome c has advanced through the contributions of disciplines as different as, for example, physical chemistry and microbial physiology. A comprehensive review must therefore deal with this diversity of interest and be intelligible to the chemist and biologist alike. To achieve such a synthesis is a daunting task but we feel it is an important one in providing a basis for developing research.

This book is divided into two volumes which correspond roughly to a biologist's view of cytochrome c and a chemist's view. However, this division was largely dictated by considerations of size and the two volumes should be seen as a single review. The book is directed at the graduate student and the research worker. To this end, it is detailed in its treatment of the recent research literature and incorporates much methodological material. However, it is our hope that within the complexity of cytochrome c, we have found unifying themes which will be of interest to the more general reader.

We would like to thank all our colleagues who critically read sections of the manuscript and Maxine Pettigrew who helped in its preparation.

GRAHAM W. PETTIGREW
GEOFFREY R. MOORE

Contents

Chapter 1 Resolution, Characterisation and Classification of c-Type Cytochromes

Chapter 2 The Role of Mitochondrial Cytochrome c in Electron Transport

Chapter 3 The Function of Bacterial and Photosynthetic Cytochromes c

Chapter 4 The Biosynthesis of Cytochrome c

Chapter 1 Resolution, Characterisation and Classification of c-Type Cytochromes

Functional studies of cytochromes fall into one of two groups. In the first, the intact respiratory system is examined using methods which allow the resolution of individual components. In the second, components are isolated and characterised and their role studied in reconstituted partial systems. Relatively few studies incorporate both approaches and we suggest that this has led to problems in the interpretation of results.

A good example is the investigation of the role of cytochrome c_2 in bacterial photosynthesis (Chap. 3). Models of electron transport were proposed based on the light-induced absorbance changes in the intact photosynthetic system assuming that cytochrome c_2 was the single c-type cytochrome present. Only with the demonstration by Wood (1980) of the presence of a cytochrome c_1 could the kinetic results of the whole system be correctly re-interpreted. Conversely, studies on isolated components without reference to the intact system can be misleading. Thus the study of electron transfer between purified cytochrome c_3 and ferredoxin of the sulfidogenic bacteria has no physiological relevance because the two proteins are separated by the cell membrane and cannot interact in vivo (Chap. 3).

In the following we discuss methods for resolution of the whole system and methods for characterisation of individual purified components. In the final section we describe a classfication scheme for the cytochromes c.

1.1 Resolution

1.1.1 The Location of Cytochrome c

Gram-negative bacteria are surrounded not only by the cell membrane but also by a peptidoglycan cell wall and an outer membrane. Proteins which are secreted into the space between the cell membrane and the outer membrane are termed periplasmic. They can be selectively released by treatment with lysozyme and EDTA in a sucrose medium which provides osmotic support for the spheroplasts as they are formed. The fragile spheroplasts can then be osmotically shocked to release the cytoplasmic contents. Thus periplasmic, cytoplasmic and membrane fractions can be obtained (Fig. 1.1) and their composition investigated.

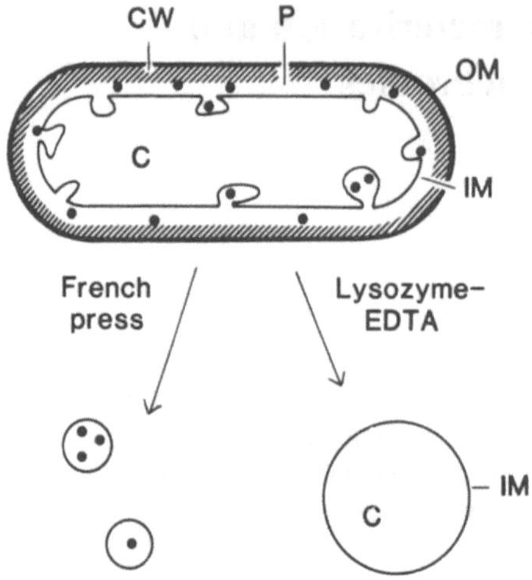

Fig. 1.1. Spheroplast formation in gram-negative bacteria. Gram-negative bacteria contain a periplasmic compartment (*P*) situated between the outer membrane (*OM*) and inner (plasma) membrane (*IM*). The cell wall (*CW*) is denoted by *hatching*. Passage of bacteria through a French pressure cell yields membrane vesicles which may contain periplasmic proteins (●) depending on the way the inner membrane fragments and reseals. In some bacteria, the inner membrane is highly invaginated, increasing the likelihood of trapping the periplasmic proteins in vesicles. Treatment with lysozyme and EDTA removes the outer membrane and cell wall and the spheroplasts that are formed contain an intact cytoplasmic compartment (*C*) if osmotically supported

Table 1.1. The cellular location of cytochrome c in *Pseudomonas stutzeri*

	Cyt. c (%)	ICDH (%)
Periplasmic	52	2
Cytoplasmic	8	98
Membrane	40	0

The amounts of total cytochrome c and isocitrate dehydrogenase (ICDH) present in each fraction were expressed as a percentage of the total present in the three fractions. (G. W. Pettigrew, unpublished results)

Wood (1983) has proposed that all c-type cytochromes are either periplasmic or are bound to the periplasmic side of the cell membrane. This model, which receives increasing experimental support, has important implications for the energy conserving mechanisms of bacterial oxidative phosphorylation (Chap. 3). The criterion for periplasmic location is the release of a protein without release of cytoplasmic contents and this is shown in Table 1.1 for *Pseudomonas stutzeri* where the cytochrome c is measured spectrophotometrically and isocitrate dehydrogenase is used as a cytoplasmic marker.

Less is known of gram-positive bacteria which lack the outer membrane. It may be that, in these organisms, cytochromes c are more tightly bound to the surface of the cell membrane (Jacobs et al. 1979) so that they are not lost through the relatively porous cell wall.

1.1.2 SDS Gel Electrophoresis and Heme Detection

The simple spectrophotometric method used in Table 1.1 cannot distinguish the individual cytochromes c that are present. This however is possible using SDS gel electrophoresis followed by heme detection.

One method of heme detection is based on the peroxidase activity of heme using 3 3′ 5 5′ tetramethylbenzidine as an oxidisable substrate (Thomas et al. 1976). The method was developed, and has been widely used, for proteins such as cytochrome P450 and cytochrome b which contain protoheme IX but, because this heme is not covalently bound, only a small fraction is retained after denaturation and this varies depending on the precise conditions employed.

On the other hand, c-type cytochromes contain covalently bound heme and are therefore ideally suited to the application of the peroxidase activity method after SDS gel electrophoresis (Goodhew et al. 1986). In Fig. 1.2, the method is applied to the fractions obtained during spheroplast preparation of aerobic and denitrifying *Pseudomonas stutzeri*. This experiment allows several important conclusions to be made.

First, individual cytochromes can be identified using purified markers. Second, the study of the general location of cytochrome c summarised in Table 1.1 is extended to define the location of the individual cytochromes. The figure of 8% total cytochrome appearing in the cytoplasmic fraction of Table 1.1 can be seen from Fig. 1.2 to be due mainly to membrane material which is difficult to fully sediment from the viscous spheroplast lysate. Third, membrane-bound c-type cytochromes are resolved and shown to be distinct from the soluble cytochromes. With few exceptions such cytochromes are rarely purified and are poorly characterised. Fourth, the effect of growth conditions on the complement of c-type cytochromes can be defined; cytochrome cd_1 (nitrite reductase) and a band of M_r 30 K are induced in denitrifying conditions, while a membrane-bound cytochrome c of M_r 32 K is characteristic of aerobic growth.

Sample preparation for electrophoresis usually involves the addition of a sulfhydryl reducing agent but this must be avoided if heme staining is to be performed because the ferrous iron is readily lost from the heme to give the porphyrin. However, although porphyrins have no peroxidase activity, they are fluorescent and this property is used in an alternative and equally sensitive method for the detection of c-type cytochromes in the presence of reducing agents (Wood 1981).

The heme-staining method is also usefully applied during purification of c-type cytochromes. Chromatographic peaks may contain single cytochromes

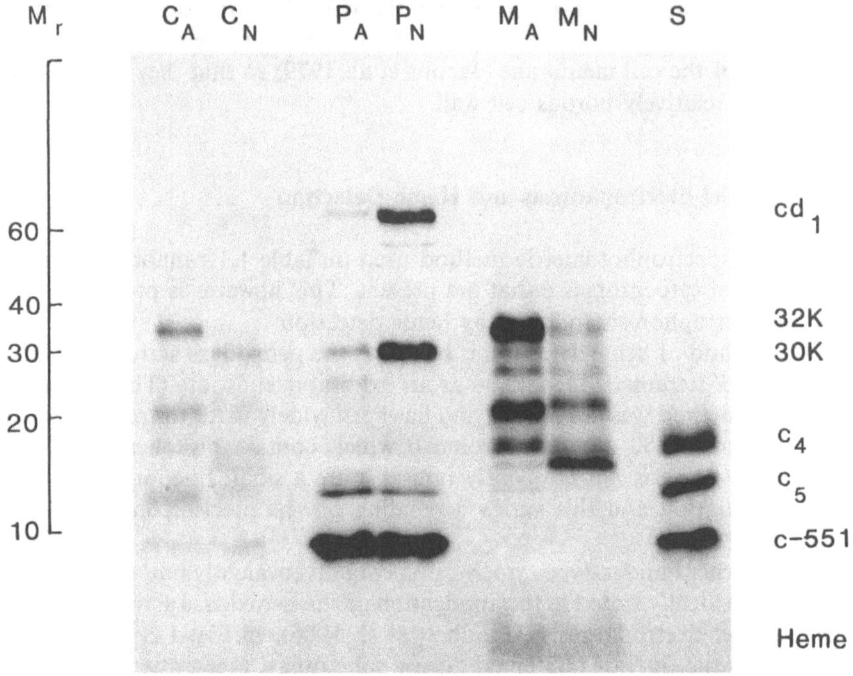

Fig. 1.2. The location and induction of the c-type cytochromes of *Pseudomonas stutzeri.* Cytoplasmic (*C*), periplasmic (*P*) and membrane (*M*) fractions obtained from aerobic (*A*) and nitrate-grown cells (*N*) cells of *Ps. stutzeri* stain 224 were subjected to SDS gel electrophoresis and then heme peroxidase staining for the detection of c-type heme. The positions of bands were compared with the purified standards (*S*) containing cytochromes c_4, c_5 and c-551 of *Ps. stutzeri* and to a set of molecular weight marker proteins (not shown) which were used to construct the scale of M_r (in kilodaltons). The channels were scanned at 560 nm and the following areas (in arbitrary units) were obtained for individual bands

	P_N	P_A
c-551	251	243
30 K	197	40
cd_1	184	12

or unresolved mixtures as shown in Fig. 1.3. In this molecular exclusion chromatography of the periplasmic fraction of denitrifying *Ps. stutzeri* 224, the cytochrome cd_1, the band of M_r 30 K and cytochrome c-551 are each separated from other cytochromes c but cytochromes c_4 and c_5 are unresolved. This mixture would be difficult to observe by visible spectroscopy but is readily detected by SDS gel electrophoresis of selected fractions followed by heme staining.

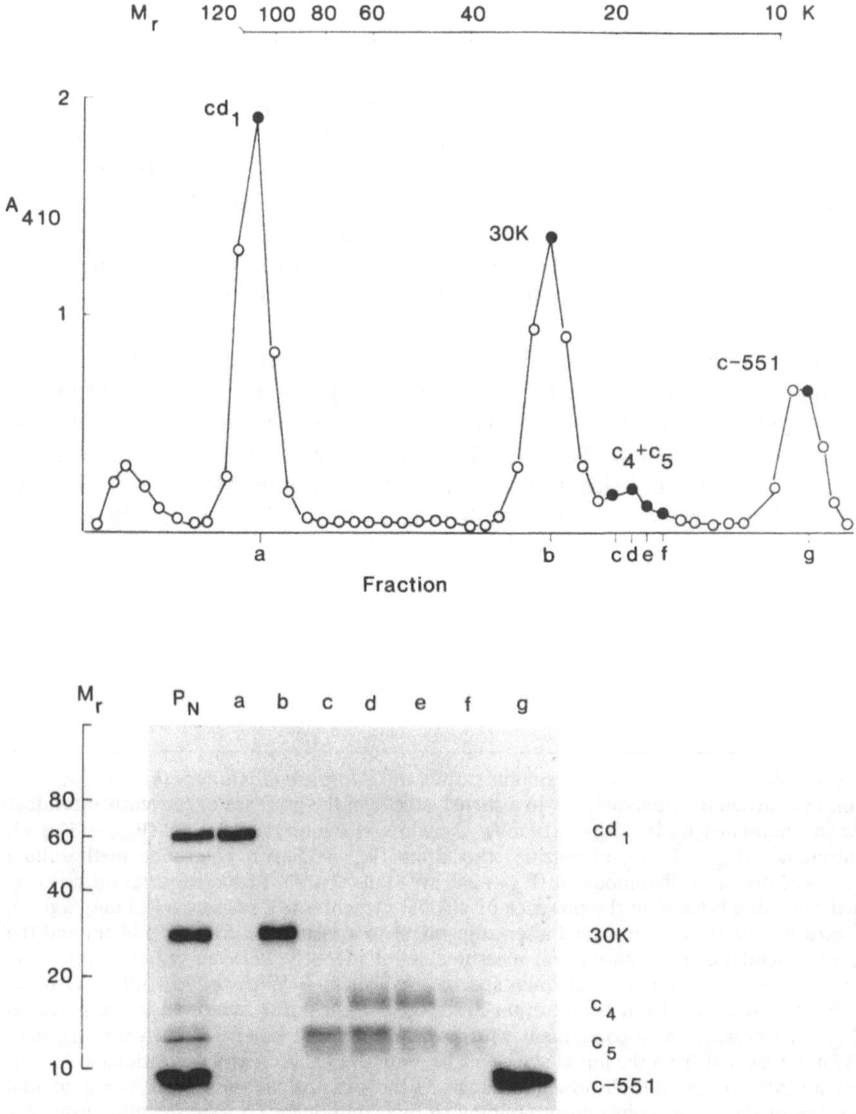

Fig. 1.3. Molecular exclusion chromatography of the periplasmic cytochromes c of denitrifying *Pseudomonas stutzeri.* The periplasmic fraction from a spheroplast preparation of denitrifying *Ps. stutzeri* was concentrated on DEAE cellulose and the concentrated eluate was applied to a Sephadex G-150 superfine column (1.5×90 cm). Portions of selected fractions ($a-g$) were subjected to SDS gel electrophoresis and heme staining in parallel with the periplasmic concentrate before chromatography (P_N). Both the molecular exclusion column and the SDS gel were calibrated using known molecular weight marker proteins and the logarithmic scales of their relative exclusion and relative mobility are shown

1.1.3 Redox Potentiometry and Spectroscopy

At room temperature the visible spectrum of a mixture of c-type cytochromes will contain featureless, composite α-peak which cannot be directly resolved. Although lower temperatures may sharpen individual α-components and allow their partial resolution, each α-component may also be split leading to an increase in complexity which may be difficult to interpret. However, stepwise reduction of the mixture may reveal spectral components not evident from the fully reduced α-band and, if the redox potential of the solution is measured at each step, even spectroscopically similar components may be potentiometrically resolved.

A relatively simple example is the redox titration of the soluble cytochromes of *"Chloropseudomonas ethylica"* (Fig. 1.4A). This consortium of two organisms synthesises a cytochrome c-555 of $E_{m,7}$ +103 mV (from the green sulfur bacterium, Shioi et al. 1972) and a triheme cytochrome c_3 of $E_{m\ average,\ pH7}$ −194 mV (from the sulfidogenic bacterium, Meyer et al. 1971). It is clear from the spectra of the soluble extract that although the fully reduced

Fig. 1.4. A Redox titration of the soluble extract of *Chloropseudomonas ethylica.* The titration was carried out anaerobically in a stirred cuvette in the presence of the redox mediators flavin mononucleotide ($E_{m,7}$ −219 mV), 2-hydroxy-1,4-naphthoquinone ($E_{m,7}$ −152 mV), duroquinol ($E_{m,7}$ +5 mV), phenazine ethosulfate ($E_{m,7}$ +55 mV), phenazine methosulfate ($E_{m,7}$ +80 mV) and diaminodurol ($E_{m,7}$ +240 mV) (all 20 μM). Reductive titration was carried out using NADH in the presence of NADH cytochrome c reductase (0.5 mg, Sigma). Successive spectra were recorded after adjustment to an isobestic point of 544 nm and the redox potential of the solution was measured simultaneously. Selected values of the redox potential corresponding to individual spectra are shown. **B** Values of log (total ox)/(total red) (●) calculated from the titration are plotted against the measured redox potential (E_h). The presence of a component with positive midpoint potential and with a_{max} near 555 nm is evident from the initial stages of the reductive titration and was estimated to contribute 19% to the total absorbance change on the basis of the position of the sigmoidal portion of the curve. Points corresponding to this component (I) were replotted using the new value of fully reduced leaving a second component (II) which was also replotted using the new value for fully oxidised. The *broken line* of component I has a slope of 60 mV indicating a good fit to a single heme. The *broken line* of component II has a steeper slope (100 mV), indicating the presence of more than one heme contribution. Further arithmetical analysis allows the construction of a theoretical curve (*solid line*) containing contributions from three heme components – component I (19%, E_m +80 mV), component IIa (26%, E_m −150 mV) and component IIb (55%, E_m −230 mV). Thus component II contains contributions from two components in proportions of 2:1 and separated by 80 mV. Published values for the purified cytochromes c from *Chloropseudomonas ethylica* are cytochrome c-555, +103 mV (Shioi et al. 1972); cytochrome c_3 (3 heme), −194 mV (average) (Meyer et al. 1971)

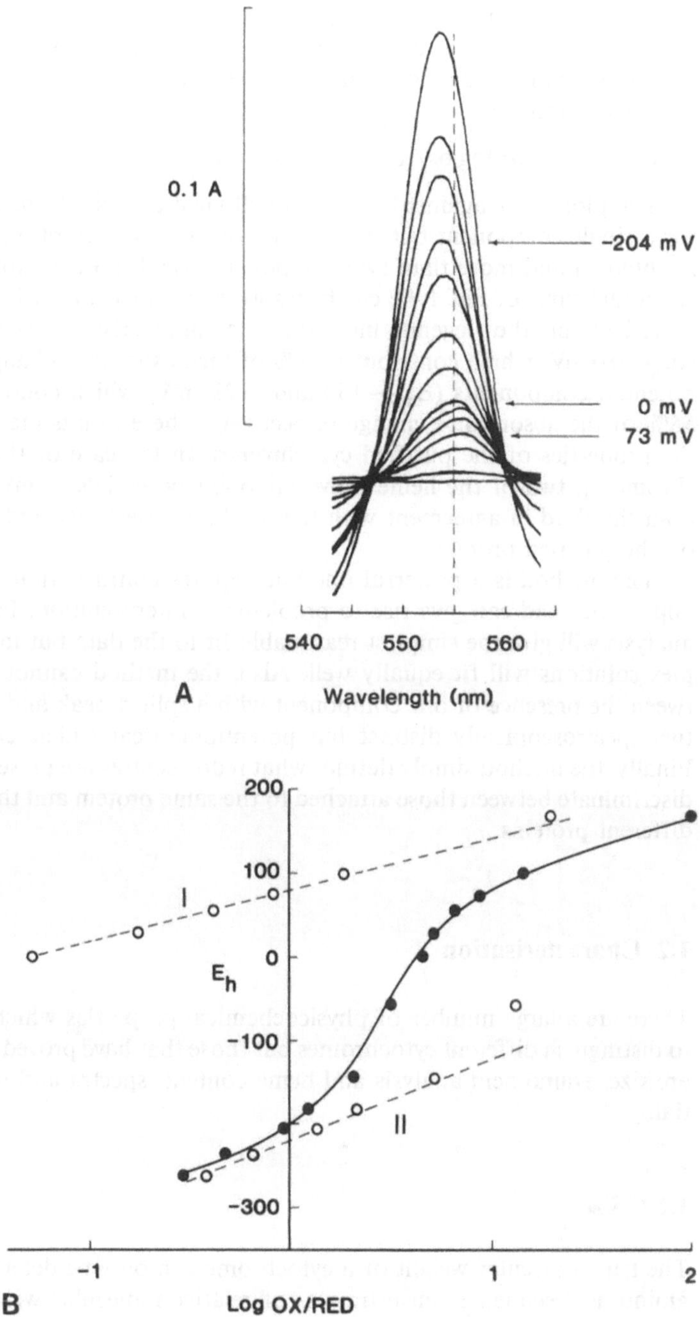
0.1 A
-204 mV
0 mV
73 mV
540
550
560
A
Wavelength (nm)
200
100
0
-100
-300
E_h
I
II
-1
1
2
B
Log OX/RED

α-peak is symmetrical at 553 nm, the initial stages of reduction produce an α-peak at 555 nm. This is consistent with prior reduction of the heme of most positive potential and is confirmed by the potentiometric analysis based on the Nernst relationship:

$$E_h = E_m + 0.06 \log \text{ox/red} \quad (n = 1,\ 30\,°C).$$

The plot of E_h against log ox/red will yield a straight line of slope 60 mV for a single component but if two components are present a sigmoidal curve is obtained and more than two components yield more complex curves. The sigmoidal curve of Fig. 1.4B can be resolved arithmetically using the procedure described in the legend into three components: a cytochrome c-555 ($E_{m,7}$ +80 mV) which contributes 19% of the absorbance change and two low potential components ($E_{m,7}$ −150 and −230 mV) which contribute 26% and 55% of the absorbance change respectively. These results are consistent with the properties of the purified cytochromes. In the case of the triheme cytochrome c_3, two of the hemes have midpoint potentials 80 mV more negative than the third in agreement with the results of Fiechtner and Kassner (1979) on the purified protein.

The method is a powerful one but requires confirmation by independent approaches and can give rise to problems in interpretation. For example, the analysis will give the simplest reasonable fit to the data but many more complex solutions will fit equally well. Also, the method cannot distinguish between the presence of one component with a split α-peak and the presence of two spectroscopically distinct but potentiometrically identical components. Finally, the method simply defines what redox centres are present but does not discriminate between those attached to the same protein and those attached to different proteins.

1.2 Characterisation

There are a large number of physicochemical properties which could be used to distinguish different cytochromes but those that have proved the most useful are size, amino acid analysis and heme content, spectra and midpoint potential.

1.2.1 Size

The true molecular weight of a cytochrome can only be determined from the amino acid sequence but estimates of relative molecular weight (M_r) can be obtained from a variety of methods of which SDS gel electrophoresis and molecular exclusion chromatography are the most popular. These two methods should give concordant results for cytochromes composed of a single polypeptide chain but cytochromes are known which are aggregates of the same or different polypeptide chains and these will dissociate to the individual subunits in SDS but remain as the aggregate under the native conditions of molecular exclusion chromatography. For example, in Fig. 1.3, cytochrome cd_1 and c_5 appear as dimers on the molecular exclusion column while cytochrome c-551 is monomeric.

For both methods, there are instances of anomalous behaviour. For example, non-spherical molecules such as the cytochrome c′ dimer (Meyer and Kamen 1982), appear larger on molecular exclusion chromatography than their actual molecular weight. Also, a cytochrome may appear intermediate in M_r between monomer and dimer if a rapid equilibrium is present. In the case of SDS gel electrophoresis, anomalous mobilities are seen for hydrophobic membrane proteins and, in the case of cytochrome c_4, the holoprotein migrates more rapidly in SDS than the apoprotein (Pettigrew, G. W., unpublished results).

1.2.2 Amino Acid Analysis and Heme Content

If amino acid analysis is performed on a pure sample of known heme content, then a minimum molecular weight per mole heme can be obtained by summing the constituent amino acids. If this is then compared with the value of M_r obtained from SDS gel electrophoresis, the heme content per mole protein can be found by division. An example of such an analysis is shown in Table 1.2 and Fig. 1.5. On the basis of SDS gel electrophoresis, the apocytochrome c_4 of *Azotobacter vinelandii* has a M_r of 20400. A figure of 19347 is obtained from the amino acid composition associated with 2 mol heme indicating that the cytochrome is a diheme protein. In this case, the amino acid sequence is known (Ambler et al. 1984) with a molecular weight of 19652 thus confirming the analysis.

Cysteine is usually underestimated after hydrolysis of a holocytochrome and a more reliable figure can be obtained after removal of the heme group with mercuric chloride. Tryptophan is destroyed by hydrolysis in HCl and as with the results of Table 1.2 is often not reported in amino acid compositions. It can be determined after hydrolysis in mercaptoethane sulfonic acid (Penke et al. 1974) or by its contribution to the ultraviolet spectrum of the protein. The problem with the latter approach is that the heme absorbs in the same region and therefore a method such as that of Goodwin and Morton (1946) which analyses the dual contributions of tyrosine and tryptophan to the composite UV spectrum can only be performed on the apoprotein. Alternatively, an approximate estimate of tryptophan may be found by subtracting the tyrosine and heme contribution from the holoprotein extinction at 280 nm as shown in Table 1.2.

Table 1.2. The amino acid composition and heme content of *Azotobacter vinelandii* cytochrome c_4

	nmol[a]	mol/2 mol heme	Sequence
ASP	48.3	23.6	21
THR	16.4[b]	8	9
SER	15.6[b]	7.6	10
GLU	38.5	18.8	18
PRO	21.7[c]	10.6	11
GLY	47.3	23.1	25
ALA	60.1	29.3	29
VAL	11	5.4	5
MET	11	5.4	6
ILE	12.3	6	6
LEU	26.6	13	14
TYR	11.9	5.8	6
PHE	8.3	4	4
HIS	8.7	4.2	4
LYS	20.7	10.1	11
ARG	14.7	7.2	7
CYS	8[c]	3.9	4
TRP	ND[d]	–	0[e]
		19347	19652

[a] This sample contained 4.1 nmol heme as determined by the pyridine hemochrome method. The sample was hydrolysed for 70 h in 6 M HCl at 105 °C.
[b] Not corrected for breakdown during the 70 h hydrolysis.
[c] Cysteine as cysteic acid and proline were estimated after removal of the heme group, performic acid oxidation and acid hydrolysis for 20 h.
[d] Tryptophan was not determined by amino acid analysis. An estimate of the tryptophan content can be made by subtracting the tyrosine and heme contribution from the absorbance of the holoprotein at 280 nm. This calculation assumes $E_{Tyr, 280 nm}$ 1.1 $mM^{-1} cm^{-1}$ and $E_{heme, 280 nm}$ 13.85 $mM^{-1} cm^{-1}$, the latter being calculated from cytochromes of known tryptophan content. This calculation for *A. vinelandii* cytochrome c_4 is

	$mM^{-1} cm^{-1}$ (280 nm)
Heme	27.7 (2 heme)
Tyr (6)	6.6
	34.3
Holocytochrome	36.7
Remainder	2.4 } Trp content 0.46
Trp	5.2 }

[e] The sequence analysis found no tryptophan. The fractional value obtained above may be due to differences in heme absorption at 280 nm between different cytochromes or to differences in tyrosine absorption due to protein environment. (G. W. Pettigrew, unpublished results)

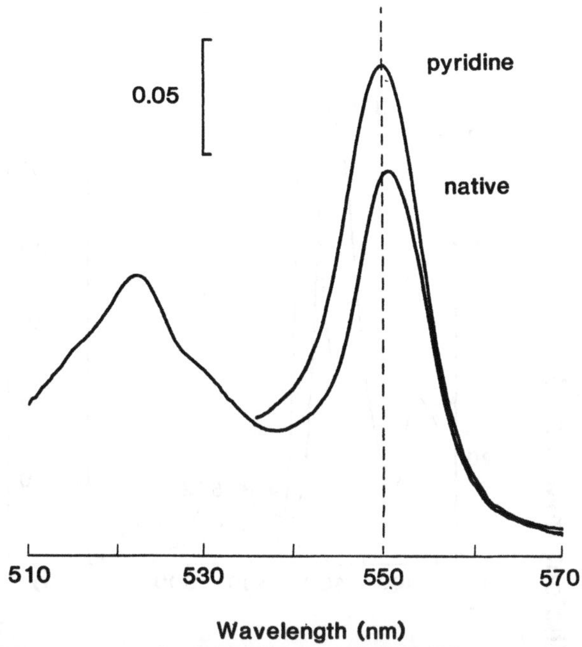

Fig. 1.5. The native and pyridine ferrohemochrome spectra of cytochrome c_4 from *Azotobacter vinelandii.* The native spectrum of ferrocytochrome c_4 in the region of the α- and β-bands was obtained by addition of solid sodium dithionite to the cytochrome in 0.1 M phosphate pH 7. The pyridine ferrohemochrome of the same concentration of cytochrome was formed in 0.15 M NaOH, 0.2 M pyridine with the addition of dithionite. Application of the extinction coefficient of the cytochrome c heme to the latter spectrum ($E_{550\,nm}$ 31.18 $mM^{-1}\,cm^{-1}$; Barstch 1971) allows the calculation of a heme concentration of 164 µM for the parent cytochrome c_4 solution. From this, 4.1 nmol heme were taken for amino acid analysis (Table 1.2)

1.2.3 Spectra

Cytochromes function by a reversible change in the redox state of their heme group between the levels of Fe II and Fe III, the only known exception being the peroxidatic centre of *Pseudomonas* cytochrome c peroxidase which can adopt an Fe IV state. Apart from the single thioether attachment of some protozoan cytochromes c (Vol. 2, Chap. 3), the heme group of cytochrome c is believed to be structurally constant and the visible spectrum is influenced only by the extraplanar ligands provided by the protein and also by the general protein environment.

The spectra of Figs. 1.6 and 1.7 illustrate the former effect. The cytochromes c′ have a pentacoordinate heme giving rise to a high-spin spectrum (Figs. 1.6a and 1.7). Other cytochromes c have low-spin spectra (Fig. 1.6b, c and d) but a major subdivision of these is based on the presence of a band at 695 nm in the ferricytochrome (Fig. 1.7). This is present in those cyto-

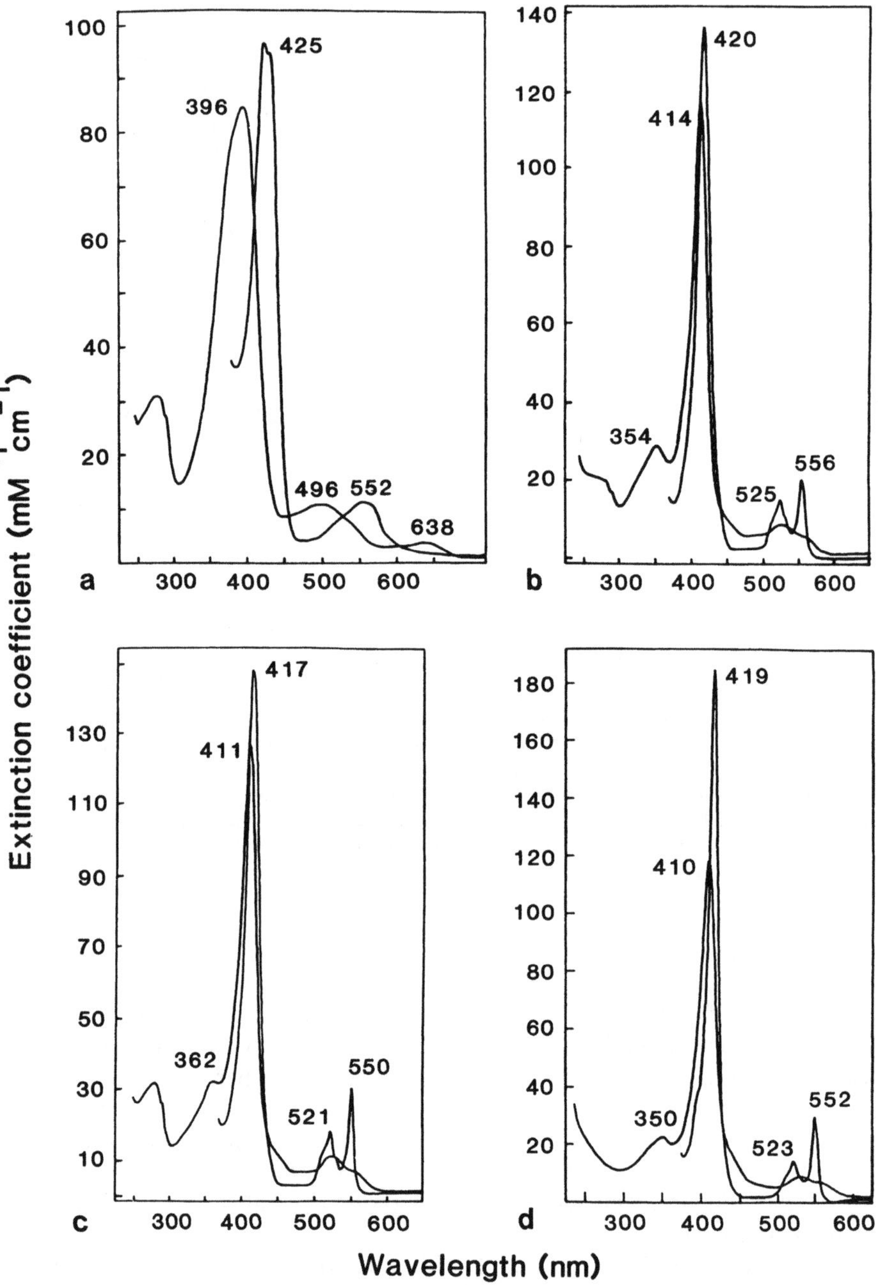

Fig. 1.6 a–d. UV-visible absorption spectra of selected c-type cytochromes. **a** cytochrome c′ from halotolerant *Paracoccus* sp.; **b** cytochrome c-556 from *Rhodopseudomonas palustris*; **c** cytochrome c_2 from *Rhodopseud. sphaeroides;* **d** cytochrome c_3 from *Desulfovibrio vulgaris.* In each case the ferri- and ferrocytochrome spectra are obtained by addition of potassium ferricyanide and sodium dithionite respectively to solutions of the cytochrome in 0.1 M phosphate pH 7. (Meyer and Kamen 1982). The ferrocytochrome spectra are those showing a band near 550 nm which is often used to designate a particular cytochrome

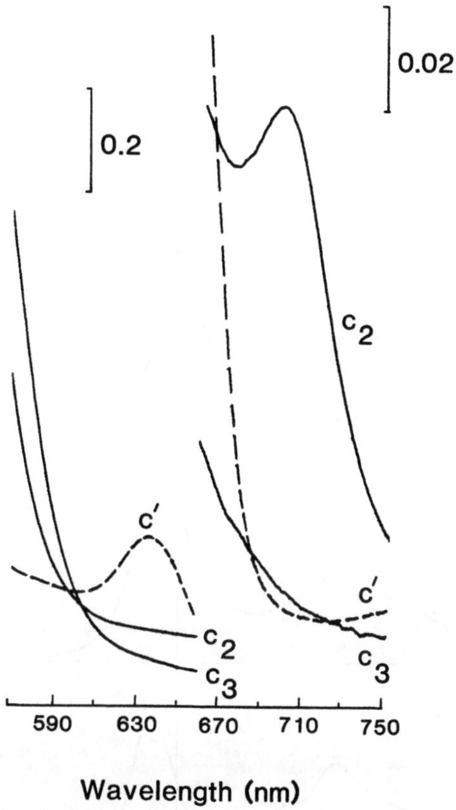

Fig. 1.7. Near-infrared absorption spectra of selected c-type ferricytochromes. Cytochrome c_2 is from *Rhodopseud. viridis* (118 μM), cytochrome c′ from *Rhodopseud. capsulata* (100 μM) and cytochrome c_3 from *Desulfovibrio gigas* (153 μM)

chromes with a histidinyl-methionyl-Fe coordination while those with a bis-histidinyl coordination lack a band at this wavelength.

Rather more subtle spectroscopic differences further distinguish the different low-spin cytochromes c. These are presumably based on variations in the general environment provided by the protein but the detailed structural basis for these minor spectroscopic changes is not known. Illustrative differences in the symmetry and wavelength maximum of the α-band and the α/β ratio are shown in Fig. 1.8. Asymmetry in the α-band can be an intrinsic property of a single heme as in the case of *Porphyra* cytochrome c-553 (Fig. 1.8F) or it may be due to the presence of two hemes as in the case of *Pseudomonas stutzeri* cytochrome c_4 (Fig. 1.8I). With both cytochrome c-551 of *Ps. aeruginosa* (Fig. 1.8G) and cytochrome c-555 of *Chlorobium limicola* (Fig. 1.8H), the asymmetry is dependent on pH, being more pronounced above pH 7. An extreme case of asymmetry is found in the split α-band of cytochrome c-554 (548) from halotolerant *Paracoccus* (Fig. 1.8J).

The position of the α-band can vary between 549 and 556 nm. The exceptions are certain protozoan cytochromes (e.g. *Crithidia oncopelti* cytochrome

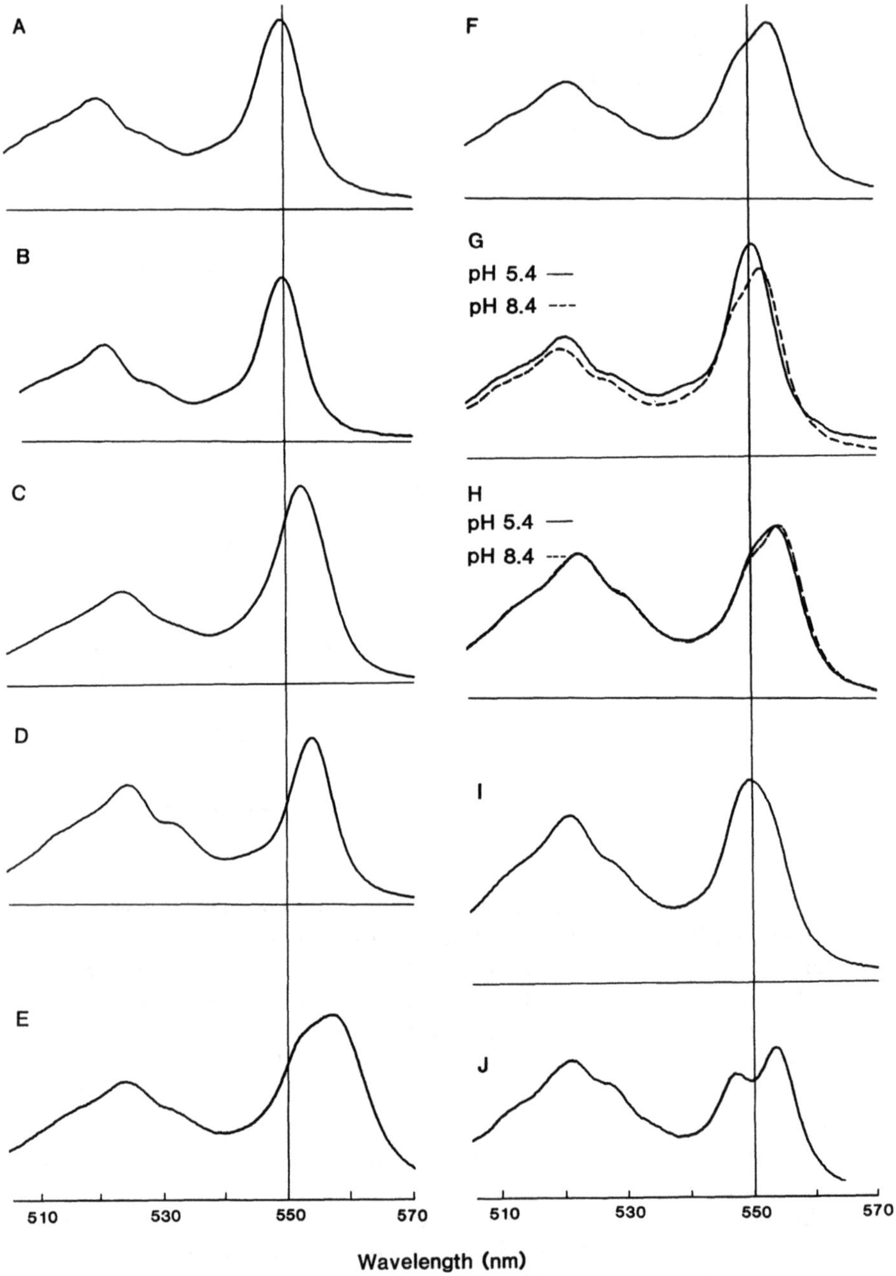

Fig. 1.8 A–J. Absorption spectra of selected ferrocytochromes c in the region of the α- and β-bands. Spectra were recorded after addition of sodium dithionite for **A**, horse heart cytochrome c; **B** *Rhodopseudomonas capsulata* cytochrome c_2; **C** *Desulfovibrio gigas* cytochrome c_3; **D** *Ps. mendocina* cytochrome c_5; **E** *Crithidia oncopelti* cytochrome c-557; **F** *Porphyra tenera*, cytochrome c-553; **G** *Ps. aeruginosa* cytochrome c-551; **H** *Chlorobium*

Table 1.3. Quantitative features of the absorption spectra of selected c-type cytochromes

	α_{max}	$E_{\alpha max}$ (mM^{-1})	β_{max}	$E_{\beta max}$ (mM^{-1})	α/β	Reference
Horse heart c	550	29	521	15.5	1.87	Pettigrew et al. 1975
Rps. capsulata c_2	550	29	521	17.2	1.69	This work
D. gigas c_3 (4 heme)	552.5	124.4	524	58.4	2.13	This work
Ps. mendocina c_5	554	26.7	524	19.7	1.36	This work
C. oncopelti c-557	557	24.7	524	15	1.65	Pettigrew et al. 1975
P. tenera c-553	553	23	521	15.2	1.51	This work
Ps. aeruginosa c-551						
pH 5.4	550.5	31.8	521	17.7	1.79	This work
pH 8.4	551.5	28.1	520	15.9	1.77	This work
Chl. limicola c-555						
pH 5.4	555	19.8	523	16.5	1.2	Meyer (1970)
Ps. stutzeri c_4	550	44.4	521	36.6	1.21	This work
Paracoccus sp. c-554 (548)	554	18.1	521	16.6	1.09	This work

Spectra in 0.1 M phosphate pH 7 were recorded after addition of dithionite. Extinction coefficients were calculated on the basis of pyridine hemochrome spectra ($E_{mM, 550 nm}$ 31.18, Bartsch 1971).

c-557, Fig. 1.8E) which contain a single thioether linkage. The α/β peak ratio can be as low as 1.1 (Fig. 1.8J) and as high as 2.1 (Fig. 1.8C) (Table 1.3).

Together, these spectroscopic differences are an important basis for the identification and classification of cytochromes discussed in Section 1.3.

1.2.4 Midpoint Oxidation Reduction Potentials

As noted earlier in Section 1.1.3, the Nernst plot will give a slope of 60 mV for a pure monoheme system. This is illustrated in Fig. 1.9 for the redox titration of cytochrome c-554 (548) of halotolerant *Paracoccus* sp. Such titrations can be performed either with the aid of a redox electrode and mediators to report the ambient potential of the solution or by the method of mixtures with a known redox couple. They should be performed in both the oxidative and reductive directions to demonstrate reversibility.

Redox titrations of cytochromes having more than one heme can be complex and difficult to analyse and interpret (Fig. 1.4). A discussion of this problem is reserved for Vol. 2, Chap. 7.

◄

limicola, f. *thiosulfatophilum,* cytochrome c-555; **I** *Ps. stutzeri* cytochrome c_4; **J** halotolerant *Paracoccus* sp. cytochrome c-554 (548). Cytochromes were dissolved in 0.1 M phosphate pH 7 except in the cases of *Ps. aeruginosa* cytochrome c-551 and *Chlorobium limicola* cytochrome c-555 where 0.1 M acetate pH 5.4 and 0.1 M Tris Cl pH 8.4 were used

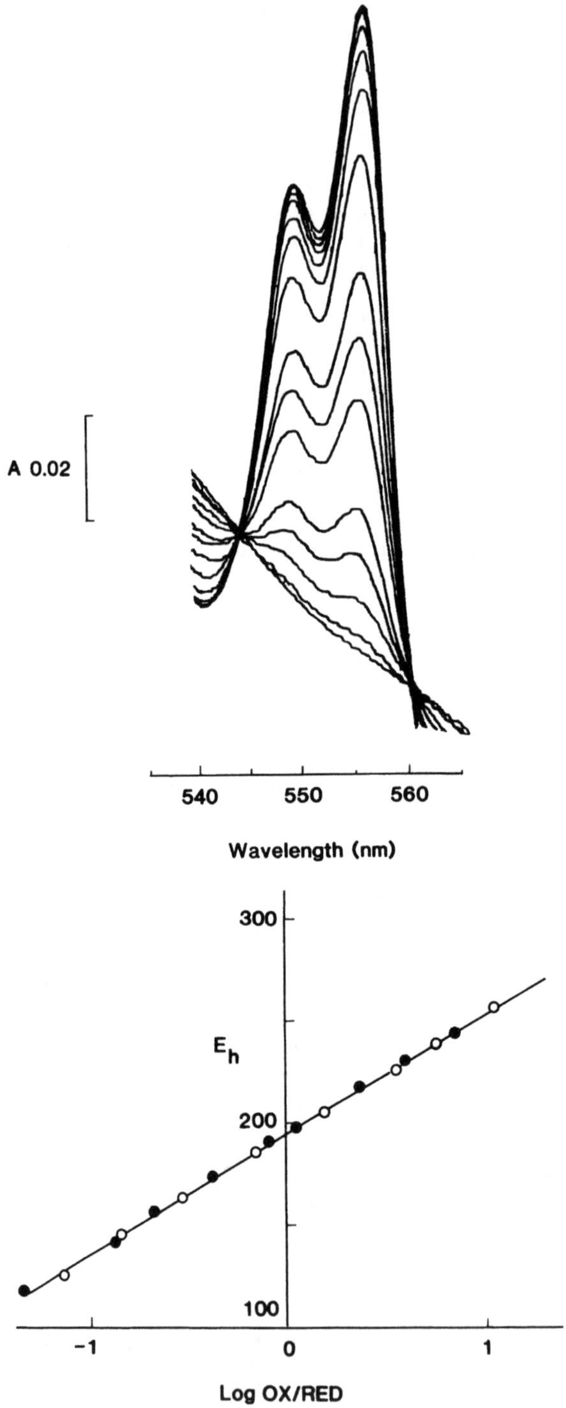

Fig. 1.9. The redox titration of cytochrome c-554 (548) from halotolerant *Paracoccus* sp. Cytochrome spectra were recorded in a stirred anaerobic cuvette containing 20 μM phenazine methosulfate, diaminodurol and ferric ammonium sulfate with 0.4 mM EDTA. Each spectrum corresponds to a particular ambient potential (not shown) and is used to calculate log $cyt._{ox}/cyt._{red}$. ○, ● Experimental points from oxidative and reductive titrations respectively. The *line* has a slope of 60 mV

1.3 Classification

Classification in biology is both a fascinating and a frustrating exercise. Definitions are based on incomplete knowledge and so must be regarded as temporary and subject to modification (if not complete change) with the appearance of new information. In spite of this, many biologists feel that a good classification scheme approaches a natural order – a phylogeny – which has a historical validity and which maps the evolutionary relationships of the defined groups.

With whole organisms we have no difficulty in exclusively defining a group such as the insects (although we have more difficulty with a group such as the Pseudomonads). Such classification has only been achieved by the accumulation of observations on species, thus building up the natural history of the group. But with proteins we are just at the beginning of defining their natural history (Ambler 1977) and because information is sparse, our present minimum definition of a group will become inadequate with future discoveries.

The classification of cytochromes into the types a, b, c and d is, however, a quite straightforward and accepted procedure. The associated prosthetic heme groups (Fig. 1.10) are sufficiently different to give rise to distinctive optical absorption spectra (although those of the b- and c-type cytochromes do overlap). In Fig. 1.11 we have illustrated these spectral distinctions by showing three complex systems each of which has a c-type cytochrome tightly associated with another heme. C-type cytochromes are further distinguished by the covalent attachment of their heme and can be quantified in complex mixtures after extraction of the non-covalently bound hemes of other cytochromes into acidified acetone.

Further subdivision of the c-type cytochromes is important if we are to understand the relatedness of the electron transport chains and organisms of which they are a part. It might be argued that we could use existing whole organism groups to classify their constituent cytochromes. However, this assumes congruence of the history of a single gene with that of the whole genome. As we shall see, present evidence suggests that this may be valid for eukaryotes but not necessarily so for prokaryotes. Thus mitochondrial cytochrome c appears to be a sensible grouping but *Pseudomonas* cytochrome c-551 is more of a problem. Not all Pseudomonads have this cytochrome and some non-Pseudomonads have structurally close relatives of the cytochrome (Ambler 1977). The latter observation, with others, suggests that cross-species gene transfer may have given rise to bacterial genomes which are historical mosaics rather than of single divergent origin (Ambler et al. 1979a; Vol. 2, Chap. 6).

For the following three reasons we are on no stronger ground if we attempt classification on the basis of function. Firstly, we often have very sketchy information on the function of bacterial electron transport proteins. This is partly because of the complexity of their terminal electron transfer processes in comparison to those of the mitochondrion. For example, in the case of *Pseudo-*

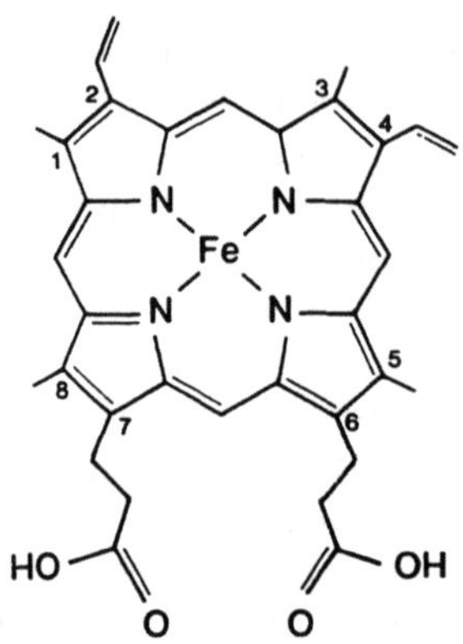
Fe
N
1
2
3
4
5
6
7
8
HO
OH
O

a

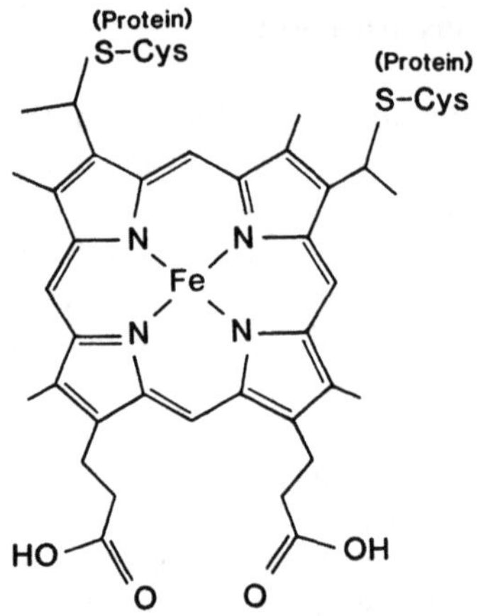
(Protein)
S-Cys
Fe
N
HO
OH
O

b

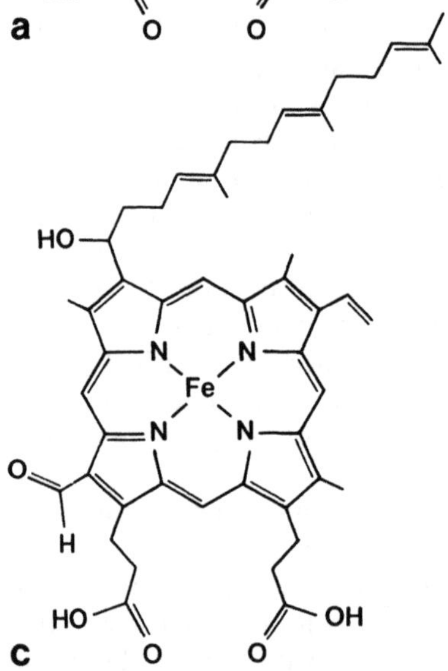
HO
Fe
N
O
H
OH

c

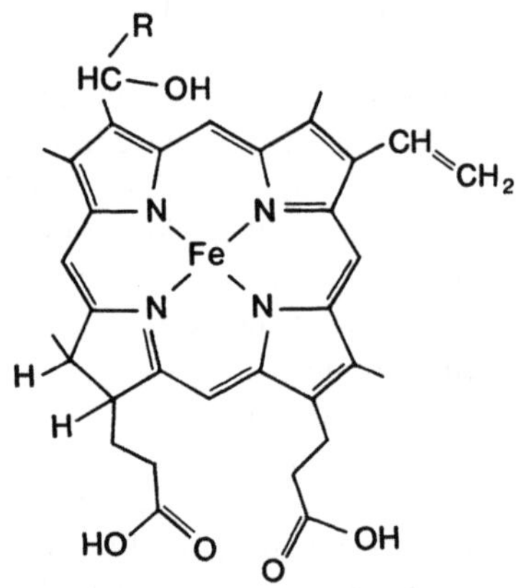
R
HC
OH
CH
CH2
Fe
N
H
HO
O

d

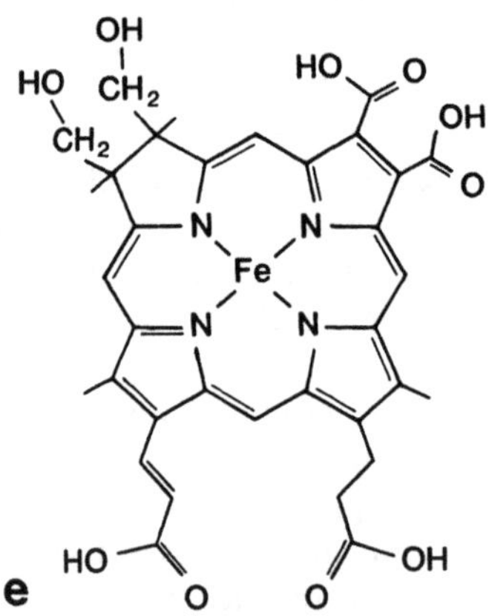
OH
HO
CH2
O
Fe
N

e

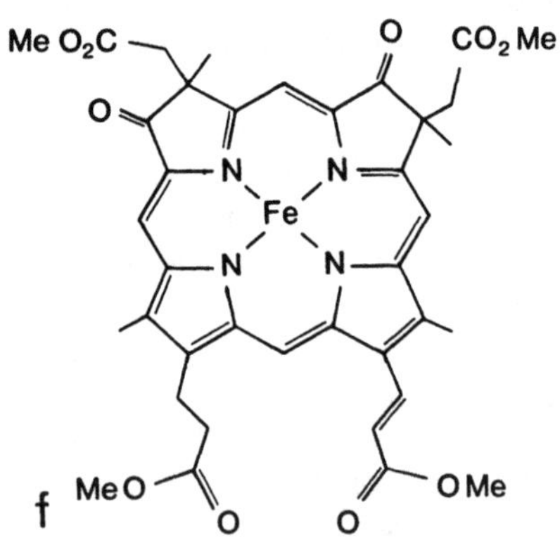
Me O2C
O
CO2 Me
Fe
N
MeO
OMe

f

monas aeruginosa, we find cytochrome c-551, c_4 and c_5 as well as the copper protein azurin which, as a group, are presumed to act as e^- donors to the terminal reductases for oxygen, hydrogen peroxide and the oxides of nitrogen.

Secondly, there is a relative lack of specificity of electron transfer in vitro. Thus *Pseudomonas* nitrite reductase (cytochrome cd_1) will react with a wide variety of bacterial cytochromes c and will accept electrons with similar efficiency from c-551, c_4, c_5 and azurin (Yamanaka 1972, 1975). The small cytochromes themselves show cross-reactivity, for example mitochondrial cytochrome c will quite rapidly reduce *Pseudomonas* cytochrome c-551, so that to deduce electron transport sequences and functional homology from the reactivity of purified components is of doubtful validity. This lack of specificity in electron transport between the small c-type cytochromes themselves can lead to problems in the use of oxidase particle preparations where some endogenous c-type cytochrome may still be present. In such cases the endogenous protein may catalyse an electron transfer to the oxidase which would not otherwise occur.

Thirdly, there are clear examples of close structural relatives performing different functions and of distant relatives performing the same function. Thus cytochrome c_2 of *Rhodopseudomonas capsulata* is the e^- donor to the photo-oxidised reaction centre, yet is very similar in amino acid sequence to cytochrome c-550 from *Paracoccus denitrificans,* a cytochrome which donates electrons to a cytochrome oxidase. On the other hand, cytochrome c-553 of *Rhodospirillum tenue* probably performs the same function as cytochrome c_2 but is not closely related in structure (Ambler et al. 1979a, b).

1.3.1 Cytochrome Classes I, II and III

The amino acid sequence work on the bacterial cytochromes, principally by R. P. Ambler and colleagues, has revolutionised our understanding of the relationships between different cytochromes and between the organisms of which they are a part. The ubiquitous small bacterial cytochromes divide into three groups on the basis of amino acid sequence (Ambler 1980) and the distinct and exclusive structure of these groups has been confirmed by X-ray crystallography (Almassy and Dickerson 1978; Korszun and Salemme 1977; Weber et al. 1980; Haser et al. 1979). We call these groups class I, II and III after Ambler (1980) and define them in Table 1.4.

Apart from the Cys-X-Y-Cys-His heme attachment, there is no sequence similarity between the three classes and thus no ambiguity in assigning a particular sequence to a particular class. However, not all amino acid sequences

Fig. 1.10a–f. The hemes of cytochromes. **a** protoheme IX; **b** substituted protoheme IX (heme c); **c** heme a (Caughey et al. 1975; Thompson et al. 1977); **d** proposed structure of heme d (Barrett 1956; Lemberg and Barrett 1973); **e** proposed structure of heme d_1 (Timkovich et al. 1984); **f** proposed structure of heme d_1 (methylester derivative) (Chang 1985)

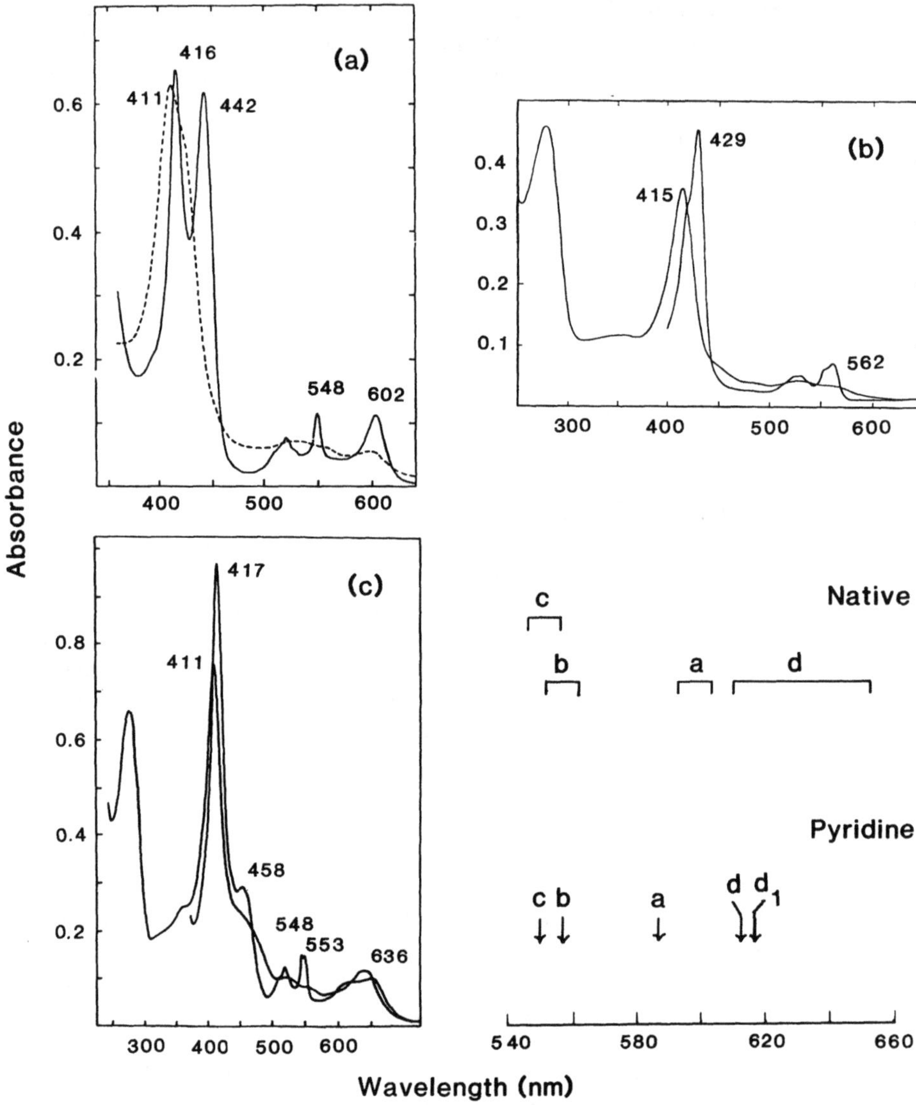

Fig. 1.11a–c. UV-visible absorption spectra of complex cytochromes containing heme c. The ferrocytochrome spectra are those showing a band near 550 nm. **a** The caa_3 complex of *Thermus thermophilus* (Hon-nami et al. 1980); **b** the cytochrome bc_1 complex of mitochondria (Yu et al. 1974); **c** cytochrome cd_1 of *Ps. aeruginosa* (Meyer and Kamen 1982). The *lower right* part of the figure shows the range of the α-peak maximum for native ferrocytochromes (┌──┐) and for the alkaline pyridine ferrohemochromes (↓) (Yamanaka and Okunuku 1974; Falk 1964; Lemberg and Barrett 1973). The range for the native ferrocytochrome d_1 is large because the α-band maximum is dependent on the nature of the reductant (635 nm with dithionite, 655 nm with ascorbate) (Parr et al. 1974). A few protozoan cytochromes have only one thioether linkage to the heme group and show pyridine ferrohemochrome spectra intermediate between cytochromes b and c (Pettigrew et al. 1975)

Table 1.4. Classification of c-type cytochromes: major structural divisions

Class	Features	Subdivision	Features	Examples
I	Low-spin, His + Met heme coordination,	(a) Large	Loop of residues closes bottom of heme crevice	Mitochondrial cyt.c, cyt.c_2
	heme near N-terminus, 80 – 120 amino acids	(b) Small	Lacks loop of (a). Left side folds downward to close bottom of heme crevice	*Pseudomonas* cyt.c-551 *Chlorobium* cyt.c-555 (Algal cyt.c-553) Cyt.c_4, Cyt.c_5 (*Desulfovibrio* cyt.c-553)
II	Heme near C-terminus	(a) High-spin	His only heme coordination	Cyt.c′
		(b) Low-spin	His + Met heme coordination	(Cyt.c-556)
III	Multi-heme, one heme per 30 – 40 amino acids bis-Histidyl coordination	On basis of heme content		[Cyt.c_3 (3 heme)] Cyt.c_3 (4 heme) [Cyt.c_3 (8 heme)]

Cytochrome groups in parentheses have not yet been fully characterised by X-ray crystallography although preliminary studies are in progress in most cases.
The original class IV of Ambler (1980) included complex proteins with prosthetic groups other than heme c. Other c-types, such as cytochrome c_1 and cytochrome c peroxidase were unclassified. We suggest that this is not a useful arrangement and propose that class IV and subsequent classes be reserved for homogeneous structural groups. In Volume 2, Chapter 5 we propose that the reaction centre cytochrome c of *Rhodopseudomonas viridis* be the prototype of a new class IV. It is possible that complex cytochromes such as cytochrome cd_1 and flavocytochrome c may be shown in future to have domains resembling existing classes such as class I. If this is so, we propose that these domains be included in the existing class rather than given the status of a new class.*

will be determined and it is useful to ask whether exclusive definitions can be formulated on the basis of more immediately accessible properties of the molecules such as spectra, molecular size, redox potential and reactivity with model oxidase and reductase preparations. An important point to bear in mind when attempting classification on these bases is that for a given cytochrome, a property is not an absolute parameter but depends on the conditions of measurement. We cannot solve the problem by measuring the property under "physiological conditions" because usually we have little or no reliable information about what these conditions are. Thus the reactivity of different mitochondrial cytochromes c with cytochrome oxidase was originally claimed to be highly variable (Yamanaka and Okunuki 1968), then constant (Smith et al. 1973) and more recently variable (Ferguson-Miller et al. 1978), the results depending on the exact conditions used. Similarly, midpoint redox potentials are often pH-dependent and, because the cytochromes are located at the acid side of bioenergetic membranes, it is not at all clear that pH 7 affords the most

* See appendix note 1

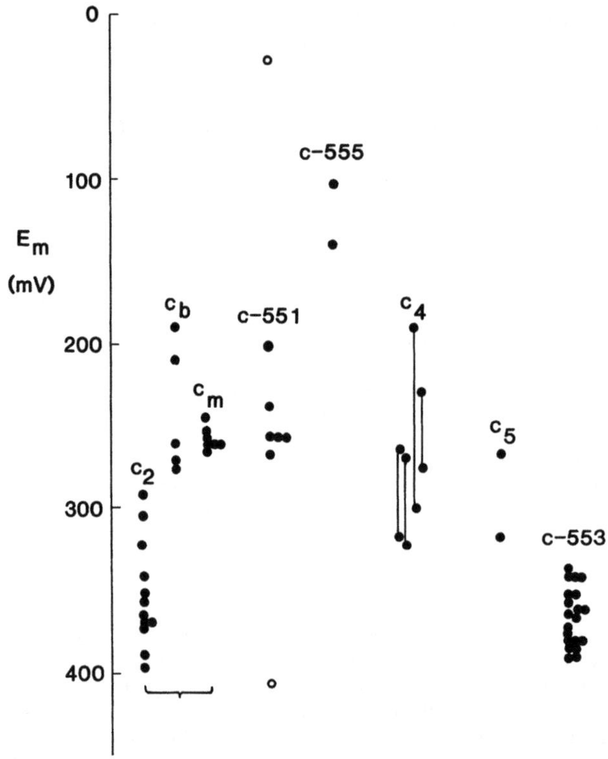

Fig. 1.12. The midpoint oxidation reduction potentials of the class I cytochromes c. Midpoint potentials are compiled from values given elsewhere in the book. Each *circle* represents a measurement of the midpoint potential of one member of the cytochrome group named above the columns. The *brackets* indicate that cytochrome c_2 (c_2), mitochondrial cytochromes c (c_m) and the bacterial cytochromes c, known to be donors to bacterial cytochrome aa_3 (c_b), are structurally homologous. The *open circles* of the cytochrome c-551 group correspond to measurements for *Rps. gelatinosa* cytochrome c-550 (+20 mV) and *Rsp. tenue* cytochrome c-553 (+405 mV). These proteins show remote but convincing sequence similarity to the *Pseudomonas* c-551 group (Vol. 2, Chap. 3). c-555 are the cytochromes c-555 of the Chlorobiaceae; c-553 are the cytochromes c-553 from algae. The *lines* joining values for cytochrome c_4 indicate the presence of two potentiometrically distinct hemes per molecule

valuable and valid point of comparison. Nonetheless physicochemical properties do greatly assist classification.

The class I cytochromes all have the 695-nm band in the ferric state which is dependent on the presence of a histidinyl-methionyl-Fe coordination as noted in Section 1.2.3. A further spectroscopic indicator of such coordination is the pronounced upfield shift of the methionine methyl resonance in ^{1}H-NMR spectra (Vol. 2, Chap. 2). Class I cytochromes contain 1 heme per 70–120 residues (a definition that includes the diheme cytochrome c_4 of M_r 20 K). They act as immediate electron donors to a terminal oxidising centre which may be a photo-oxidised chlorophyll species or a terminal respiratory enzyme. As a consequence they tend to exhibit redox potentials near the positive end of the biological scale, but since the terminal oxidising centres show considerable variation in operating potential, the range covered by their cytochromes c is correspondingly large (0 to +500 mV) (Fig. 1.12).

The class II cytochromes originally included only the cytochromes c′ which were characterised by high-spin absorption spectra (Fig. 1.6a). The discovery of low-spin homologues has complicated any definition of the group, particularly as some of these have a 695-nm band and contain methionine as a sixth heme ligand. These low-spin cytochromes have pronounced red shifts in most of their absorption bands (Meyer and Kamen 1982; Fig. 1.6b) but it is doubtful whether they can be reliably identified on this basis. This, so far, small subgroup probably constitutes the main risk of misclassification of a structurally uncharacterised cytochrome.

Cytochromes of class III are distinguished by their multi-heme nature and very low redox potentials. Even in the absence of analytical information of this sort they can be identified by the presence of a distinctive shoulder on the trailing edge of the reduced Soret peak, a high Soret (red.)/Soret (ox.) ratio and the absence of a 695-nm band in the oxidised form. These features serve to distinguish the class III cytochromes from those of other groups (Figs. 1.6d and 1.7).

Thus the three main structural groups of c-type cytochromes can be defined on exclusive lines in terms of their properties, and a novel cytochrome can usually be reliably designated. Problems arise when subdivision of class I is attempted.

1.3.2 Subdivision of Class I

Attempts to subdivide the cytochromes of class I on the basis of size (Ambler 1980) and three-dimensional structure (Dickerson 1980) have been made and will be discussed more fully in Vol. 2, Chaps. 3 and 4. We feel that some differences in size may reflect rather minor genetic changes in otherwise related proteins. For example, the cytochromes c_2 form a quite homogeneous group of proteins with greater than 30% identity in amino acid sequence (Meyer and Kamen 1982), yet subdivision on the basis of size splits this group into two.

In this case, size differences involve short loops of chain on the surface of the molecule. A more major size difference is observed between the cytochromes c_2/mitochondrial cytochrome c group on the one hand, and a group of smaller class I cytochromes, on the other. Here the size difference is due to the presence or absence of a large loop of chain which closes off the bottom of the heme crevice (Almassy and Dickerson 1978; Korszun and Salemme 1977) and it is possible that this represents a useful basis for subdivision. Implicit in this proposal is the belief that such a major change could have occurred only once in evolution and that it gave rise to two coherent lines of descent.

Table 1.5. Subdivision of class I cytochromes c; distinguishing features of subclasses

Subclass	Size	Spectra	$E_{m,7}$	Structure	Function	Comments and Anomalies
Mitochondrial cytochrome c	*103 – 112 aa* Size variation at N-terminus	*α (550 nm), sym.* α/β 1.8	*244 – 264 mV*	40% minimum pairwise sequence identity	*Reduced by cytochrome c_1, oxidised by cyt.aa_3*	*Tetrahymena* cyt.c does not react with mammalian oxidase. Other protozoan cytochromes have red-shifted absorption bands due to a single cysteine heme attachment site
Rhodospirillum cyt.c_2	*97 – 124 aa* Size variation at surface loops	*α (550 nm), sym. α/β 1.6*	*290 – 390 mV*	30% minimum pairwise sequence identity	Probably reduced by cyt.c_1, *oxidised by reaction centre* and in some species by a cyt.c oxidase	By simple sequence comparison, cyt.c_2 cannot be distinguished from mitochondrial cyt.c yet no cytochrome c_2 reacts well with mammalian oxidase. *Paracoccus denitrificans* cyt.c-550 is a close homologue of *Rps. capsulata* cyt.c_2 (54% identity) yet occurs in a non-photosynthetic organism and has an $E_{m,7} = 260$ mV
Pseudomonas cyt.c-551	*81 – 82 aa*	*α (551 – 553 nm)* sym. or asym., *$\alpha/\beta = 1.7 – 2.1$*	*200 – 265 mV*	64% minimum pairwise sequence identity within limited groups. *Lacks 13 residue loop at base of heme crevice*	Perhaps reduced by cyt.c_1, possibly oxidised by cyt.cd_1 and/or an o-type oxidase	A homologue with 64% identity is found in the aerobe *Azotobacter vinelandii*. More distant homologues are also found in the photosynthetic bacteria (*Rsp. tenue* cyt.c-553, 50% identity; *Rps. gelatinosa* cyt.c-550, 35% identity)
Chlorobiaceae cyt.c-555	*86,99 aa* Size variation mostly at N-terminus	*α (555 nm), asym., $\alpha/\beta = 1.2$*	*154, 103 mV*	54% sequence identity. *Lacks loop at base of heme like cyt.c-551*	Probably involved in electron transfer from sulfide and thiosulfate to reaction centre	

Algal cyt. c-553	*83 – 89 aa*	*α (552 – 554 nm), usually asym., $\gamma_r/\alpha = 5.7 - 7.7$*	*335 – 390 mV*	38% minimum pairwise sequence identity. *Lacks loop at base of heme like cyt.c-551*	*Interchangeable with plastocyanin in algae. Acts between cyt.f and photo-oxidised chlorophyll*	*Euglena* cyt.c-552 has a symmetrical α-peak and $\gamma_r/\alpha = 5.2$. It is also the most divergent sequence
Cytochrome c_4	*181, 190 aa*	*α (550 nm), sym. $\alpha/\beta = 1.2$. Low 260 – 280 nm absorption, no shoulder at 290 nm*	*190 – 320 mV*	77% sequence identity. *Two-domain structure* with one heme per domain, *gene doubling,* each domain lacks loop at base of heme. *No Trp*	Not known but reactive with *Pseudomonas* cyt.cd_1 in vitro	Hemes probably have different redox potentials with variable degree of separation. Cytochrome c-554 (548) from halotolerant *Paracoccus* is 50% identical to cytochrome c_4 and is isolated as a dimer of mol.wt.18000
Cytochrome c_5	*87 aa*	*α (554 nm) $\alpha/\beta = 1.4$*	*265, 320 mV*	*Contains a second Cys-X-Y-Cys site with no heme bound.* Ragged N-terminus	Not known, suggested to be a fragment of larger protein	
Desulfovibrio cyt.c-553	*82 aa*	*α (553 nm)*	*0 mV*	Little similarity to other sequences. *No Trp*	Not known	The amino acid sequence of the cytochromes c-553 from the Hildenborough and Miyazaki strains have been determined and show little resemblance in spite of the similarity in their cytochromes c_3 (4 heme). The E_m for the Miyazaki strain was quoted as −280 mV yet the cytochrome was reducible with ascorbate

Principal discriminating features are indicated by italic script

If cytochromes are found which lack this portion of chain and yet are close relatives of the large cytochromes on the basis of the rest of the amino acid sequence such a proposal would have to be discarded.

With this doubt about classification on the basis of detailed structural information it is not surprising that classifications on other bases are not rigorous. We offer in Table 1.5 some tentative definitions of subgroups of class I on the basis of size, spectrum, redox potential and reactivity with oxidase and reductase preparations. Together, these properties allow a composite definition which distinguishes one group of cytochromes from another but often there are anomalies which must be taken into account.

We have reached a point where the amount of available structural information should encourage a re-examination of the commonly used subdivisions of class I and we offer a partial reorganisation and renaming of the cytochrome subclasses in Table 1.6. In doing this two main problems have to be faced. One is whether to group together or to subdivide extensively, i.e. what degree of structural difference merits a separate grouping? The second is whether to

Table 1.6. A cytochrome c nomenclature

Names in common usage	Suggested[a] name
Cytochrome f Cytochrome c_1	c_1
Mitochondrial cyt.c Cytochrome c_2	c_2
Cytochrome c_3[b] Cytochrome cc_3 Cytochrome c_7	c_3 (4 heme) (8 heme) (3 heme)
Cytochrome c_4	c_4
Cytochrome c_5	c_5
Cytochrome c_6 Algal cytochrome f Algal cytochrome c-553	c_6
Pseudomonas cyt.c-551[c]	c_7
Cytochrome c′[d] Cytochrome c-556	c_8

[a] We should emphasise that these are personal recommendations. Cytochrome nomenclature remains in the hands of an International Committee.

[b] The cytochrome c_3 group has also been subdivided on the basis of M_r (c_3-14 K, c_3-26 K, c_3-9 K). We have not adopted this nomenclature because of variation of M_r within a subdivision.

[c] Cytochromes c from *Azotobacter vinelandii, Rhodospirillum tenue* and *Rhodopseudomonas gelatinosa* are homologous to the cytochromes c-551 of the Pseudomonads.

[d] The cytochromes c′ were originally given a variety of names, e.g. cytochromoid, cryptocytochrome c, cytochrome cc′, RHP which are no longer in use.

employ functional as well as structural criteria. Our revised scheme is conservative in its degree of subdivision and essentially ignores functional considerations. An alternative scheme based on such considerations could be constructed but would be sketchy through lack of knowledge and would incorporate both convergent and divergent effects.

References

Almassy RJ, Dickerson RE (1978) Pseudomonas cytochrome c-551 at 2.0 Å resolution: enlargement of the cytochrome c family. Proc Natl Acad Sci USA 75:2674–2678

Ambler RP (1977) Amino acid sequences and bacterial phylogeny. In: Matsubara H, Yamanaka T (eds) Evolution of protein molecules. Univ Tokyo Press

Ambler RP (1980) The structure and classification of cytochromes c. In: Robinson AB, Kaplan NO (eds) From cyclotrons to cytochromes. Academic Press, London New York, pp 263–279

Ambler RP, Meyer TE, Kamen MD (1979a) Anomalies in amino acid sequences of small cytochromes c and cytochromes c′ from two species of purple photosynthetic bacteria. Nature (London) 278:661–662

Ambler RP, Daniel M, Hermoso J, Meyer TE, Bartsch RG, Kamen MD (1979b) Cytochrome c_2 sequence variation among the recognised species of purple non-sulfur photosynthetic bacteria. Nature (London) 278:559–560

Ambler RP, Daniel M, Mellis K, Stout CD (1984) The amino acid sequence of the diheme cytochrome c_4 from *Azotobacter vinelandii.* Biochem J 222:217–227

Barrett J (1956) The prosthetic group of cytochrome a_2. Biochem J 64:626–639

Bartsch RG (1971) Bacterial cytochromes. In: San Pietro A (ed) Methods in enzymology, photosynthesis, part A. Academic Press, London New York, pp 344–363

Caughey WS, Smythe GA, O'Keefe DH, Maskasy JE, Smith ML (1975) Heme A of cytochrome oxidase – structure and properties – comparisons with hemes B, C and S and derivatives. J Biol Chem 250:7602–7612

Chang CK (1985) On the structure of heme d_1. J Biol Chem 260:9520–9522

Dickerson RE (1980) Cytochrome c and the evolution of energy metabolism. Sci Am 242, 3:98–109

Falk JE (1964) Porphyrins and metalloporphyrins. Elsevier, Amsterdam

Ferguson-Miller S, Brautigan DL, Margoliash E (1978) Correlation of the kinetics of electron transfer activity of various eukaryotic cytochromes c with binding to mitochondrial cytochrome c oxidase. J Biol Chem 251:1104–1115

Fiechtner MD, Kassner RJ (1979) Redox properties and heme environment of cytochrome c-551.5 from *Desulfuromonas acetoxidans.* Biochim Biophys Acta 579:269–278

Goodhew C, Brown K, Pettigrew GW (1986) Haem staining in gels, a useful tool in the study of bacterial c-type cytochromes. Biochim Biophys Acta 852:288–294

Goodwin TW, Morton RA (1946) The spectrophotometric determination of tyrosine and tryptophan in proteins. Biochem J 40:628–632

Haser R, Pierrot M, Frey M, Payan F, Astier JP, Bruschi M, Le-Gall J (1979) Structure and sequence of the multiheme cytochromes c_3. Nature (London) 282:806–810

Hon-nami K, Oshima T (1980) Cytochrome oxidase from an extreme thermophile, *Thermus thermophilus* HB8. Biochem Biophys Res Commun 92:1023–1029

Jacobs AJ, Kaha VK, Cavari B, Brodie AF (1979) Purification, physicochemical properties and localisation of cytochrome c in energy transducing membranes from *Mycobacterium phlei.* Arch Biochem Biophys 194:531–541

Korszun ZR, Salemme FR (1977) Structure of cytochrome c-555 from *Chlorobium thiosulfatophilum:* primitive low-potential cytochrome c. Proc Natl Acad Sci USA 74:5244–5247

Lemberg R, Barrett J (1973) The cytochromes. Academic Press, London New York

Meyer TE (1970) Comparative studies on soluble iron containing proteins in photosynthetic bacteria. PhD thesis, Univ Cal

Meyer TE, Kamen MD (1982) New perspectives on c-type cytochromes. Adv Protein Chem 35:105–212

Meyer TE, Bartsch RG, Kamen MD (1971) Cytochrome c_3 – a class of electron transfer proteins found in both photosynthetic and sulfate reducing bacteria. Biochim Biophys Acta 245:453–464

Parr SR, Wilson MT, Greenwood C (1974) The reaction of *Pseudomonas aeruginosa* cytochrome c oxidase with sodium metabisulfite. Biochem J 139:273–276

Penke B, Ferenczi R, Kovacs K (1974) A new acid hydrolysis method for determining tryptophan in peptides and proteins. Anal Biochem 60:45–50

Pettigrew GW, Leaver JL, Meyer TE, Ryle AP (1975) Purification, properties and amino acid sequence of atypical cytochrome c from two Protozoa *Euglena gracilis* and *Crithidia oncopelti*. Biochem J 147:291–302

Shioi Y, Takamiya K, Nishimura M (1972) Studies on energy and electron transfer systems in the green photosynthetic bacterium, "*Chloropseudomonas* 2K". J Biochem 71:285–291

Smith L, Nava ME, Margoliash E (1973) In: King TE, Mason HS, Morrison M (eds) Oxidases and related redox systems, vol. 2. Univ Park Press, Baltimore, pp 629–638

Thomas PE, Ryan D, Levin W (1976) An improved staining procedure for the detection of the peroxidase activity of cytochrome P450 on sodium dodecyl sulfate polyacrylamide gels. Anal Biochem 75:168–176

Thompson M, Barrett J, McDonald E, Battersby AR, Fookes CJR, Chaudhry IA, Clezy PS, Morris HR (1977) Cytochrome c oxidase – isolation, crystallisation and synthesis of porphyrin A-dimethylester. J Chem Soc Chem Commun 278–279

Timkovich R, Cork MS, Taylor PV (1984) Proposed structure for the non-covalently associated heme prosthetic group of dissimilatory nitrite reductases. J Biol Chem 259:15089–15093

Weber PC, Bartsch RG, Cusanovich MA, Hamlin RC, Howard A, Jordan SR, Kamen MD, Meyer TE, Weatherford DW, Xuong NH, Salemme FR (1980) Structure of cytochrome c′: a dimeric high-spin heme protein. Nature (London) 286:302–305

Wood PM (1980) Interrelationships of the two c-type cytochromes in *Rhodopseudomonas sphaeroides* photosynthesis. Biochem J 192:761–764

Wood PM (1981) Fluorescent gels as a general technique for characterising bacterial c-type cytochromes. Anal Biochem 111:235–239

Wood PM (1983) Why do c-type cytochromes exist? FEBS Lett 164:223–226

Yamanaka T (1972) Evolution of the cytochrome c molecule. Adv Biophys 3:229–276

Yamanaka T (1975) A comparative study on the redox reactions of cytochromes c with certain enzymes. J Biochem 77:493–499

Yamanaka T, Okunuki K (1968) Comparative studies on reactivities of cytochrome c with cytochrome oxidases. In: Okunuki K, Sekuzu I, Kamen MD (eds) Structure and function of cytochromes. Univ Park Press, Baltimore, pp 390–403

Yamanaka T, Okunuki K (1974) Cytochromes. In: Microbial iron metabolism. Academic Press, London New York

Yu CA, Yu L, King TE (1974) Soluble cytochrome bc_1 complex and the reconstitution of succinate cytochrome c reductase. J Biol Chem 249:4905–4910

Chapter 2 The Role of Mitochondrial Cytochrome c in Electron Transport

2.1 Introduction

The location and role of mitochondrial cytochrome c are represented diagrammatically in Fig. 2.1, a scheme which can be applied in a broader way to many bacterial systems (Chap. 3). The intermembrane location was established by several lines of evidence (reviewed in Nicholls 1974) but the means by which cytochrome c conducts electrons between its membrane reductase and oxidase remains controversial and forms the main subject of this chapter. Also controversial is the relative importance of the reactions of cytochrome c with soluble redox systems. This subject will be considered in Section 2.8.

The central problem is represented in Fig. 2.2. Does cytochrome c act as a static electron conducting link (Fig. 2.2a), as a carrier which diffuses in two dimensions only on the membrane surface (Fig. 2.2b) or as a freely diffusible carrier, available, at least in part, in the intermembrane space (Fig. 2.2c). Evidence relevant to this problem includes equilibrium and kinetic studies governing the reactions of cytochrome c and investigations of the binding surface on cytochrome c for its reaction partners. This evidence will be reviewed.

2.2 Reaction of Cytochrome c with the Inner Membrane Electron Transfer System

2.2.1 General Features of the Inner Membrane Cytochrome Complexes

2.2.1.1 Coenzyme Q-Cytochrome c Oxidoreductase (Complex III)

Ubiquinol-cytochrome c oxidoreductase catalyses the reduction of cytochrome c by ubiquinol and the translocation of protons across the membrane from the matrix to the intermembrane space (Wikstrom et al. 1981). This region of the electron transport chain can be isolated as a multi-subunit complex (Hatefi et al. 1962; Hatefi and Rieske 1967; Rieske 1967; Yu et al. 1977; Azzi et al. 1982). Relatives of the mitochondrial complex have been isolated from plants and bacteria (Chap. 3.2).

Complex III contains at least eight different subunits, three of which are known to carry redox centres (Fig. 2.3). Because of the covalently bound heme,

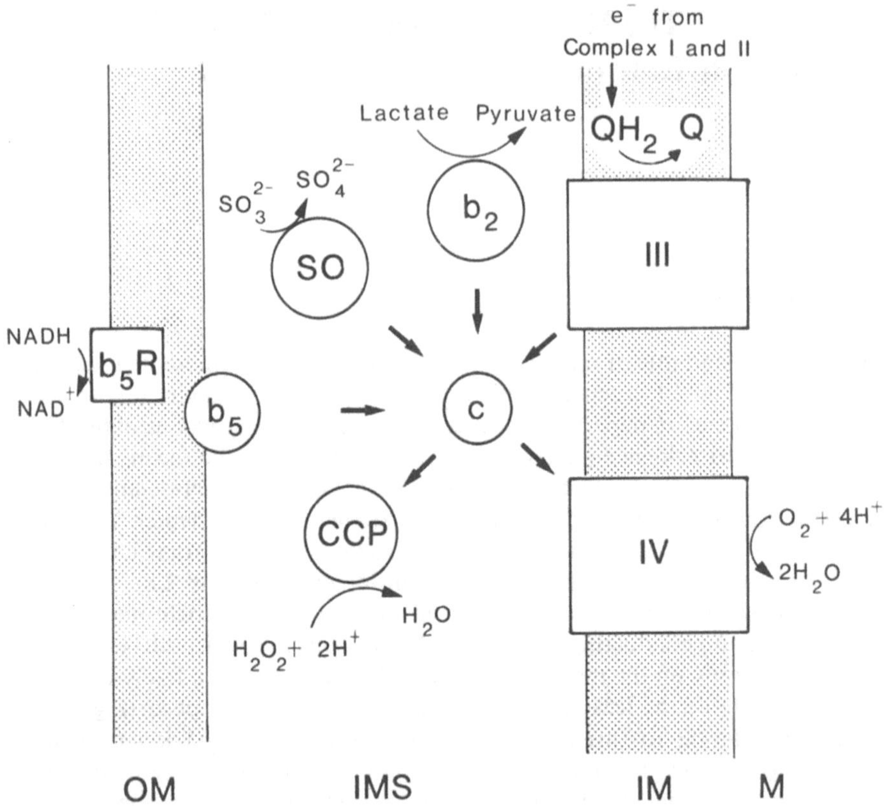

Fig. 2.1. The role of cytochrome c in mitochondrial electron transport; redox partners in the inner membrane and intermembrane space. The respiratory chain is located in the inner mitochondrial membrane (*IM*). It is organised into multi-protein complexes of which complexes *I* and *II* accept reducing equivalents from matrix (*M*) substrates and reduce coenzyme Q. Complex *III* is a coenzyme Q-cytochrome c oxidoreductase while complex *IV* is a cytochrome c oxidase. Cytochrome c (*c*) therefore acts as a link between complex *III* and *IV* and it also interacts with intermembrane (*IMS*) redox partners. These include cytochrome b_5 (situated in the outer membrane, *OM*) and sulfite oxidase (*SO*) in animals, and cytochrome c peroxidase (*CCP*) and cytochrome b_2 (a lactate dehydrogenase) in yeasts. The *heavy arrows* indicate the direction of election flow to and from cytochrome c

cytochrome c_1 is readily recognisable on SDS gel electrophoresis of the complex, and the Fe-S protein and cytochrome b have been purified (Rieske et al. 1964; Weiss 1976) and shown to correspond to the 25-K and 27-K polypeptides respectively.

Mapping the topography of this complex in terms of the relationships of the subunits to each other and to the membrane is of importance in providing a structural framework for models of electron transport. Here, a most exciting development has been the calculation of shape and dimensions from tilted images of electron micrographs (Henderson and Unwin 1975). Membrane crys-

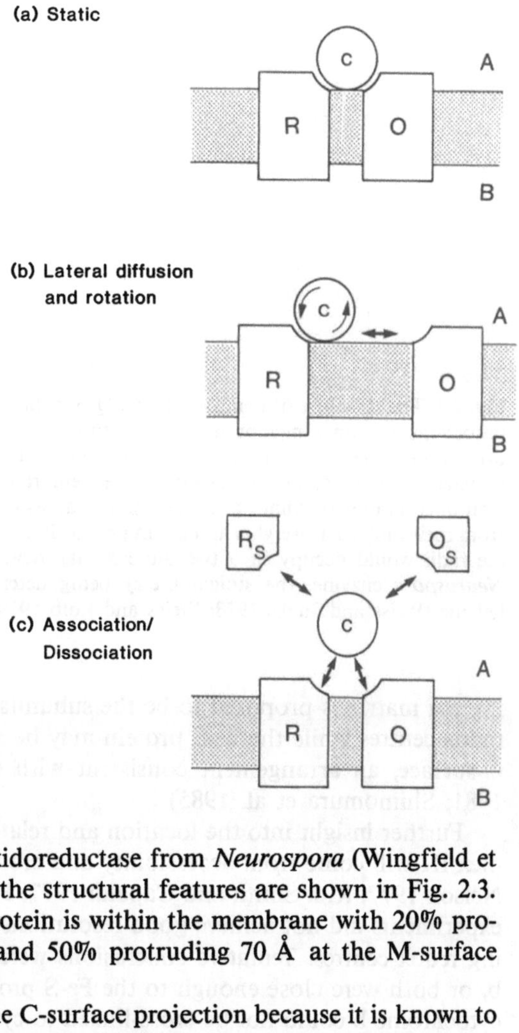

Fig. 2.2a–c. The mode of action of cytochrome c in electron transport. The diagram summarises three ways in which the peripheral membrane protein, cytochrome c, might transfer electrons between the integral membrane proteins, CoQ cytochrome c oxidoreductase (*R*) and cytochrome c oxidase (*O*). In the case of the mitochondrial inner membrane, compartment *A* represents the intermembrane space and *B* the matrix. For bacterial respiration or photosynthesis *A* represents the periplasmic space and *B* the cytoplasm. In model (a) cytochrome c requires two interaction surfaces. In model (b) the *arrows* indicate rotational and lateral diffusion of the cytochrome on the membrane. In model (c) R_s and O_s represent soluble reductases and oxidases and the *arrows* represent free diffusion of cytochrome c to the alternative sites

tals of the CoQ-cytochrome c oxidoreductase from *Neurospora* (Wingfield et al. 1979) have been studied and the structural features are shown in Fig. 2.3. Remarkably, only 30% of the protein is within the membrane with 20% protruding 30 Å at the C-surface and 50% protruding 70 Å at the M-surface (Leonard et al. 1981).

Cytochrome c_1 is placed in the C-surface projection because it is known to be the donor to cytochrome c. Also, if *Neurospora* are grown in the presence of chloramphenicol (an inhibitor of mitochondrial protein synthesis), the inner mitochondrial membrane contains cytochrome c_1 inserted without association with other proteins (Yun et al. 1981). Protease treatment of such membranes released a heme-containing water-soluble fragment of Stokes radius 35 Å and M_r 24 K. This suggests that cytochrome c_1 is an amphiphilic polypeptide anchored to the membrane by a cleavable hydrophobic tail (see also Chap. 3.2.3). The Fe-S protein also falls into this category.

A cytochrome bc_1 complex lacking the Fe-S protein and the large subunits can be prepared by mild salt treatment (Hovmoller et al. 1981) and electron microscopy reveals a less bulky, 30-Å C-surface extension and a loss of the 70-Å M-surface extension (Li et al. 1981). Thus the large mass of protein fac-

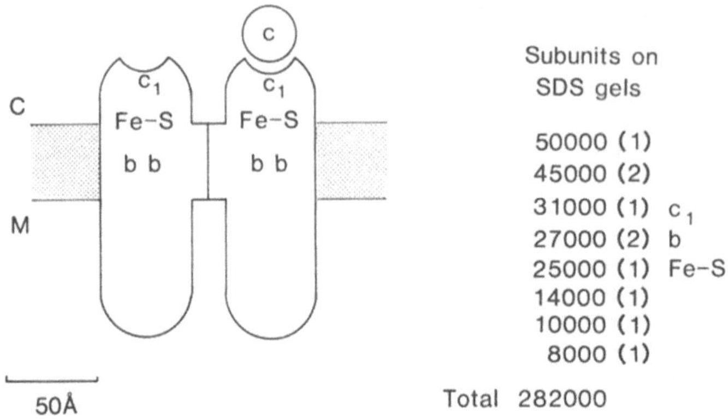

Fig. 2.3. The structure of complex III (CoQ cytochrome c oxidoreductase). A portion of the mitochondrial inner membrane is represented diagrammatically in transverse section. *C* cytoplasmic side; *M* matrix side. The CoQ cyt.c oxidoreductase is represented as a dimer and dimensions within and on either side of the membrane are those obtained by analysis of electron micrographs (Leonard et al. 1981). The areas covered by the molecules are calculated from their molecular weights and are in proportion to the surfaces that spheres of appropriate radii would occupy in a two-dimensional view. The subunit composition is for the *Neurospora* enzyme, the stoichiometry being determined after growth of cells on ^{3}H-leucine (Weiss and Juchs 1978; Weiss and Kolb 1979; Weiss and Wingfield 1979)

ing the matrix is proposed to be the subunits of M_r 45 K and 50 K having no redox centres while the Fe-S protein may be adjacent to cytochrome c_1 at the C-surface, an arrangement consistent with electron transfer (Bowyer et al. 1981; Shimomura et al. 1985).

Further insight into the location and relationship of redox centres in complex III will come from accessibility and cross-linking studies (Gellerfors and Nelson 1977; R. J. Smith and Capaldi 1977; H. T. Smith et al. 1977) but initial experiments did not achieve good resolutions of those three subunits containing redox centres. Tentative conclusions were that eighter cytochromes c_1, or b, or both were close enough to the Fe-S protein to allow cross-linking while cytochrome b could not be cross-linked to cytochrome c_1. A detailed account of the internal electron transfer properties of complex III is beyond the scope of the present review and the reader is referred to Hauska et al. (1983) and Rich (1984).

Cytochrome c₁. Cytochrome c_1 has been purified from bovine, *Neurospora* and bakers' yeast mitochondria. Most isolation methods are based on those of Yu et al. (1972) and King (1978) which involve cleavage of the strong interaction between the cytochrome and complex III by treatment with 5–15% mercaptoethanol in the presence of bile salts and ammonium sulfate. The ingenious device of growth in chloramphenicol, mentioned above, can be used in the case of *Neurospora* to produce a mitochondrial membrane deficient in cytochrome b (Yun et al. 1981) thus simplifying the purification of the Triton-solubilised cytochrome c_1 by affinity chromatography on immobilised cytochrome c.

Purified preparations of cytochrome c_1 have a heme content of 32 nmol mg^{-1} consistent with M_r 31 K observed on SDS gel electrophoresis. This is also the molecular weight observed in a crude cell extract of yeast grown on ^{14}C-δ-amino laevulinic acid (Ross and Schatz 1976) and is in reasonable agreement with the formula weight of 27924 derived from the amino acid sequence (Wakabayashi et al. 1982a). In the absence of SDS, aggregates of cytochrome c_1, probably containing five molecules, are formed (King 1978). A major point of disagreement between King and co-workers and other groups is the presence and significance of a colourless polypeptide of M_r 15 K associated with the purified protein. Although Ross and Schatz (1976) observed this band in yeast cytochrome c_1 preparations they found it in variable amount and it was absent in the cytochrome c_1 immune-precipitated from crude extracts (Ross and Schatz 1978) and in the preparation of Konig et al. (1980). Chromatography under denaturing conditions was required in order to isolate the single heme-containing polypeptide (Trumpower and Katki 1975; King 1982) but Robinson and Talbert (1980) found that high concentrations of bile salts during extraction and purification yielded a native single chain cytochrome c_1. This unresolved dispute regarding the significance of the colourless peptide is important for our understanding of the cytochrome c_1: cytochrome c interaction (Sec. 2.2.2).

The occurrence of cytochrome c_1 in bacteria and the relationship between cytochrome c_1 and cytochrome f are discussed in Chap. 3.

2.2.1.2 Cytochrome Oxidase

Cytochrome oxidase is one of the most complex and most extensively studied redox enzymes. It not only catalyses the four-electron transfers from the single electron donor, cytochrome c, to dioxygen but is proposed to do so across the mitochondrial inner membrane thus acting as a charge transfer system within the Mitchell chemiosmotic model of the oxidative phosphorylation process (Mitchell 1980). Recent results (Wikström et al. 1981; Wikström 1984) suggest that it also acts as a redox-linked proton pump so that its action can be expressed as:

$$4\,c_r + nH_i^+ + O_2 \rightarrow 4\,c_o + 2\,H_2O + (n-4)H_o^+ \,, \qquad (2.1)$$

where H_i^+ and H_o^+ refer to protons within and outside the matrix and n has a value between 8 and 12 (Reynafarge et al. 1982).

This functional complexity is matched by a complexity of structure, the details of which remain controversial (Saraste 1983; Kadenbach et al. 1983). For many years eukaryotic oxidase preparations were believed to contain seven or eight equimolar subunits. More resolving gel electrophoresis has revealed 12 or 13 bands in the case of the bovine heart enzyme (Fig. 2.4B) and the amino

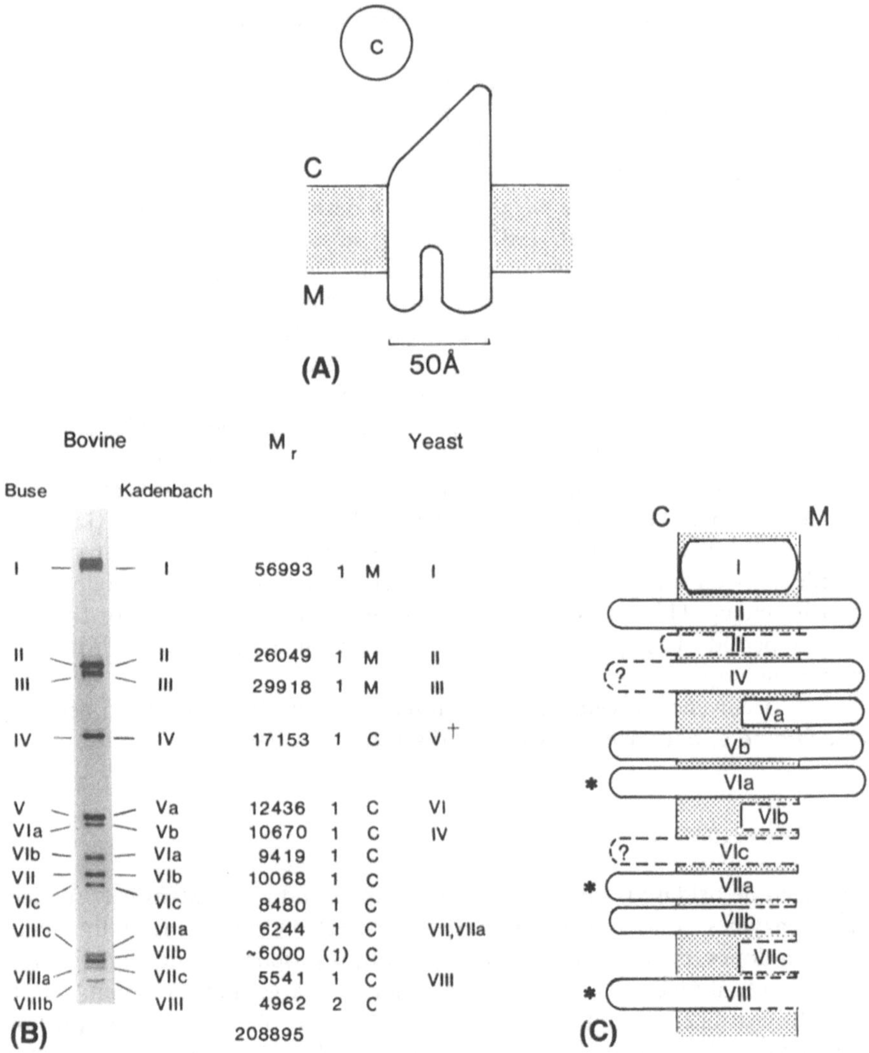

Fig. 2.4A–C. The structure of cytochrome c oxidase. **(A)** The areas covered by cytochrome c **(C)** and the oxidase are calculated from their molecular weight and are in proportion to the surface that spheres of appropriate radii would present in a two-dimensional view. **(B)** The SDS gel electrophoresis pattern of cytochrome c oxidase (Kadenbach et al. 1983) with naming of corresponding bands and stoichiometry according to Buse et al. (1985). Molecular weights are from the amino acid sequences except in the cases of *I* and *III* which are based on the DNA sequences, and *VIIb* (Kadenbach), the sequence of which has not been determined. The site of synthesis is indicated by *M*, mitochondrial or *C*, cytoplasmic. The yeast enzyme (Power et al. 1984a) has a different pattern of electrophoretic separation, and subunits corresponding to the bovine enzyme are indicated. † = Two genes. **(C)** Representation of the topology studies for rat liver oxidase using proteolysis of mitoplasts and sonicated particles (Jarausch and Kadenbach 1985a) or subunit-specific antisera (Kuhn-Nentwig and Kadenbach 1985b). Subunit *I* was poorly susceptible to proteolysis and was heavily labelled

acid sequences of 12 subunits have been determined (Buse et al. 1985). In the yeast enzyme, nine subunits have been identified (Power et al. 1984a). These yeast subunits are homologous to counterparts in the bovine enzyme (Power et al. 1984b, c) but their relative electrophoretic mobility in SDS differs in some cases and an accepted naming scheme has become essential (Fig. 2.4B). Some authors contest the validity of this "stoichiometric" definition of a cytochrome oxidase subunit and propose that peptides which can be proteolytically removed without loss of enzyme activity arise from fortuitous copurification (Capaldi et al. 1983).

This dispute and the markedly anomalous behaviour of subunits I and III on SDS gel electrophoresis have led to uncertainties in the literature regarding the molecular weight of the monomer. In Fig. 2.4B we accept the stoichiometric analysis of Buse et al. (1985) which gives M_r 209 K and 9.57 nmol heme mg^{-1}.

An interesting finding and one potentially important to our understanding of function is the tissue specificity of some of the smaller subunits. Bands VIa, VIIa and VIII (after Kadenbach) migrate differently in SDS gel electrophoresis of the purified enzymes from bovine heart and liver and this may be correlated with differences in turnover and apparent K_m values (Merle and Kadenbach 1982). In the rat, all nine nuclear-encoded subunits show immunological differences in different tissues unlike the three mitochondrial-encoded subunits (Kuhn-Nentwig and Kadenbach 1985a). The largest differences are again seen in subunits VIa and VIII (after Kadenbach). In yeast, Cumsky et al. (1985) have shown that the counterpart to mammalian subunit IV is encoded by two genes with 62% homology which may be differentially expressed depending on the physiological status of the cell. Thus a regulatory role for the nuclear subunits is a possibility and one form that this could take is the modulation of the affinity for cytochrome c. This is further discussed in Sect. 2.5.

The overall dimensions of cytochrome oxidase are known from electron diffraction studies of two crystal forms: dimeric enzyme in flattened vesicle bilayers (Henderson et al. 1977; Frey et al. 1978; Deatherage et al. 1982) and monomeric enzyme in detergent-rich sheets (Fuller et al. 1979, 1982). Results from the two forms are in agreement in showing the oxidase monomer as a lop-sided Y-shape inserted across the membrane with the arms facing the matrix (Fig. 2.4A).

by "deep probe" arylazido phospholipids (Bisson et al. 1979; Prochaska et al. 1980; Cerletti and Schatz 1979; Gutweniger et al. 1981). Inconclusive results (*broken lines*) for subunit *III* were obtained in the studies of Ludwig et al. (1979) and Jarausch and Kadenbach (1985a) but it is heavily labelled by deep probe phospholipids (references above) and is susceptible to C-side proteolysis (Zhang et al. 1984). *IV* was accessible from the C-surface in the studies of Jarausch and Kadenbach (1985a) and Kuhn-Nentwig and Kadenbach (1985b) in contradiction to previous labelling work (Ludwig et al. 1979) and proteolysis (Zhang et al. 1984). *VIc* gave a weak antibody reaction and was poorly digestible in mitoplasts. Less is known of M-surface topology because the small subunits were not studied. This is indicated by *open-ended oblongs*; * indicates that these subunits show tissue specificity (Kuhn-Nentwig and Kadenbach 1985a; Merle and Kadenbach 1982) and are protected by cytochrome c (Kadenbach and Stroh 1984)

The topology of cytochrome oxidase within the membrane and the nearest neighbour relationships of subunits has been a difficult analytical problem and earlier investigations were additionally hampered by incomplete resolution of subunits. The accessibility of subunits to water-soluble, membrane-impermeable labelling reagents such as diazobenzene sulfonate, subunit-specific antisera and lipid-soluble labelling reagents such as the arylazidophospholipids has been studied in mitoplasts, submitochondrial particles and reconstituted vesicles. In the latter, oxidase insertion appears to be asymmetric giving rise to predominantly right-side-out vesicles (Zhang et al. 1984) and these can be further purified using a cytochrome c affinity column (Madden and Cullis 1985). Nearest neighbours have been investigated using bifunctional cross-linking reagents (Briggs and Capaldi 1978; Jarausch and Kadenbach 1985b).

All these studies suffer from the problem that for a positive result to be obtained not only must a region of a particular subunit be accessible to protease, antibody or chemical reagent but the region must also be susceptible to attack or binding. Thus negative results can be difficult to interpret. In Fig. 2.4C we have summarised the studies of Jarausch and Kadenbach (1985a) and Kuhn-Nentwig and Kadenbach (1985b) on the rat liver enzyme. Cross-linking studies are difficult to incorporate in such a two-dimensional view and remain problematic in interpretation because of incomplete knowledge of the topology.

Because of the susceptibility to dissociation of the prosthetic groups, the crucial question of the location of the redox centres in cytochrome oxidase has been controversial. Subunit III does not contain redox centres because it can be lost from the complex without affecting the oxidase activity (Saraste et al. 1981). It is proposed to form a proton channel across the membrane (Casey et al. 1981; Senior 1983).

Direct evidence that subunit I and II contain redox centres came from the work of Winter et al. (1980), who found 90% of the Cu and 50% of the heme associated with subunit II after dissociation of the complex in 0.1% SDS, 0.025% Triton X-100. The remaining heme was associated with subunit I. Similar results were obtained by Corbley and Azzi (1984). Indirect evidence supports these conclusions. Thus the oxidase from *Paracoccus denitrificans* contains two hemes and two Cu but bound to only two subunits (Ludwig and Schatz 1980)*. These are of similar molecular weight and show immunological cross-reactivity with subunits I and II of yeast cytochrome oxidase (Ludwig 1980). Also, arguments based on chemical modification (reviewed by Capaldi et al. 1983) and sequence homology (Steffens and Buse 1979) support the proposal that at least one copper is located in subunit II. Thus in mitochondrial cytochrome oxidase all four redox centres may be located in subunits I and II.

Kinetic and physical studies of the enzyme have led to the concept of an "electron-accepting pole" containing heme a and Cu and a "ligand-binding pole" containing a magnetically coupled binuclear centre of heme a_3 and Cu (Brunori and Wilson 1982). Since heme a is considered to be the redox centre kinetically proximal to cytochrome c (Sect. 2.2.2) and since cytochrome c can be cross-linked to subunit II (Sect. 2.5.1) we may speculate that this subunit

* See appendix note 2

contains the "electron-accepting pole". If the assignment of redox centres to subunits found by Winter et al. is correct then subunit II must also contribute to the ligand-binding pole jointly with subunit I.

Further discussion of the role of the redox centres of cytochrome oxidase is beyond the scope of this book and the reader is referred to the reviews of Caughey et al. (1976); Malmström (1979); Wikström and Krab (1979); Wikström et al. (1981); Brunori and Wilson (1982) and Vol. 23 of *J. Inorg. Biochem.* We will return to cytochrome oxidase when we consider the bacterial relatives of the mitochondrial enzyme (Chap. 3).

2.2.1.3 The Stoichiometry of the Cytochrome System

Mitochondria from different species and from different tissues within the same species show considerable variation in their cytochrome c content (Table 2.1). In vertebrate sources there is usually less than 1 mol cyt. c/mol cyt. aa_3 and in intestinal tissue the ratio can be as low as 0.1 mol cyt. c/mol cyt. aa_3. On the other hand, *Neurospora* mitochondria contain a considerable excess of cytochrome c.

Most research on mitochondria has been performed on those isolated from rat liver, pigeon heart or bovine heart, tissues that contain approximately equimolar cytochrome c and cytochrome aa_3. When such mitochondria are washed in 0.15 M KCl after hypotonic pretreatment they lose their cytochrome c and all respiratory activity (Jacobs and Sanadi 1960). Oxidative phosphorylation can be restored by addition of an amount of cytochrome c equal to that removed. It is important to emphasise the difference between this intact, phosphorylating system and similar depletion/reconstitution experiments performed with non-phosphorylating Keilin Hartree particles

Table 2.1. Relative concentrations of cytochromes in mitochondria

Source	mol/mol aa_3		Reference
	c	c_1	
Rat liver[a]	0.8	0.6	Williams 1968
Rat intestine[a]	0.1	0.2	Williams 1968
Chick liver	0.5	0.7	Williams 1968
Chick intestine	0.2	0.4	Williams 1968
Bovine heart[b]	1.1	0.6	Nicholls 1976
Neurospora[c]	2.9	1.0	Sebald et al. 1979

[a] Rat kidney, brain and heart mitochondria fall between these extremes.
[b] 0.5 nmol aa_3 mg^{-1} mitochondrial protein.
[c] 0.24 nmol aa_3 mg^{-1} mitochondrial protein.

(KHP) where a fraction (10–20%) of the cytochrome c originally present will allow maximum respiration (Nicholls 1964b; Jacobs et al. 1964; Nicholls et al. 1969). Nicholls (1974) proposed that this striking difference may be due to the presence of branching or interchain electron transfer in the KHP that is absent in the intact mitochondrion.

2.2.2 Direct Binding Studies

Cytochrome c forms tight complexes with both its reductase and oxidase on the mitochondrial membrane and in the purified state. The present consensus is that these complexes are productive and their strength has been inferred from kinetic studies and measured directly by binding studies.

Most binding studies are analysed by the graphical method of Scatchard (1949) according to Eq. (2.2):

$$\text{bound/free} = K_A n - K_A \text{ bound} \ , \tag{2.2}$$

where bound is the mol cytochrome c bound/mol binding partner, free is the concentration of free cytochrome c, K_A is the association constant and n is the number of binding sites. An example of such an analysis is shown in Fig. 2.5. In principle, such studies can give information on dissociation constants ($1/K_A$) and the numbers of sites for cytochrome c binding. In practice, reliable data at low cytochrome c concentrations can be difficult to obtain due to binding to glass or Sephadex or other materials. Also Scatchard plots may be curved, rather than linear or biphasic – an indication of interacting sites.

Scatchard analysis requires measurement of the concentrations of free and bound cytochrome c and these can be obtained in a number of different ways. Binding to particulate systems can be assessed by simple centrifugation followed by spectrophotometry or radioactive counting (Nicholls 1964b; Williams

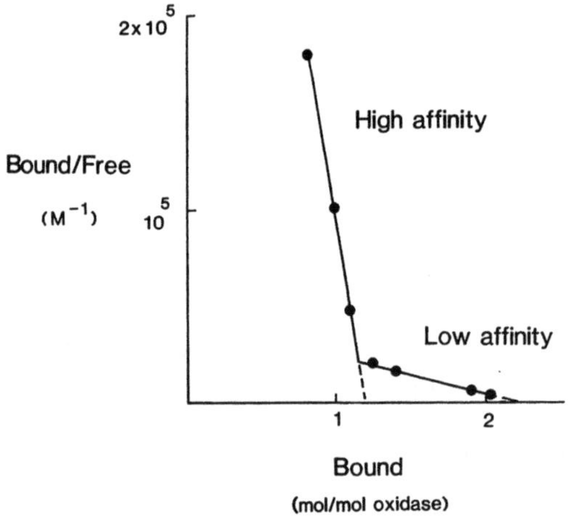

Fig. 2.5. The binding of cytochrome c to cytochrome oxidase (Data from Diggens and Ragan 1982). Oxidase preparation was from bovine heart, cytochrome c was from horse heart

and Thorp 1970; Vanderkooi et al. 1973a, b; Diggens and Ragan 1982). Binding to mitochondria can also be detected without centrifugation by the tendency of cytochrome c to quench the fluorescence of 12-(9-anthroyl) stearic acid incorporated into the membranes (Vanderkooi et al. 1973a) or by EPR spectroscopy of immobilised, spin-labelled cytochrome c (Vanderkooi et al. 1973b).

For soluble systems the method of choice is equilibrium molecular exclusion chromatography (Hummel and Dreyer 1962) using a material such as Sephadex G-100 equilibrated with a known concentration of cytochrome c. Bound cytochrome c is calculated either from the peak or the trough of cytochrome c concentration in the elution profile. Binding of high affinity ($K_D < 10^{-7}$ M) may be detected by molecular exclusion chromatography in the absence of an equilibrating concentration of cytochrome c (Ferguson-Miller et al. 1976; Rieder and Bosshard 1978). Under such conditions Eq. (2.3) applies:

$$r - 1 - \ln r = K_D v / c a \ , \qquad (2.3)$$

where r is the fractional saturation, K_D is the dissociation constant, v the volume of buffer through which the complex is filtered, c the concentration of binding sites and *a* the volume applied to the column (Dixon 1976).

An alternative to molecular exclusion chromatography was devised by Petersen (1978) who analysed the partition of cytochrome c:cytochrome oxidase mixtures between polyethylene glycol (which preferentially takes up the oxidase, with or without bound cytochrome c) and dextran (which preferentially takes up the free cytochrome c).

These studies provide quantitative information on the affinity and number of sites. Qualitative estimates of the relative affinities of cytochrome c for its redox partners can also be obtained from the ionic strengths required to elute complex III or cytochrome oxidase from cytochrome c affinity columns (Bill and Azzi 1984).

Mitochondria. Two classes of binding sites were found for cytochrome c in pigeon heart mitochondria (Vanderkooi et al. 1973a, b) (Table 2.2). Those of higher affinity ($K_D = 5 \times 10^{-8}$ M) are found at a concentration of 1 nmol mg^{-1} mitochondrial protein and probably represent the sum of the sites on cytochrome c_1 and cytochrome aa_3 which are each expected to be present at approx. 0.5 nmol mg^{-1}. Those of lower affinity ($K_D = 5 \times 10^{-7}$ M) are found in larger concentration (5 nmol mg^{-1}) and probably involve phospholipid in the membrane (Table 2.2).

Cytochrome Oxidase. Within rather precise, narrow (and unphysiological!) conditions of ionic strength and ionic composition, bovine cytochrome oxidase contains one high affinity (K_D approx. 10^{-7} M) and one low affinity site (K_D approx. 10^{-6} M) for horse cytochrome c (Table 2.2). The low affinity site disappears under conditions of high ionic strength or in the presence of phosphate or ATP (Ferguson-Miller et al. 1976; Petersen 1978). It also disappears

Table 2.2. Binding studies on cytochrome c

Ligand	System	Method[e]	Ionic conditions	K_{DH}[a]	K_{DL}
Horse ferri-cyt.c	Cyt.c depleted mitochondria	F	0.05 M MOPS pH 7.2	$\sim 5\times10^{-8}$ M[f]	5×10^{-7} M
Horse ferro-cyt.c	Cyt.c depleted mitochondria	F	0.05 M MOPS pH 7.2		2×10^{-6} M
Horse ferri-cyt.c	Bovine oxidase	EM, C	0.025 M Tris-acetate pH 7.8	$\sim 10^{-7}$ M (1)	$\sim 10^{-6}$ M (1)[b,g]
Horse ferri-cyt.c	Bovine oxidase	P	0.02 M phosphate pH 7.4	1.3×10^{-7} M[h]	
Horse ferro-cyt.c	Bovine cyano-oxidase	P		3.5×10^{-7} M	
Yeast ferri-cyt.c	Bovine oxidase	EM	0.025 M Tris-acetate pH 7.8	5×10^{-7} M[i] (2)	
Horse ferri-cyt.c	*Neurospora* complex III (oxidised)	EM	0.02 M Tris-acetate + 0.02 M NaCl, pH 7	1.2×10^{-7} M[j] (1)	[c]
Horse ferro-cyt.c	*Neurospora* complex III (oxidised)	EM	0.02 M Tris-acetate + 0.02 M NaCl, pH 7	6×10^{-7} M (1)	
Horse ferri-cyt.c	Bovine complex III (oxidised)	EM	0.05 M Tris Cl pH 7.5	1.9×10^{-7} M[k]	[d]
Horse ferro-cyt.c	Bovine complex III (reduced)	EM	0.05 M Tris Cl pH 7.5	6.4×10^{-7} M	[d]

[a] Figures in parentheses are the number of binding sites. K_{DH} and K_{DL} are dissociation constants for high affinity and low affinity sites.
[b] Binding to the low affinity site was abolished by the presence of ATP or phosphate.
[c] At lower ionic strength, curved Scatchard plots were obtained indicating more than two binding sites. Purified cytochrome c_1 gave similar results (Yun et al. 1981).
[d] At $I<0.05$ M, low affinity phospholipid binding sites were detected.
[e] Methods: F fluorescence, EM equilibrium molecular exclusion chromatography, P partition, C centrifugation (see text).
[f] Vanderkooi et al. (1973a, b).
[g] Ferguson-Miller et al. (1976); Dethmers et al. (1979); Rieder and Bosshard (1978); Diggens and Ragan (1982).
[h] Petersen (1978).
[i] Dethmers et al. (1979).
[j] Weiss and Juchs (1978).
[k] Speck and Margoliash (1984).

if tightly bound phosphatidylglycerol is removed from cytochrome oxidase by exchange with Triton X-100 and it can be restored by adding back this phospholipid (Vik et al. 1981). In contrast, yeast iso-1 cytochrome c bound to two sites with equal affinity (K_D approx. 5×10^{-7} M) on both bovine and yeast cytochrome oxidase (Ferguson-Miller et al. 1976; Dethmers et al. 1979).

These studies are of importance in the debate as to whether one or two binding sites exist on cytochrome oxidase (Sect. 2.8) but they do not necessarily define catalytically active sites. In contrast, Michel and Bosshard (1984) found only one binding site by measuring the small perturbations that occur in the spectra of both heme c and heme a on complex formation. These authors argued that such perturbations will occur only for closely and, by implication, functionally apposed redox centres and that non-catalytic sites therefore will not be observed by their method. Thus cytochrome oxidase probably has one catalytic site of high affinity and a second lower affinity site formed in part by tightly bound phospholipid. The role of this low affinity site is the subject of much debate and is further discussed in the following sections.

CoQ Cytochrome c Oxidoreductase and Cytochrome c_1. Direct binding studies by Weiss and Juchs (1978) with the *Neurospora* reductase, and using equilibrium molecular exclusion, found a single site with $K_D = 1.2 \times 10^{-7}$ M for ferricytochrome c at pH 7 in 20 mM Tris-acetate, 20 mM NaCl. Additional weaker binding sites were apparent at lower ionic strength (Table 2.2). Similar results were obtained by Speck and Margoliash (1984) for the bovine enzyme.

Cytochrome c also forms a complex with purified cytochrome c_1 which is stable to non-equilibrium molecular exclusion and which can be detected by changes in circular dichroism and kinetics of ascorbate reduction (Chiang et al. 1976; Bosshard et al. 1979; Yun et al. 1981). The binding affinity is similar to that observed for the intact reductase (Yun et al. 1981). Some controversy surrounds the function of the colourless polypeptide of M_r 9175 in the cytochrome c_1 preparations of King's group. Kim and King (1981, 1983) claimed that no cytochrome c binding can be detected with a preparation of cytochrome c_1 lacking this polypeptide, which they therefore called "hinge protein" to indicate its mediating role (Wakabayashi et al. 1982a). Hinge protein is highly acidic (Wakabayashi et al. 1982a) and may escape detection on SDS gels. However, the cytochrome c_1 preparation of Konig et al. (1980), which does not appear to contain this polypeptide, was competent in electron transfer (Konig et al. 1981) and could be covalently cross-linked to cytochrome c (Broger et al. 1983). A very acidic subunit of M_r 17 K was also found in complex III of yeast (van Loon et al. 1984).

Thus there is one high affinity site for cytochrome c on CoQ cytochrome c oxidoreductase which is formed by cytochrome c_1. An acidic colourless polypeptide has been proposed to mediate the binding but its status is uncertain.

Redox State Dependent Binding. The studies in Table 2.2 agree that ferricytochrome c binds three to four times more strongly than the ferrocytochrome to both cytochrome oxidase and CoQ cytochrome c oxidoreductase. However, Bill and Azzi (1984) found that immobilised yeast ferri- and ferrocytochrome c had the same affinity for the reductase while the ferricytochrome showed a *lower* affinity for the oxidase than the ferrocytochrome c. These conflicting results make it difficult to come to conclusions about three important areas of cytochrome c function which are influenced by redox state-dependent binding.

Firstly, the lowering of redox potential on binding of cytochrome c to mitochondria or to purified redox proteins suggests that the ferricytochrome binds more tightly than the ferrocytochrome (Vol. 2, Chap. 7). This is consistent with most of the binding studies discussed above but not the finding of Bill and Azzi (1984). Secondly, and in contradiction to the first point, the accepted kinetic analysis of cytochrome oxidase requires equal binding of substrate and product (Sect. 2.2.3). Thirdly, an aesthetically satisfying model for cytochrome c function might be that the ferricytochrome has a high affinity for the reductase and low for the oxidase since it is the substrate for the former and the product of the latter. The ferrocytochrome would show the converse pattern. However, there is little experimental justification for such a view.

In summary, cytochrome c binds to both its oxidase and reductase at single sites of high affinity (K_D approx. 10^{-7} M). The affinity of these sites decreases with rising ionic strength, indicating electrostatic interactions and, at lower ionic strength, low affinity sites (K_D approx. 10^{-6} M) can also be detected which probably involve phospholipid associated with the enzymes. Binding to the high affinity sites may be influenced by the redox state of the cytochrome.

2.2.3 Kinetics

2.2.3.1 Cytochrome Oxidase

The oxidation of cytochrome c by cytochrome oxidase has been studied using particulate preparations, detergent-solubilised membranes and purified enzyme. The observed turnover number is increased by treatment of particles by certain non-ionic detergents but is generally decreased by further purification (L. Smith et al. 1979b). The relatively poor activity of the purified enzyme can be improved by adding back phospholipid preparations, particularly those containing cardiolipin (Roberts and Hess 1977; Thompson and Ferguson-Miller 1983; Speck et al. 1983).

Lauryl maltoside appears to be the detergent of choice for solubilisation of the oxidase (Rosevear et al. 1980; Thompson et al. 1982; Thompson and Ferguson-Miller 1983). Its effectiveness was originally thought to be due to its ability to stabilise the dimeric state (Rosevear et al. 1980) as had been demonstrated for Triton X-100 (Robinson and Capaldi 1977). However, after a reconsideration of the molecular exclusion behaviour of the enzyme, taking into account the contribution of the detergent micelle and assuming a 12–13 subunit monomer of M_r 210 K, Thompson and Ferguson-Miller (1983) concluded that the enzyme is monomeric in lauryl maltoside. This question of aggregation state is of importance in the kinetic and structural description of cytochrome oxidase to be considered in the following sections.

We consider the kinetics of the cytochrome c reactions in some detail, even when it is clear that certain postulated mechanisms are not applicable to the mitochondrial system. In this way we hope to clarify some of the confusion surrounding this subject and illustrate the kind of mechanisms that might be applicable to related bacterial enzymes. One kinetic model emerges as compatible with all the kinetic evidence and its structural significance will be further discussed at the end of this chapter.

Steady-State Kinetic Measurements. There exist two approaches to the steady-state kinetic investigation of the cytochrome c oxidase reaction which differ in principle and may also differ in what they tell us about the reaction. One is the spectroscopic assay, in which the time course of the reaction is monitored by absorption changes due to falling ferrocytochrome c concentration. The second is the polarographic assay, in which cytochrome c catalyses the electron transfer between the ascorbate-TMPD (tetramethyl-p-phenylene diamine) reduction system and the oxidase, the time course being monitored by falling oxygen concentration. The two will be dealt with separately.

The Spectroscopic Assay. Smith and Conrad (1956) showed that the oxidation of cytochrome c followed strictly first-order kinetics at all concentrations of ferrocytochrome tested (Fig. 2.6a). Furthermore, the derived pseudo-first-order rate constants decreased with rising cytochrome c and were a function of total cytochrome c present $[C_t]$, not ferrocytochrome c alone $[C_r]$ (in this text, $[C_o]$ and $[C_r]$ are the concentrations of oxidised and reduced cytochrome c and $[C_t] = [C_o] + [C_r]$). Nevertheless, an apparent hyperbolic relationship between initial velocity (k_{obs} $[C_r]$) and substrate concentration was obtained from which apparent K_M and V_{max} values could be derived by conventional kinetic plots.

At very low substrate concentrations, deviations from linearity in such plots are observed (Errede et al. 1976), suggesting the presence of a high affinity kinetic phase.

However, the incompatibility between first-order kinetics and simple Michaelis-Menten analysis of the enzyme substrate complex led Smith and Conrad to reject the derived Michaelis parameters as mechanistically meaningless. As Errede et al. (1976) pointed out, deviations from first-order kinetics should be detectable at substrate concentrations 10% of K_M while cytochrome oxidation remained first-order up to cytochrome concentrations at least ten times greater than K_M.

The attempts by L. Smith and Conrad (1956) and Minnaert (1961) to provide a mechanistic interpretation of the kinetic data has provided the framework for most subsequent analyses. Such mechanisms are considered after a description of the polarographic assay and the pre-steady-state kinetics.

The Polarographic Assay. The principle of this assay is outlined in Fig. 2.6b and involves cytochrome c acting in a catalytic rather than a stoichiometric role. Initial velocities are available from the time course of oxygen consumption and conventional linearising plots are used to analyse the relationship be-

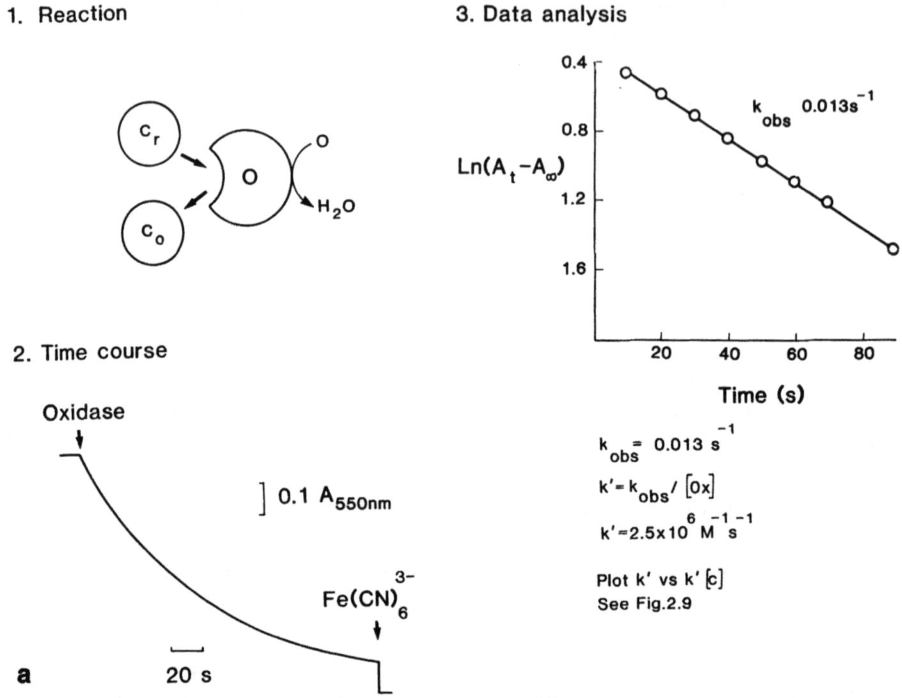

Fig. 2.6a. The spectroscopic assay of cytochrome c oxidase

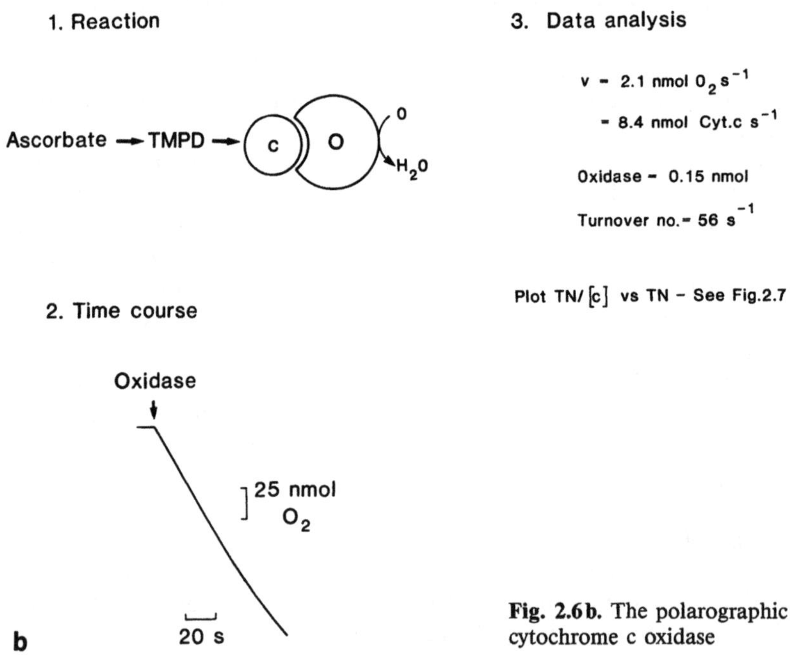

Fig. 2.6b. The polarographic assay of cytochrome c oxidase

Fig. 2.7. Kinetic data from the polarographic assay (Data from Rosevear et al. 1980)

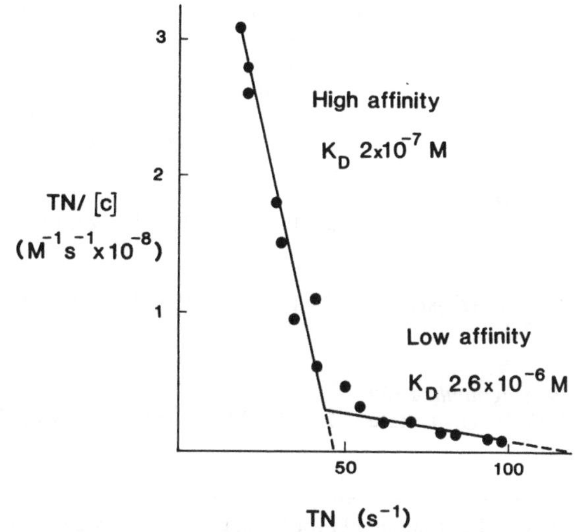

tween these velocities and cytochrome concentration. Early indications of complexity in such plots consistent with the operation of two apparent K_M values were noted by Nicholls (1964b), and have been extensively analysed by Ferguson-Miller et al. (1976, 1978a) (Fig. 2.7). Thus, superficially, the polarographic assay has produced results similar to the pattern seen with the spectroscopic assay and the two apparent K_M values obtained from the Eadie-Hofstee plot have been interpreted as estimates of dissociation constants for high and low affinity sites. The good agreement between the derived K_M values and the K_D values obtained from direct binding studies support this interpretation (Ferguson-Miller et al. 1976; Vanderkooi et al. 1973a and b; Sect. 2.2.2).

Cytochrome c in the high affinity site is proposed to turn over without dissociation because the overall rate of oxygen consumption is related to the level of reduction of cytochrome c bound to the oxidase, not to that of free cytochrome c. Thus the limiting rate under these circumstances must be direct reduction of the cytochrome c : oxidase complex [k_{TMPD} in Eq. 2.4)], a process which is only important when the rate of dissociation (governed by k_2 and k_4) is considerably less than the rate of reduction by TMPD.

$$E+S \underset{k_2}{\overset{k_1}{\rightleftharpoons}} ES \xrightarrow{k_3} EP \underset{k_5}{\overset{k_4}{\rightleftharpoons}} E+P \quad (EP \xrightarrow{k_{TMPD}} ES) \; . \tag{2.4}$$

These observations have two important implications. Firstly, they suggest that the cytochrome c : cytochrome oxidase complex is kinetically competent. Secondly, the possibility arises that cytochrome c_1 might reduce cytochrome c in the same way as TMPD without dissociation of the cytochrome c from its complex with the oxidase (Fig. 2.2a). This mechanism will be considered further in Section 2.8.

Table 2.3. Kinetic parameters derived from the spectroscopic and polarographic assays

	High affinity reaction		Low affinity reaction		Reference
	K_D	TN	K_D	TN	
Spectroscopic assay					
0.1 M MES pH 6.0	2×10^{-7} M	40 s^{-1}	2×10^{-5} M	100 s^{-1}	Calculated from Errede and Kamen 1978
0.025 M Tris-acetate pH 7.9	8×10^{-8} M	10 s^{-1}	3.5×10^{-6} M	70 s^{-1}	Calculated from Rosevear et al. 1980
Polarographic assay					
0.025 M Tris-acetate pH 7.9	2×10^{-7} M	40 s^{-1}	2.7×10^{-6} M	120 s^{-1}	Calculated from Rosevear et al. 1980

Notes: The optimum pH for the spectroscopic assay is approximately 6.0 while that for the polarographic assay is 7.8. Comparison between the two assays should be done under the same conditions, e.g. the results of Rosevear et al. (1980)

Comparison of the Two Methods. The K_D values derived by the spectroscopic and polarographic methods are in reasonable agreement (Table 2.3) even though the two assays almost certainly have different rate-limiting steps. Thus conditions which maximise activity in the latter are low ionic strength which permits tight binding of cytochrome c to the oxidase and direct reduction of the complex by TMPD. Increasing ionic strength leads to continuously falling rates. The turnover numbers, at least for the high affinity site, are functions of k_{TMPD} and [TMPD] and thus are of no physiological interest. On the other hand, the rate obtained in the spectroscopic assay is limited at low ionic strength by dissociation of the enzyme-product complex (k_{off}) and at high ionic strength by the association of enzyme and substrate. Thus at values below the optimal ionic strength (approx. 0.1 M phosphate at pH 6, Waino et al. 1960), turnover numbers of $2-40\ s^{-1}$ for the high affinity reaction (L. Smith et al. 1979a; Rosevear et al. 1980; Errede and Kamen 1978) are probably measures of k_{off}.

At higher concentrations of cytochrome c the operation of the low affinity site allows much higher turnover on the enzyme (Table 2.3). This may be due either to an intrinsically greater activity at a second independent site or to a facilitated release of the first cytochrome by the second (see later). In the latter situation, the high turnover number of the low affinity phase would again correspond to a k_{off} but one that is enhanced by the effect of the second molecule.

Pre-Steady-State Kinetics. The interpretation of the spectroscopic changes that occur in cytochrome oxidase on reduction with cytochrome c is complicated by the joint contribution of redox centres to particular absorption bands.

Under anaerobic conditions or with heme a_3 blocked with cyanide, the appearance of the band at 603 nm (Gibson et al. 1965) and the disappearance of the "visible" low-spin ferric heme EPR signal (Hartzell et al. 1973) on reaction with ferrocytochrome c are believed to reflect reduction of heme a as the primary electron acceptor from cytochrome c. Accompanying this reduction is a bleaching at 830 nm (Gibson and Greenwood 1965) and the disappearance of the visible Cu EPR signal (which originates from Cu_A; Hartzell et al. 1973; Greenwood et al. 1976) at a rate independent of cytochrome c concentration (Wilson et al. 1975). This suggests that the Cu_A centre accepts the second electron via heme a. The stoichiometry of approx. 2 mol cytochrome c oxidised per mole heme a reduced is in agreement with this interpretation (Antalis and Palmer 1982). Further electron transfer into the oxidase is inhibited by the anaerobic conditions under which these experiments have been carried out. Flash flow experiments using cytochrome c bound to oxidase allow further electron transfer within the oxidase to be studied by the rapid replacement of the inhibitory CO by O_2. Such experiments reveal heterogeneity of cytochrome c oxidation that may be due to the presence of more than one electron transfer pathway within the oxidase (Hill and Greenwood 1984).

Three types of reaction have been studied with the anaerobic system. One is the reaction of ferrocytochrome c with an excess of oxidised cytochrome oxidase – the high affinity reaction. This reaction is so fast that it cannot be adequately observed at low ionic strength (Veerman et al. 1980). Extrapolation of studies at ionic strengths greater than 130 mM suggest that the second-order rate constant k_{on}, may be as large as $10^{10}-10^{11}\,M^{-1}\,s^{-1}$ (Veerman et al. 1982). The value of the first-order rate constant k_{off} (the rate of dissociation of ferricytochrome c) was estimated by Wilms et al. (1981a) to be $2.5\,s^{-1}$ from the reaction of cytochrome c_1 with the cyt. c: cyt. oxidase complex and this is in reasonable agreement with the maximum turnover number of $10\,s^{-1}$ obtained for the high affinity site in the spectroscopic steady-state assay (Rosevear et al. 1980) where dissociation of the product is believed to be rate-limiting (see earlier).

The second type of experiment that can be done is reduction of the cytochrome c: cytochrome oxidase complex by ferrocytochrome c – the low affinity reaction. Here we have a slower process with a second-order rate constant k_{on} extrapolated to $I = 0$ of $2.3 \times 10^8\,M^{-1}s^{-1}$ and a $k_{off} = 300\,s^{-1}$ (van Gelder et al. 1975). This reaction was originally considered to take place at a second independent site on the oxidase (van Gelder et al. 1982) that was equated with the low affinity site of the steady-state experiments. However, several recent lines of evidence suggest that the low affinity reaction involves binding of a second cytochrome c at a single, already occupied site. The essential feature of this model is that the negative interactions between the two molecules cause the first to be more readily lost (i.e. k_{off} raised) and the second to exhibit a weaker binding. Such a mechanism is supported by the finding that polylysine (and by implication, other polycations such as cytochrome c itself) causes release of cytochrome c from the complex at rates at least as high

Table 2.4. Rate constants for the low affinity reaction of cytochrome c with cytochrome oxidase; evidence for interaction between high and low affinity reactions

k_{on} for donor	Cytochrome in complex with bovine oxidase	
	Human	Horse
Horse cyt.c	$7.4 \times 10^6\ M^{-1}\ s^{-1}$	$3.5 \times 10^7\ M^{-1}\ s^{-1}$
Human cyt.c	$10^7\ M^{-1}\ s^{-1}$	$3.5 \times 10^7\ M^{-1}\ s^{-1}$

as $250\ s^{-1}$ (Wilms et al. 1981a). Also, in contrast to previous proposals (Ferguson-Miller et al. 1978a), Veerman et al. (1983) have found that the family of monosubstituted Lys derivatives of cytochrome c (Sect. 2.4) showed the same order of reactivity under high and low affinity conditions indicating that the binding surface on cytochrome c is the same for both. Finally, the rate of the low affinity reaction was found to depend on the nature of the occupancy of the high affinity site (Table 2.4, Osheroff et al. 1983) clearly demonstrating that the sites are not independent.

The third type of experiment is that of van Gelder et al. (1979) and Veerman et al. (1982) in which electron transfer was observed between ferrocytochrome c and a cytochrome oxidase in which the high affinity site was blocked by the redox-inactive porphyrin cytochrome c. Although the results were taken as evidence for an active and separate low affinity site, the observed electron transfer could also be explained by rapid dissociation of porphyrin cytochrome c caused by the incoming molecule (Veerman et al. 1980). This mechanism might also explain the complete replacement of native horse cytochrome c by a modified cytochrome when the horse cytochrome c : oxidase complex is passed through a gel filtration column equilibrated with the modified cytochrome (Ferguson-Miller et al. 1978a). Osheroff et al. (1983) and Veerman et al. (1980) argued that the rate of the low affinity reaction cannot be limited by a dissociation event because the rate constant would then be expected to be a true first-order constant independent of cytochrome c concentration. However, the present mechanism proposes a dissociation enhanced by the incoming cytochrome molecule and therefore dependent on its concentration.

The dissociation constant for the high affinity reaction is strongly dependent on ionic strength and when this is high, the distinction between high and low affinity sites disappears. In contrast, the dissociation constant of the low affinity reaction is independent of ionic strength. Both van Gelder et al. (1975) and Antalis and Palmer (1982) found that the matching fall of k_{on} and k_{off} with rising ionic strength led to a constant value of $K_D = 10^{-6}\ M$ for this reaction. We interpret the fall in k_{off} with rising ionic strength as due to a decreasing effectiveness of the incoming cytochrome in expelling the occupant of the site.

In conclusion, the pre-steady-state kinetics of ferrocytochrome c with cytochrome oxidase are best interpreted by a single-site model in which a high affinity reaction can occur with the first molecule of cytochrome c and a low affinity reaction involves binding of a second molecule and displacement of the first. The dissociation constant associated with the high affinity reaction is approx. 10^{-8} M and the complex turns over at 2.5 s^{-1} in the absence of excess cytochrome c. In contrast, the second molecule binds with a weaker affinity ($K_D = 10^{-6}$ M) but induces rapid turnover (250 s^{-1}).

Kinetic Models. A variety of kinetic models have been proposed to describe the steady-state kinetics of the cytochrome c oxidase reaction. In the following review, we will discuss their evolution from the original single-site models of L. Smith and Conrad (1956) and Minnaert (1961) to the four more complex models proposed by Errede and Kamen (1979) in the light of their extended kinetic analysis.

L. Smith and Conrad (1956) proposed that the two key features of the spectroscopic assay, namely first-order kinetics and a hyperbolic relationship between initial velocity and substrate concentration might be reconciled by a model involving inactive complex formation. This suggestion and several alternative models were examined by Minnaert (1961). All models were required to obey the general equation:

$$v = \frac{\alpha_1}{\beta_1 [C_t] + \beta_2} [C_r] [\text{OXIDASE}] \; , \qquad (2.5)$$

where α_1, β_1, β_2 are constants. Equation (2.5) predicts:

1. a first-order reaction at constant $[C_t]$ ($v = k_{obs} [C_r]$),
2. diminishing k_{obs} as $[C_t]$ increases,
3. a straight line Lineweaver-Burke or Eadie-Hofstee plot.

Models were then assessed for simplicity, and the viability of assumptions used in their derivation. Three models survived which were kinetically indistinguishable: mechanism IV, which involved productive complex formation with ferrocytochrome c and inhibitory complex formation by ferricytochrome c; mechanism V, which involved self-exchange between free and bound cytochrome c, and an extended "dead-end" complex model of the type originally suggested by L. Smith and Conrad (1956) which required that complex formation with either ferro- or ferricytochrome c be non-productive while electron transfer took place via transient collisions. These models have formed the basis for interpretations of the more complex kinetics which became evident as the substrate concentration range was extended.

[A note of caution concerning the models described below is that they all assume that the two redox states of cytochrome c bind with equal affinity to the oxidase. This was supported by the results of Yonetani and Ray (1965) who showed that K_I for the binding of the ferricytochrome c as an inhibitor was equal to the apparent K_M for the ferrocytochrome under conditions where first-order kinetics were observed. However, as discussed in Section 2.2.2, direct binding studies and redox potentiometry have suggested that the ferricytochrome binds three to four times more tightly than the ferrocytochrome to both crude and purified oxidase preparations. This conflict worried Nicholls (1974) who suggested that these effects were due to binding to non-specific sites (phospholipid) rather than specific ones (cytochrome oxidase). However, Ferguson-Miller et al. (1978b) offered a simpler resolution of the conflict. They noted that binding and redox potential studies have tended to be performed at higher pH values than those of the kinetic studies and under these conditions, deviations from first-order kinetics and a fall in K_I relative to K_M have been observed (Yonetani and Ray 1965). Repeating the binding studies at lower pH values would be of considerable interest.]

L. Smith and Conrad (1956) and Nicholls (1964b) noted that Lineweaver-Burke plots for the steady-state spectroscopic assay of ferrocytochrome c oxidation deviated from linearity at low concentrations of substrate and this led to a detailed re-examination of the reaction by Errede et al. (1976) using a wider range ($0.7-160\times10^{-6}$ M) of ferrocytochrome c concentrations.

To fit the results, Eq. (2.5) was extended to give:

$$v = \frac{\alpha_1 + \alpha_2 [C_t]}{1 + \beta_1 [C_t] + \beta_2 [C_t]^2} [C_r] [OXIDASE] \tag{2.6}$$

or

$$v = k' [C_r] [oxidase] \ , \tag{2.7}$$

where α_1, α_2, β_1, β_2 are constants which vary in nature depending on the proposed mechanism. The quadratic term indicated that under certain conditions, two molecules of cytochrome c may be bound (however, see Sect. 2.3.2). Again there was no unique mechanistic basis for the interpretation of Eq. (2.6) and four possible models were presented (Errede and Kamen 1978) which were extensions of the original one-site models of Minnaert (1961).

Although each of these four possible models equally well describe the steady-state kinetic data, the dependent-site model seems the most compatible with other lines of evidence and is the one described most fully here. This model (Fig. 2.8) involves reaction at higher substrate concentrations of a second cytochrome c with an oxidase molecule already productively associated with a cytochrome c. As with the Minnaert IV model the non-saturation kinetics are explained by the combined effect of increasing product inhibition at the two sites and decreasing substrate saturation. The difference in binding affinity at the two sites in such that they can be recognised by varying the cytochrome c concentration.

A number of graphical methods can be used to analyse the results (Errede et al. 1976; Errede and Kamen 1978). We have chosen here the Eadie-Hofstee-Scatchard analysis (Fig. 2.9) to facilitate comparison with the polarographic (Fig. 2.7) and binding studies (Fig. 2.5). Since $v = k' [C_r]$ [oxidase] and turnover number (TN) = v/[oxidase], the plot of k' vs $k' [C_r]$ is

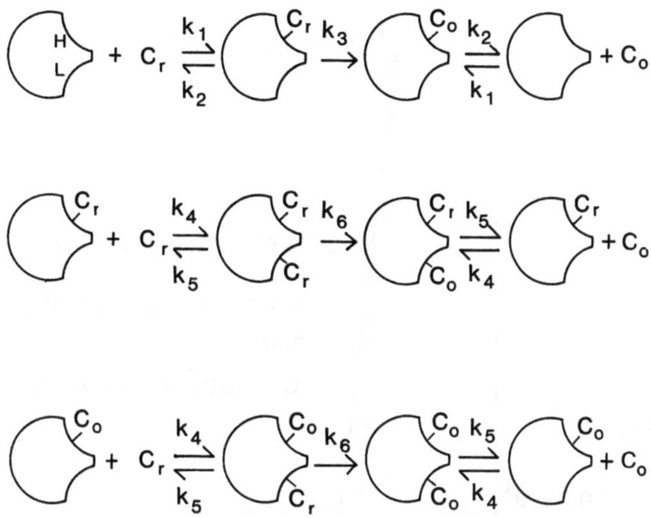

Fig. 2.8 a–c. Kinetic models for cytochrome oxidase: the dependent-site model (After Errede and Kamen 1978). Two active sites of high (*H*) and low (*L*) affinity are represented on the oxidase. Occupancy of the *H*-site by either c_o or c_r is required for binding to the L-site. Assumptions in the derivation of the rate equation: reverse reaction (reduction of c_o) is negligible; reactions of c_r with oxidase: c_o and oxidase: c_r are equivalent; k_{on} (k_1 and k_4) and k_{off} (k_2 and k_5) for binding are independent of the redox state of cytochrome c. The rate equation is then:

$$v = \frac{k_H + K_H k_L [C_t]}{1 + K_H [C_t] + K_H K_L [C_t]^2} \text{[OXIDASE]} [C_r] \qquad (2.27)$$

	Limiting rate of reaction	Affinity
At high affinity site	$k_H = k_1 k_3 / (k_2 + k_3)$	$K_H = k_1 / k_2$
At low affinity site	$k_L = k_4 k_6 / (k_5 + k_6)$	$K_L = k_4 / k_5$

equivalent to a plot of TN/[C_r] vs TN and closely related to the plot of v/S vs v. From the negative slopes of the plot in Fig. 2.9, binding constants were derived which, according to the model, represent productive complex formation at two sites of different affinity (K_H and K_L).

The limiting rates of reaction at the two sites k_H and k_L are complex constants. They are given by the intercepts on the y-axis (Fig. 2.9). For the first site $k_H = k_1 k_3/(k_2 + k_3)$, where k_1 and k_2 are the on/off rate constants for binding and k_3 is the rate constant for electron transfer (Fig. 2.8). Ferguson-Miller et al. (1978b) noted that the observed value for k_H ($2.5 \times 10^8\ M^{-1}\ s^{-1}$) is close to values measured for k_1 (k_{on}) by pre-steady-state methods and this would occur if $k_3 \gg k_2$. The diagram of Fig. 2.8 shows two separate binding sites for cytochrome c but Errede and Kamen (1979) pointed out that the sites referred to need not necessarily be physically separate locations on the oxidase molecule. A mechanism in which binding of the first cytochrome to a single surface reduced the affinity of the second cytochrome for the same surface is compatible with the model (see Sect. 2.8).

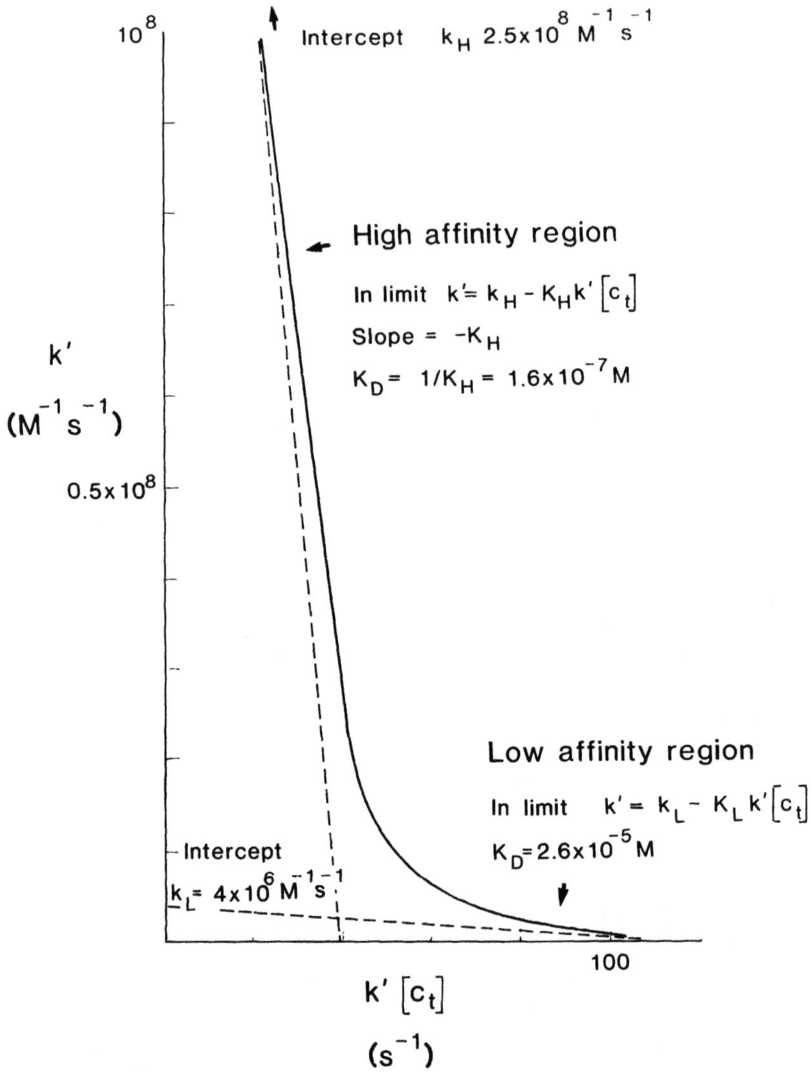

Fig. 2.9. Analysis of the kinetics of cytochrome oxidase according to the dependent-site model (Data from Errede and Kamen 1978). Eadie-Hofstee-Scatchard analysis of the rate equation:

$$v = \frac{k_H + K_H k_L [C_t]}{1 + K_H [C_t] + K_H K_L [C_t]^2} [\text{OXIDASE}] [C_r] \tag{2.27}$$

$$v = k' [\text{OXIDASE}] [C_r] \tag{2.28}$$

The *broken lines* denote the limiting cases defined by Errede et al. (1976). The *solid curve* is defined by the above equation.

In this guise, the dependent-site model resembles the single catalytic site model proposed on the basis of the pre-steady-state kinetic analysis in which the second cytochrome enhances the dissociation of the first, thus increasing turnover. We will find later that this model is also compatible with most of the chemical evidence for the binding of cytochrome c to the oxidase. A similar model has been proposed by Speck et al. (1984) involving a catalytic and a non-catalytic site which interact.*

In the independent site-productive complex model, binding and electron transfer can take place at the second site without occupancy of the first. Thus in this model, the values of the constants k_H, k_L, K_H and K_L, which in the dependent-site model are determined by the occupancy of the oxidase, define the reaction at two independent sites. Because of this, the rate equation contains a different arrangement of these constants and the limiting cases do not give rise to simple determination of parameters from the Eadie-Hofstee plot as they do for the dependent-site model.

The self-exchange model (Fig. 2.10A) involves electron transfer between a second cytochrome c and an oxidised cytochrome c bound to the oxidase. It is related to mechanism V of Minnaert. The limiting-rate constant $k_B = k_1k_3/(k_2+k_3)$ is given by the y-intercept on the Eadie-Hofstee plot but other features of the plot are complex functions of rate constants and binding constants. Errede and Kamen considered this model to be incompatible with other known kinetic data. Thus, if this model is assumed, $k_4/k_2 = 2.6\times 10^5\,M^{-1}$ can be calculated from the experimental data at 4 mM ionic strength. If k_4 is taken as $5\times10^2\,M^{-1}\,s^{-1}$ for this ionic strength (Gupta et al. 1972), k_2 would be $2\times10^{-3}\,s^{-1}$. However, measured values of k_2 in the presence of excess cytochrome c are probably higher than $100\,s^{-1}$ (van Gelder et al. 1975; Wilms et al. 1981a).

Self-exchange could be enhanced by the binding of the first and second cytochromes c to a common surface but probably not by enough to explain the discrepancy between the observed and calculated values of k_2. For example, the self-exchange of cytochromes c bound to phosvitin was enhanced by only two orders of magnitude (Yoshimura et al. 1981). The self-exchange model is also not compatible with the observation that the rate of ferrocytochrome c ox-

◄———

High affinity region at limit (low $[C_t]$)

$$k_H \gg K_H k_L [C_t] \quad \text{and}$$
$$(1+K_H[C_t]) \gg K_H K_L [C_t]^2$$
$$k' = k_H - K_H k'[C_t]$$

Low affinity region at limit (high $[C_t]$)

$$K_H k_L [C_t] \gg k_H \quad \text{and}$$
$$K_H[C_t] + K_H K_L [C_t]^2 \gg 1$$
$$k' = k_L - K_L k'[C_t]$$

Note that the *broken lines* are for the two limiting cases only. Due to intermixing of the two phases the equation gives rise to a *curved line* in the region of overlap where limiting assumptions do not apply

---

* See appendix note 3

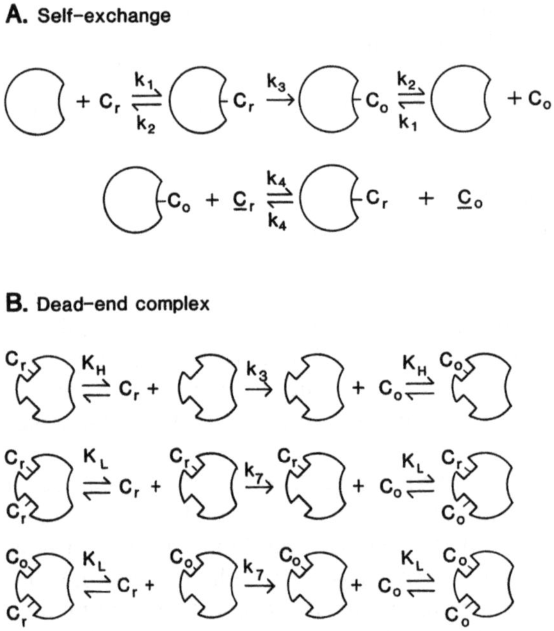

Fig. 2.10 A, B. Kinetic models for cytochrome c oxidase: the self-exchange and dead-end complex models. **A** Self-exchange. One active site of high affinity is represented on the oxidase by a *concave arc* and if this site is occupied by c_o it can receive electrons by a self-exchange mechanism. The rate equation is

$$v = \frac{k_B + K_B k_3 k_e [C_t]}{1 + [K_B + k_e][C_t] + K_B k_e [C_t]^2} \text{ [OXIDASE] } (C_r) \tag{2.29}$$

where the limiting rate constant

$k_B = k_1 k_3/(k_2 + k_3)$
$K_B = k_1/k_2$ and
$k_e = k_4/(k_2 + k_3)$.

B Dead-end non-productive complex. In this representation the oxidase contains three sites, two being non-productive bindings sites (*square sections*) having different affinity for cytochrome c while the third (*concave arc*) is involved in electron transfer via transient collision. The rate equation is:

$$v = \frac{k_3 + K_H k_7 [C_t]}{1 + K_H [C_t] + K_H K_L [C_t]^2} \text{ [OXIDASE] } [C_r] \tag{2.30}$$

idation is unaffected by the presence of the redox-inactive porphyrin cytochrome c in the high affinity site (Veerman et al. 1980).

The dead-end, non-productive complex model is represented in Fig. 2.10B. Although the rate equation is identical to that for the dependent-site model and the Eadie-Hofstee plot yields binding constants directly, their significance is totally different. In the dead-end model, binding of cytochrome c in either redox state to give a complex with the oxidase is inhibitory while electron transfer takes place via transient collision. The binding constants K_H and K_L simply act to limit the amount of free oxidase available for reaction while k_3 and k_7 are the true electron transfer rates for free oxidase and for oxidase with one non-productive cytochrome attached respectively. Therefore, the crucial distinction between this model and the productive complex models is whether the complex of cytochrome c with the oxidase is kinetically competent. Studies with the polarographic assay considered above suggest that this is so and thus favour a productive-complex model.

In conclusion, only the dependent-site model in which the properties of a single binding surface are influenced by its occupancy by a first cytochrome c molecule seems fully compatible with other lines of evidence. This model is more fully discussed in Section 2.8.

Monomers and Dimers and the Dependent-Site Model. In an oxidase monomer with a single catalytic site, the dependent-site model involves enhanced dissociation of the bound cytochrome c by an incoming molecule. However, a second possibility is that two separate sites in the dimer show dependent behaviour due to negative cooperativity. Both interpretations of the model would be consistent with the observed biphasic kinetics.

If the monomers showed monophasic kinetic character, the second possibility would be favoured. However, the results of this test are controversial. The monomeric forms isolated from rat liver (Thompson and Ferguson-Miller 1983), shark muscle (Georgevich et al. 1983) and from bovine heart after alkaline treatment (Georgevich et al. 1983) or after molecular exclusion in lauryl maltoside (Hakvoort et al. 1985) retain the two kinetic phases of cytochrome c oxidation suggesting that this is an intrinsic property of the monomer and not due to dimer formation. On the other hand, Nalecz et al. (1983) and Bolli et al. (1985a) found that dimers appeared in parallel with biphasic kinetics as the salt or enzyme concentration was raised. Indeed, these authors regarded the increase in activity of cytochrome oxidase up to an optimal ionic strength as due to enhanced dimerisation rather than increased product dissociation (Bolli et al. 1985b). They claimed that observations of biphasic kinetics with monomeric preparations were due to dimerisation in the assay mixture. This possibility is difficult to discount and is consistent with the shift to the dimer caused by cross-linking to cytochrome c or by turnover during cytochrome c oxidation (Hakvoort et al. 1985). Final judgement must therefore be reserved on this question. In the subsequent discussion and in Fig. 2.19 we have favoured intrinsic cytochrome c interactions at each monomer catalytic site.

Ionic Strength and Salt Dependence. The ionic strength dependence of the reaction kinetics is particularly interesting and potentially important for our understanding of cytochrome oxidation in situ. For example, the high affinity arm of the Eadie-Hofstee plot obtained from the polarographic assay (Fig. 2.7) is almost completely abolished by the presence of 25 mM phosphate or 3 mM

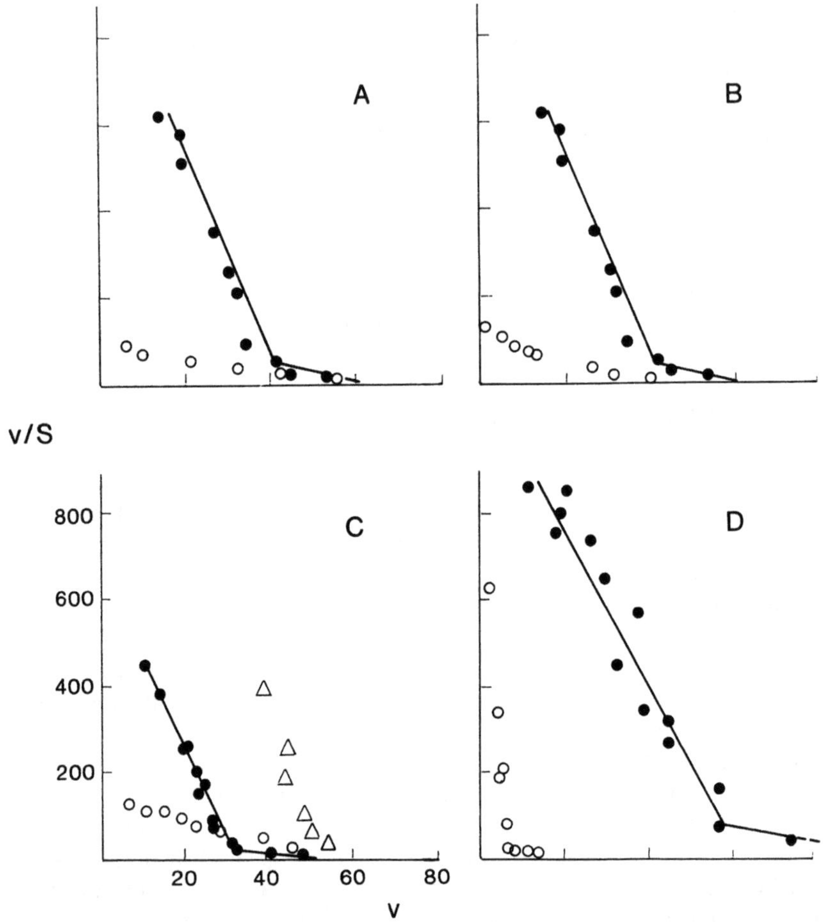

Fig. 2.11 A–D. Factors affecting the kinetics of cytochrome c oxidase (Data from Ferguson-Miller et al. 1976, 1978 b). Activities were measured polarographically using beef heart Keilin Hartree particles in the stated buffer which also contained 250 mM sucrose, 7 mM sodium ascorbate and 0.7 mM TMPD. **A** ○ 25 mM Tris phosphate, pH 7.8; ● 25 mM Tris cacodylate, pH 7.8; **B** ○ 3 mM ATP in 25 mM Tris cacodylate, pH 7.8; ● 25 mM Tris cacodylate, pH 7.8; **C** ● Horse cyt. c; ○ Euglena cyt. c; △ Yeast cyt. c. (all in 25 mM Tris cacodylate pH 7.8). **D** ● Cow cyt. c; ○ rhesus monkey cyt. c (in 25 mM Tris acetate pH 7.8). In all cases v is in nmol O_2 min^{-1}; *S* is the cytochrome concentration in units of 10^{-6} M; numerical values of the axes of C apply to all four diagrams

ATP (Fig. 2.11)*. The presence of phosphate resulted in a marked fall in the amount of cytochrome c found bound to the oxidase after non-equilibrium molecular exclusion chromatography (Ferguson-Miller et al. 1976). Since these concentrations are within the physiological range that might be expected to obtain in the natural environment of the reaction, these findings cast severe doubt on the physiological relevance of the biphasic kinetics (see Sect. 2.8).

* See appendix note 4

Comparative Kinetic Studies. Extension of the kinetic analysis to include studies, both on different eukaryotic cytochromes c and different oxidases, has yielded interesting comparative information. A comparison of the cytochromes c from *Euglena gracilis,* bakers' yeast and horse heart is shown in Fig. 2.11. Ferguson-Miller et al. (1976) interpreted this pattern as due to an oxidase with two sites per molecule, the sites being of different affinity for horse cytochrome c, of equivalent and high affinity for yeast cytochrome c and of equivalent but low affinity for *Euglena* cytochrome c. This interpretation was supported by direct binding studies on the purified oxidase which found 0.94 mol, 1.92 mol and 0.63 mol respectively of the three cytochromes per mole oxidase after non-equilibrium molecular exclusion chromatography.

Monophasic kinetics were also observed for both horse and yeast cytochrome c with yeast oxidase and a probable stoichiometry of 2 mol/mol observed in direct binding studies again suggests a two-site oxidase but with equal binding affinity at the two sites for both cytochromes (Dethmers et al. 1979).

A hope, implicit in this work and in the wider ranging studies of Errede and Kamen (1978) is that specific structural differences between cytochromes might be correlated with kinetic features. Thus, Ferguson-Miller et al. (1976) noted that *Euglena* cytochrome c is unusual in lacking a lysine at position 27 and suggested this as a cause of the low affinity with the beef oxidase. The observation that *Crithidia fasciculata* cytochrome c resembles that of *Euglena* in kinetic behaviour with beef oxidase while also lacking lysine 27 tends to support this hypothesis (Errede and Kamen 1978). However, the differences in structure are so large amongst these cytochromes that it is difficult to justify emphasis on one particular change. Comparisons restricted to the number and distribution of the surface lysines may be an oversimplification. Thus *Neurospora* cytochrome c is indistinguishable from yeast cytochrome c on this basis yet has very different kinetic properties (Dethmers et al. 1979). Other factors may influence binding such as hydrophobicity of surface areas, solvation and ion-binding properties, and flexibility of the bound molecule (Sect. 2.7). A similar comparative analysis of the related bacterial cytochromes c_2 is discussed in Chapter 3.

Cytochromes c which are much more closely related to each other yet show differences in activity offer a better chance of identifying structural bases for such differences. Thus there are only ten differences in the sequences of cow and rhesus monkey cytochromes c yet the latter (and human cytochrome) is very poorly reactive with bovine oxidase. This is not due to a trivial circumstance such as the presence of inhibitory contaminants in the preparation of the monkey cytochrome since it shows good activity with its own oxidase (Ferguson-Miller et al. 1978 b) and it is not due to defective binding or a change in redox potential (Osheroff et al. 1983). A correlation was detected between reactivity with the bovine oxidase and pK_a for the loss of the 695-nm band and it was suggested that the higher pK_a observed for primate cytochrome c implied a tighter heme crevice and a lower tendency to enter a proposed transition state on the oxidase (Osheroff et al. 1979). However, this explanation was discredited by pre-steady-state kinetic analysis which showed that human cytochrome c was more slowly reduced by TMPD than non-primate cytochromes when bound to the oxidase. Thus TMPD reduction may be rate-determining in the polarographic assay and differences in steady-state rates between different cytochromes may simply reflect subtle, but physiologically uninteresting, differences in its access to bound cytochrome (Osheroff et al. 1983).

We finish this section on a cautionary note. In 1973, Smith et al. claimed that there were "no reproducible differences in the maximal kinetics with different cytochromes c" and the observed diversity in amino acid sequences was seen as a result of the fixation of neutral mutations. Discussion now is in terms of evolutionary selection to perform subtly different physiological roles. For example, yeast cytochrome c may have evolved a weaker affinity for cytochrome oxidase so that it remains equally accessible to the peroxidase that exists in the intermembrane space (Dethmers et al. 1979).

Why are the recent results different from those obtained in 1973? The answer is that they were collected under conditions of very low ionic strength and in the absence of phosphate or nucleotides. Yet, at higher ionic strengths, the differences between cow and monkey cytochromes disappear and in the presence of a very low concentration of ATP the high affinity kinetic region is abolished. A lengthier discussion of the physiological significance of the biphasic kinetics is to be found in Section 2.8.

2.2.3.2 CoQ Cytochrome c Oxidoreductase and Cytochrome c_1

Steady-State Kinetics. CoQ cyt. c oxidoreductase (complex III) catalyses the single electron reduction of cytochrome c via its terminal redox centre, cytochrome c_1. The steady-state kinetics of this process have been studied by using KHP blocked with cyanide and purified systems in which either NADH (via complex I), succinate (via complex II) or reduced coenzyme Q analogues provide the supply of electrons to complex III.

The kinetic behaviour of these systems is not so well characterised as that of cytochrome oxidase. The progress of reduction of cytochrome c by succinate in KHP shows both a zero-order and a first-order region (L. Smith et al. 1974) although at low enough $[C_0]$, the kinetics are purely first-order (Ahmed et al. 1978). Like cytochrome oxidase the first-order rate constant diminishes with rising $[C_t]$ and kinetic mechanisms similar to those proposed for cytochrome oxidase [such as product inhibition (Minnaert type IV) and dead-end complex formation] are consistent with this behaviour.

Analysis of the reaction in terms of a Minnaert type IV mechanism is given in Fig. 2.12. The second-order rate constant of $3.4\times10^7\,M^{-1}\,s^{-1}$ in 0.1 M phosphate which was obtained (Ahmed et al. 1978) is similar to the value of k_1 derived from pre-steady-state analysis (Table 2.5) suggesting that under these conditions k_2 is small compared to k_3. However, it should be noted that

$$C_{1,r} + C_o \underset{k_2}{\overset{k_1}{\rightleftharpoons}} C_{1,r}{:}C_o \overset{k_3}{\longrightarrow} C_{1,o}{:}C_r \underset{k_5}{\overset{k_4}{\rightleftharpoons}} C_{1,o} + C_r$$

If $k_1/k_2 = k_5/k_4$ then $k_{obs} = \dfrac{k_1 k_3/(k_2 + k_3)}{1 + k_1/k_2\,[C_r + C_o]}$

Thus plots of k_{obs} $(= v/[C_o])$ vs $k_{obs}[C_o]$ yield the limiting rate constant k_v at $[C_o] = 0$.

$k_v = 3.4 \times 10^7\ M^{-1} s^{-1}$ in 0.1 M phosphate pH 7 (Ahmed et al., 1978)

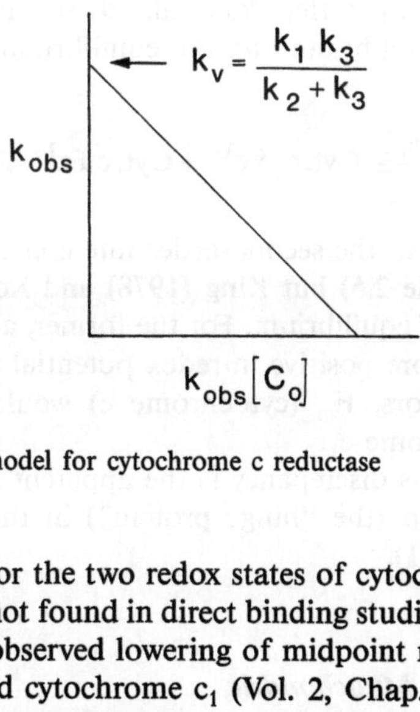

Fig. 2.12. Kinetic model for cytochrome c reductase

equal affinities for the two redox states of cytochrome c – a prerequisite for this model – is not found in direct binding studies (Sect. 2.2.2) and is not consistent with the observed lowering of midpoint redox potential on binding of cytochrome c and cytochrome c_1 (Vol. 2, Chap. 7).

The ionic strength dependence of the reaction is somewhat different from that of the oxidase, the rate being independent of ionic strength up to $I = 0.1$ M and then diminishing in a trend consistent with simple shielding of complementary charge interactions (Ahmed et al. 1978). The ionic strength independence of rate at low ionic strength may arise if k_2 were small compared to k_3 and only became comparable above $I = 0.1$ M (Ahmed et al. 1978). Speck and Margoliash (1984) described an alternative mechanism to account for the unusual ionic strength dependence. They proposed that non-productive substrate binding at low ionic strength decreases the affinity of the catalytic site for cytochrome c. The result is that K_M shows a smaller variation with ionic strength than expected for a single-site system.

Table 2.5. Pre-steady-state kinetic analysis of the reaction between cytochrome c_1 and c

Reductant	k_1 ($M^{-1} s^{-1}$)	k_2 ($M^{-1} s^{-1}$)	K_{eq}	Conditions	References
Cyt.c_1 Fe^{2+}	1.7×10^7	5×10^6	3.4	0.05 M Phosphate pH 7.4, 23 °C	King 1978
	2.4×10^5	3.3×10^5	0.73	0.5 M Phosphate pH 7.4, 10 °C	Konig et al. 1981
Succinate cyt.c reductase	2.8×10^6			0.05 M Phosphate pH 7.4, 10 °C	Yu et al. 1973
	3.4×10^7			0.1 M Phosphate pH 7, 25 °C	Ahmed et al. 1978

Pre-Steady-State Kinetics. Purified cytochrome c_1 reduces cytochrome c in a fast reaction which can be monitored spectrophotometrically because the α-peak maxima of the two proteins differ (Yu et al. 1973). This allows a study of the kinetic approach to equilibrium and the equilibrium position for the reaction:

$$\text{Cyt.}\,c_1\,Fe^{2+} + \text{Cyt.}\,c\,Fe^{3+} \underset{k_2}{\overset{k_1}{\rightleftharpoons}} \text{Cyt.}\,c_1\,Fe^{3+} + \text{Cyt.}\,c\,Fe^{2+}\ . \qquad (2.8)$$

The k_1 is found to be similar to the second-order rate constant derived from the steady-state analysis (Table 2.5) but King (1978) and Konig et al. (1981) disagreed as to the position of equilibrium. For the former, a K_{eq} of 3.4 would place cytochrome c 32 mV more positive in redox potential than cytochrome c_1 while, for the latter authors, E_m (cytochrome c) would be 8 mV more negative than that of cytochrome c_1.

One possible origin for this discrepancy is the apparent absence of a low molecular weight contaminant (the "hinge protein") in the preparation of Konig et al. (1981) (Sect. 2.2.1).

2.2.3.3 Electron Transport in Mitochondria

Models of electron transport based on reactions of purified components must be consistent with kinetic parameters derived from intact respiring mitochondria. In the ADP-stimulated state 3, the steady-state rate of electron transport is $5\ s^{-1}$ and this falls to $0.5\ s^{-1}$ in state 4 (Chance 1967).

Apparent rate constants or turnover numbers for individual oxidation reactions have been obtained using oxygen pulse rapid flow techniques and laser flash dissociation of CO from cytochrome oxidase (Chance et al. 1964). Problems associated with these measurements include partial resolution of the components under study and also the fact that the rate obtained is the difference between the oxidation rate and a reduction rate. Because of this, oxidation rates may be higher than the measured values. Chance (1974) has computed the following pattern of rate constants for succinate oxidation:

$$\text{succinate} \underset{0.01\,s^{-1}}{\overset{33\,s^{-1}}{\rightleftharpoons}} c_1 \underset{990\,s^{-1}}{\overset{1000\,s^{-1}}{\rightleftharpoons}} c \underset{460\,s^{-1}}{\overset{500\,s^{-1}}{\rightleftharpoons}} a \underset{400\,s^{-1}}{\overset{1400\,s^{-1}}{\rightleftharpoons}} a_3 \underset{0.001\,s^{-1}}{\overset{210\,s^{-1}}{\rightleftharpoons}} O_2 \,. \tag{2.9}$$

Thus it is clear that the intrinsic ability of cytochrome c to donate or accept electrons is not a limiting factor in steady-state electron transport. The value of $500\,s^{-1}$, computed for the oxidation of cytochrome c by cytochrome a is in reasonable agreement with the maximum turnover figures obtained for the low affinity reaction (Wilms et al. 1981b).

2.3 Reaction of Cytochrome c with Intermembrane Redox Systems

We group together in this section four interactions involving cytochrome c which are not directly part of the inner membrane electron transport system (Fig. 2.1) yet are believed to be of physiological significance. The four involve cytochrome b_5, sulfite oxidase, cytochrome b_2 and cytochrome c peroxidase. They are an interesting group because, although there is uncertainty as to their quantitative contribution to the flux of electrons through cytochrome c, they are the best defined systems for studying the nature of an electron transport complex of two proteins. As we shall see, in the case of the cytochrome c: cytochrome c peroxidase complex, detailed structural models have been proposed describing the route of an electron passing between the two redox centres.

From an evolutionary point of view, cytochrome b_5, sulfite oxidase and cytochrome b_2 form a distinct structural family that is characterised by possession of a common cytochrome b domain (Guiard and Lederer 1979b). Cytochrome c peroxidase has a distinctly different structure.

2.3.1 The Cytochrome b_5 Family

Cytochrome b_5. Cytochrome b_5 is an intermediate carrier in electron transfer from NADH to fatty acyl CoA desaturase on the microsomal membrane (Strittmatter et al. 1974). It is anchored to the cytoplasmic surface of the membrane by means of a hydrophobic C-terminal region which can be proteolytically removed to yield a soluble domain carrying the heme group. In the erythrocyte, a distinct, but closely related, gene product lacking the hydrophobic anchor region is a component of the soluble methemoglobin reductase system (Huttquist and Passon, 1971; Kimura et al. 1984; Abe et al. 1985). After initial controversy, the mitochondrial outer membrane has been shown to contain a third distinct cytochrome b_5 (Ito 1980; Lederer et al. 1983) which may be involved in rotenone-insensitive oxidation of cytoplasmic NADH by transfer of electrons to cytochrome oxidase via cytochrome c in the intermembrane space (Matlib and O'Brien 1976; Bernardi and Azzone 1981; Sect. 2.8). If this is so, the topological arrangement of the cytochrome b_5 and its reductase on the membrane must differ from that of their microsomal relatives.

Most work on cytochrome c interactions with cytochrome b_5 has been performed with the microsomal protein for which the X-ray crystallographic structure of the soluble domain has been determined (Argos and Mathews 1975; Mathews 1984). However, conclusions based on the microsomal cytochrome are probably broadly applicable to the mitochondrial relative because the two show 58% sequence homology and conservation of all the carboxyl groups implicated in the binding surface for cytochrome c (Sect. 2.5).

These carboxyl groups are situated around the mouth of a cleft in the protein containing the heme group and allow electrostatic association with the positively charged cytochrome c molecule. The association of the two molecules is therefore weakened by raising the ionic strength. At $I = 0.01$ M, a K_D of 10^{-5} M was obtained by monitoring the small absorption change at 416 nm associated with complex formation (Mauk et al. 1982).

The kinetics of reduction of cytochrome c by the particulate microsomal NADH cytochrome c oxidoreductase system resembled those observed for the mitochondrial CoQ cytochrome c oxidoreductase (Sect. 2.2.3) in that cytochrome c only became rate-limiting at low concentrations and the rate constants derived from the first-order progress curves decreased with rising $[C_o]$ (Ng et al. 1977). Under these circumstances, simple Michaelis-Menten analyses could not be applied and the authors used the limiting value of v/S as v tends to zero on an Eadie-Hofstee plot as an empirical parameter for comparing the effects of lysine modification of cytochrome c and of ionic strength (Sect. 2.4). If this intercept (which in Michaelis-Menten analysis is V/K_M) was divided by $[b_5]$, a second-order rate constant of 3×10^7 M^{-1} s^{-1} was obtained. This closely resembles the k_{on} of 4.6×10^7 M^{-1} s^{-1} obtained in stopped-flow measurements by Strittmatter (1964).

Sulfite Oxidase. Sulfite oxidase catalyses the terminal detoxification step in the biological degradation of sulfur-containing amino acids in animal tissues. It is a water-soluble homodimer of M_r 120 K found in the intermembrane space (Cohen et al. 1972; Johnson and Rajagopalan 1980) and transfers electrons from sulfite to cytochrome c (Macleod et al. 1961; Cohen and Fridovich 1971):

$$SO_3^{2-} + H_2O + 2c_o \rightarrow SO_4^{2-} + 2H^+ + 2c_r \ . \quad (2.10)$$

Intact mitochondria exhibit cyanide-sensitive oxidation of sulfite characterised by $SO_3^{2-}/O = 1$ and $P/O = 1$ (Oshino and Chance 1975) with electron flow to cytochrome oxidase.

On the basis of the susceptibility to "nicking" by trypsin (Ito 1977) the monomer is believed to contain two domains, one with an attached molybdenum cofactor and the other, of M_r 10 K, with protoheme IX. The visible spectrum of the latter domain is identical to that of cytochrome b_5 and a homologous relationship suggested by Kessler and Rajagopalan (1974) has been demonstrated by amino acid sequence determination (Guiard and Lederer 1979a) which found 26 identities and 34 single base changes over an 89 residue comparison.

On the basis of equilibrium molecular exclusion chromatography, sulfite oxidase contains two non-interacting binding sites of equal affinity for cytochrome c (Speck et al. 1984) which correspond to the protoheme IX domains (Johnson and Rajagopalan 1977). The ionic strength dependence of K_D is consistent with electrostatic stabilisation, rising from 10^{-7} M at $I = 0.01$ M to 1.5×10^{-5} M at $I = 0.05$ M.

The steady-state reaction kinetics are also monophasic and the apparent K_M value is markedly dependent on ionic strength (Speck et al. 1981). However, unlike the kinetics of the cytochrome oxidase reaction, K_M is consistently lower than K_D. This probably reflects a rate-limiting electron-transfer step prior to cytochrome c reduction which leads to saturation of the kinetics at concentrations of cytochrome c well below those required to saturate binding.

Flavocytochrome b_2. Flavocytochrome b_2 is a soluble L-lactate: cytochrome c oxidoreductase found in aerobically-grown yeast mitochondria that catalyses the oxidation of L-lactate to pyruvate with electrons passing to cytochrome c oxidase via cytochrome c. It contains four FMN groups and four protoheme IX groups in a tetramer of M_r 240 K (Ohnishi et al. 1966, Guiard and Lederer 1979b). It was originally isolated in a crystalline form from bakers' yeast (Appleby and Morton 1959) but subsequent work has shown that this preparation has suffered proteolytic damage and alteration in kinetic properties (Jacq and Lederer 1972, 1974). The presence of PMSF blocks this endogenous protease activity and yields a tetrameric enzyme with subunits of M_r 58 K. Cytochrome b_2 from *Hansenula anomala* (Baudras 1971) remained intact during purification and its stability has made it the protein of choice in recent investigations.

Prolonged proteolysis of flavocytochrome b_2 yields a limit peptide containing protoheme IX, the b_2 core (Labeyrie et al. 1966, 1967; Gervais et al. 1983). This fragment retains the heme spectrum and redox potential of the holoenzyme and a number of spectroscopic methods suggest a similarity to cytochrome b_5 (Labeyrie et al. 1966; Risler and Groudinsky 1973; Keller et al. 1973) which has been confirmed by sequence (Guiard et al. 1974) and X-ray analysis (Xia et al. 1986). The degree of sequence identity is 27%, similar to the figure obtained by comparing cytochrome b_5 with the heme fragment of sulfite oxidase (Guiard and Lederer 1979b).

In the overall lactate: cytochrome c oxidoreductase activity, the rate-limiting step is reduction of the prosthetic groups by lactate because these centres remain oxidised in the steady state (Morton and Sturtevant 1964). Pre-steady-state studies have revealed that an initial two electron reduction to form the flavohydroquinone is followed rapidly by a single electron transfer to protoheme IX (Capeillere-Blandin et al. 1975). Cytochrome c reduction occurs at the protoheme IX site only, this site being rapidly replenished with electrons from the flavin. Because of the rapid internal equilibration, single electron turnover kinetics of reduction of cytochrome c by fully reduced cytochrome b_2 follow strictly first-order progress curves (Capeillere-Blandin 1982). The pseudo-first-order rate constants exhibit saturation with rising $[b_2]$ indicating a reversible binding step which is strongly dependent on ionic strength and has an apparent K_D of 3×10^{-7} M at I = 0.02 M. The limiting second-order rate constant at I = 0 was found to be 2×10^{10} M^{-1} s^{-1}, comparable to figures obtained for other redox reactions of cytochrome c.

No thorough analysis of the binding of cytochrome c is available. Both Baudras et al. (1972) and Yoshimura et al. (1977) considered that the binding of cytochrome c to the bakers' yeast enzyme is anti-cooperative with only one molecule bound with high affinity. However, cytochrome c concentrates in crystals of this enzyme by diffusion up to a value of four cytochrome c per molecule and Prats (1977) found four sites for cytochrome c binding by sucrose density gradient centrifugation with the enzyme prepared from *Hansenula anomala.*

The dissociation constant, at least for the first molecule of cytochrome c, is probably less than 10^{-6} M judging from the absence of trailing effects in the ultracentrifuge, but direct measurement of K_D has been made only for the binding of the dimeric, fluorescent Zn-porphyrin cytochrome c. Thomas et al. (1983) found one Zn-porphyrin cytochrome c dimer bound per cytochrome b_2 monomer with $K_D = 10^{-7}$ M at I = 0.03 M and with no apparent anti-cooperative effects. A similar value and ionic strength dependence of K_D was obtained for binding to the proteolytic fragment of cytochrome b_2 containing only the FMN group (Thomas et al. 1983) leading to the suggestion that it is the FMN domain rather than the heme domain which specifically interacts with the cytochrome c. This is not incompatible with the kinetic evidence that protoheme rather than flavin is the electron donor to cytochrome c if the binding to the flavin domain correctly orientates the cytochrome c for heme : heme electron transfer. Such a binding mode would, however, differ from those proposed for other complexes formed by cytochrome c where the binding site and the electron transfer site are in the same surface region (see also Sect. 2.5.2 and Chap. 3, Sect. 3.6.2).

2.3.2 Cytochrome c Peroxidase

Cytochrome c peroxidase (CCP) is a water-soluble, protoheme IX-containing enzyme found in the mitochondrial intermembrane space of aerobically-grown yeast (Yonetani 1976) where it catalyses the reduction of hydrogen peroxide by cytochrome c, presumably as a detoxification mechanism:

$$H_2O_2 + 2c_r + 2H^+ \rightarrow 2H_2O + 2c_o \ . \qquad (2.11)$$

Hydrogen peroxide reacts with the free enzyme to produce "compound I" (Fig. 2.13) in which one oxidation equivalent is in the form of a relatively stable free

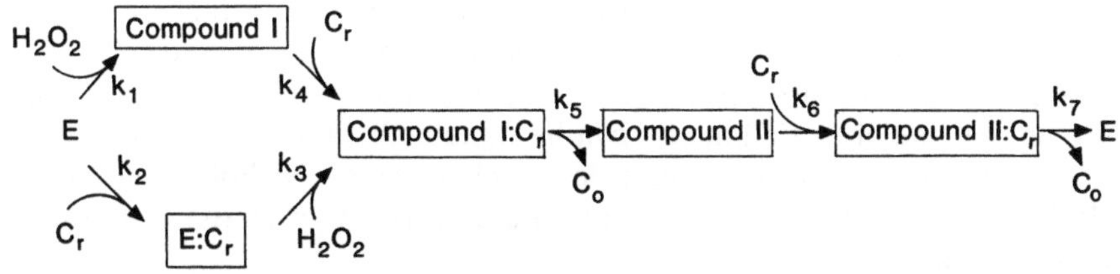

Fig. 2.13. Random order mechanism for cytochrome c peroxidase. Compounds *I* and *II* contain two and one oxidising equivalents respectively above the native ferric enzyme. k_5 and k_7 each contain electron transfer and dissociation components which cannot be distinguished by steady-state kinetics

radical of uncertain identity and the other is present as Fe IV heme (Yonetani et al. 1966; Hoffman et al. 1981; Poulos and Finzel 1984). The relatively small size (34 K) and its soluble nature, have made CCP a popular subject of study and it is one of the best characterised redox enzymes. The crystal structure has been determined (Poulos et al. 1980; Poulos and Finzel 1984) and is described in Section 2.5 in connection with the postulated structure of the peroxidase: cytochrome c complex.

Cytochrome c forms a tight complex with CCP which is detectable by ultracentrifugation (Mochan and Nicholls 1971; Dowe et al. 1984), molecular exclusion chromatography (Mochan and Nicholls 1971; Kang et al. 1978), NMR spectroscopy (Gupta and Yonetani 1973; Boswell et al. 1980) and by small absorption changes in the visible spectrum (Erman and Vitello 1980). The latter two effects titrate with a stoichiometry of one cytochrome c to one CCP and a K_D value (2×10^{-7} M at I = 0.01 M) which rises with ionic strength consistent with electrostatic stabilisation of the complex.

Ferrocytochrome oxidation follows first-order kinetics (Beetlestone 1960; Nicholls 1964a) like that for cytochrome oxidase (Sect. 2.2.3). By analogy, this is interpreted as due to product inhibition with $K_I = K_M$. Although monophasic kinetics were obtained by Yonetani and Ray (1966), later workers found curved Eadie-Hofstee plots (Kang et al. 1977; H.T. Smith and Millet 1980). Again, by analogy with cytochrome oxidase, these biphasic kinetics were thought to reflect two functional binding sites of different affinities and this hypothesis was supported by direct binding studies which found more than one binding site under conditions in which biphasic kinetics were observed. Kang et al. (1977) came to the conclusion that the two cytochromes c bound at separate sites, one associated with the free radical and the other with the Fe IV heme present in the peroxidase after reaction with hydrogen peroxide. However, several considerations suggest that this two-site model should be approached with caution.

Firstly, most studies find only one binding site for cytochrome c. Secondly, neither kinetics nor binding is biphasic in nature for either horse or yeast cytochrome c under conditions of optimal ionic strength (Kang et al. 1978). Thirdly, the region of cytochrome c implicated in binding to peroxidase is the same front-face surface found for interactions with other proteins (Sect. 2.4). In a two-site model the separate sites might have been expected to have different interactions with cytochrome c. Finally, Kang and Erman (1982) have shown that the biphasic kinetics can be explained without recourse to a two-site model. Using the random-order mechanism of Nicholls and Mochan (1971) (Fig. 2.13) they derived a rate equation which, at constant (H_2O_2) can be written as:

$$v = \frac{\alpha_1 + \alpha_2[C_r]}{\beta_o + \beta_1[C_r] + \beta_2[C_r]^2}[C_r][CCP] \ , \tag{2.12}$$

where α_1, α_2 and β_2 are collections of rate constants. Apart from the fact that $[C_t]$ does not appear because the presence of $[C_o]$ at the start of the reaction was not considered, the form of this equation is identical to that derived by Errede et al. (1976) for cytochrome oxidase [Eq. (2.6)]. Also, because it contains terms in $[C_r]$ to the second power, this equation can give rise to curved Eadie-Hofstee plots.

The kinetic model of Kang and Erman (Fig. 2.13) allows a good fit of the experimental results without the need to postulate the existence of two sites to explain the biphasic kinetics. The best fit incorporates a k_1 of 9.6×10^7 $M^{-1} s^{-1}$ and a k_3 of $4 \times 10^7 M^{-1} s^{-1}$ both in reasonable agreement with stopped-flow experiments (Loo and Erman 1975; Kang and Erman, unpublished work). The k_2 adopts a best fit value of $7.9 \times 10^{10} M^{-1} s^{-1}$ which is close to the diffusion limited rates for spheres charged to the same degree as peroxidase and cytochrome c (Eigen and Hammes 1963). In combination with a best-fit k_{-2} value of $4.6 \times 10^4 M^{-1} s^{-1}$ this yields a K_D for cytochrome c of 6×10^{-7} M similar to that measured directly (Kang et al. 1977). If the product dissociation rate resembles that of substrate then the rates defining the maximum turnover number of $380 s^{-1}$ under these conditions would appear to be electron transfer within the two types of complex.

The ionic strength dependence of the reaction is complex and it is not known whether the model and the values obtained by Kang and Erman for I = 0.01 M, pH 7 and horse cytochrome c are generally applicable. The reactions with horse and yeast cytochrome c are optimal at I = 0.05 and 0.2 M respectively and the values of apparent K_M and K_D determined under these conditions (Kang et al. 1977) are shown in Table 2.6. In both cases K_D resembles apparent K_M at the optimal ionic strength while K_D becomes larger as ionic strength increases and K_M remains relatively unchanged (Kang et al. 1978; H.T. Smith and Millet 1980). This unusual behaviour of K_M indicates that it cannot correspond in a direct way to the binding interaction (unlike the case of cytochrome oxidase) and that the agreement between K_D and K_M at the optimal ionic strengths is coincidental (H.T. Smith and Millet 1980).

The same binding constant (2×10^{-6} M) was obtained for horse and yeast cytochrome c at their kinetically optimal ionic strengths and the same amount of horse and yeast cytochrome c (0.2–0.3 mol/mol) was found bound to the

Table 2.6. Kinetic and binding properties of the cytochrome c peroxidase reaction under optimum catalytic conditions

Cytochrome	Conditions	K_M	K_D	Cyt.bound (mol/mol)	Reference
Yeast	0.2 M Tris-cacodylate pH 6	4×10^{-6} M	2×10^{-6} M	0.3	Kang et al. 1977
Horse	0.05 M Tris-acetate pH 6	4.4×10^{-6} M	2×10^{-6} M	0.2	Kang et al. 1977

Notes: There is very little variation in K_M with rising ionic strength but K_D rises markedly.

peroxidase after non-equilibrium molecular exclusion chromatography under these ionic conditions. This observation is consistent with the suggestion that the cytochrome c peroxidase reaction is a compromise between efficient association of substrate and efficient release of product.

2.4 The Binding Surface of Cytochrome c

2.4.1 Introduction

The aim of the work described below has been to characterise the interaction surface that cytochrome c presents to CoQ cyt. c oxidoreductase, cytochrome oxidase and the intermembrane redox systems with which it interacts. Two important questions can be asked. Are these surfaces identical, overlapping or separate? And do the surfaces enclose the point of electron transfer? The questions are considered below and further discussed in Sects. 2.7 and 2.8.

There are two important general methods used to investigate this problem.

The Kinetic Examination of Singly Modified Cytochrome c Derivatives. A large number of studies have shown that the loss of two to six positive charges by lysine modification results in complete loss of activity with both oxidase and reductase (reviewed in Ferguson-Miller et al. 1978b). The groups of Margoliash and Millet undertook the daunting task of preparing singly modified cytochrome c derivatives to extend the work of Wada and Okunuki (1969) with trinitrophenyl Lys 13 cytochrome c. The success of ion exchange chromatography in the separation of complex mixtures of these isoelectric species is a reflection of the asymmetric charge distribution on the molecule and it is notable that the effect of a lysine modification on the activity of the cytochrome is proportional to its effect on the affinity of the cytochrome for ion exchange material.

The modifications used were trifluoroacetylation (TFA), trifluoromethylphenylcarbamoylation (TFC) and 4-carboxy-2,6-dinitrophenylation (CDNP)

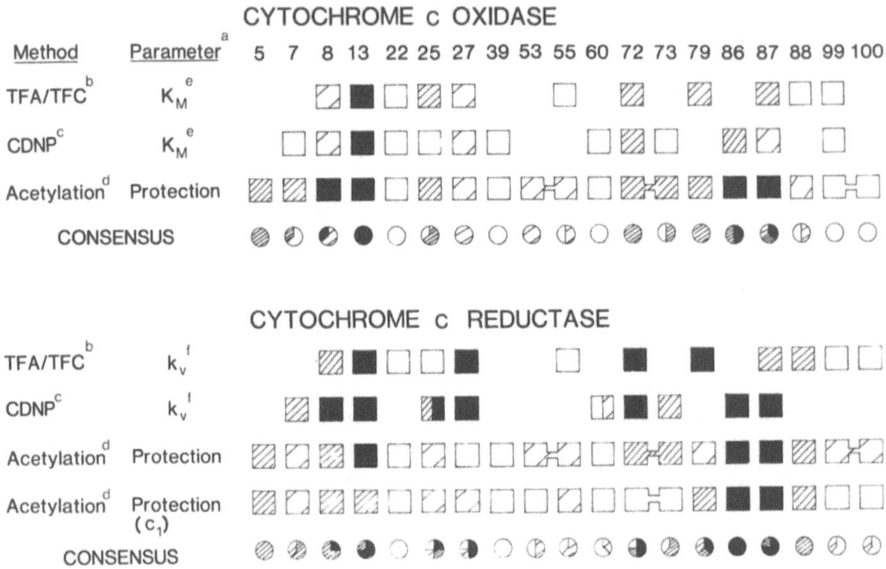

Fig. 2.14. Mapping the interaction surface on cytochrome c for cytochrome oxidase and cytochrome c_1. *a* Results are scored as a "% effect" on the various parameters measured by different authors. The lysine residue which, on modification, gave the greatest loss of activity or which was most protected was taken as the 100% effect point. Untreated cytochrome gave the 0% effect point and intermediate values of K_M, k_v or degree of protection were calculated accordingly [e.g. $(K_{M(mod.)} - K_{M(native)}]/[K_{M(max.)} - K_{M(native)}]$. These "% effect" figures were graded as

% Effect	Grading
−10 to + 9	unaffected
+10 to + 29	low
+30 to + 69	medium
+70 to +100	high

In the protection experiments *joined boxes* indicate lysines present in the same peptide. For these instances the % effect is an average one which may or may not conceal the presence of differentially affected lysines. *b* Trifluoroacetylation (*TFA*) with S-ethylthioltrifluoroacetate or trifluoromethylphenylcarbamoylation (*TFC*) using trifluoromethylphenyl isocyanate (Staudenmeyer et al. 1977; H.T. Smith et al. 1977; Ahmed et al. 1978). *c* 4-Carboxy-2,6-dinitrophenylation using 4 chloro-3,5-dinitrobenzoate (Ferguson-Miller et al. 1978a, b; Osheroff et al. 1980; Speck et al. 1979; Konig et al. 1980). *d* Acetylation using acetic anhydride (Rieder and Bosshard 1978, 1980). *e* Apparent K_M values derived from plots of v/S against v (Eadie-Hofstee). In the case of cytochrome oxidase where such plots are biphasic (Sect. 2.2.3) the "high affinity" region of the plot was used (0.01 − 1.5 μM cytochrome c). *f* k_v is a limiting pseudo-first-order rate constant at zero cytochrome c concentration given by the y-intercept of the v/S against v plot and divided by the enzyme concentration to give a second-order rate constant

(Fig. 2.18). Effects on activity do appear to be due to alterations in charge rather than to the bulk of the reagents because the TFA results were comparable to those for the much bigger TFC species (H. T. Smith et al. 1977). Unchanged visible spectra and only slightly altered NMR spectra and redox potential (Falk et al. 1981) indicate that effects are not due to major conformational changes in the derivatives.

The main problem with the approach appears to be the difficulty in obtaining a complete set of singly modified derivatives. Thus the results shown in Figs. 2.14 and 2.16 are for incomplete sets and the mapping of the interaction area suffers accordingly.

The Study of Lysines Shielded in a Protein Complex. In contrast to the kinetic method, the differential chemical modification approach of Rieder and Bosshard (1978, 1980) can accomplish the mapping of the lysines of cytochrome c in a single experiment. Cytochrome c, in the presence and absence of a redox partner, is modified with trace ^{3}H-labelled lysine-specific reagent and the lysines are then completely modified with cold or ^{14}C-labelled reagent (Fig. 2.18). Shielding factors are calculated from the ratio of the reactivity of the lysine in the free cytochrome to the reactivity of that lysine in the complex. Shielding is always observed to increase in the presence of a redox partner. Differences in shielding cannot be due to conformational changes produced by the

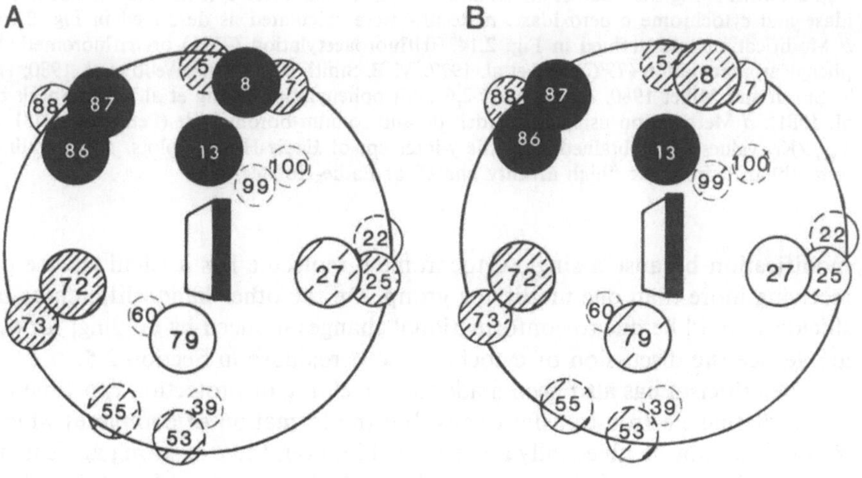

Fig. 2.15 A, B. The binding surfaces on cytochrome c involved in interaction with cytochrome oxidase and CoQ cytochrome c reductase. Results are from Rieder and Bosshard (1980) using acetylation of cytochrome c in the presence of cytochrome oxidase (**A**) or CoQ cyt.c reductase (**B**). The display of cytochrome c is from the front with the heme edge viewed side on. The approximate positions of lysine side chains are indicated by *larger circles of solid lines* at the front of the molecule and *smaller circles of broken lines* at the rear of the molecule. The degree of protection of a lysine in the presence of the oxidase or the reductase is indicated by the *degree of shading* of a circle according to the ranks established in the legend to Fig. 2.14. Thus ● 70–100% shielding; ⊘ 30–69%; ⊘ 10–29%; ○ −10 to +9%

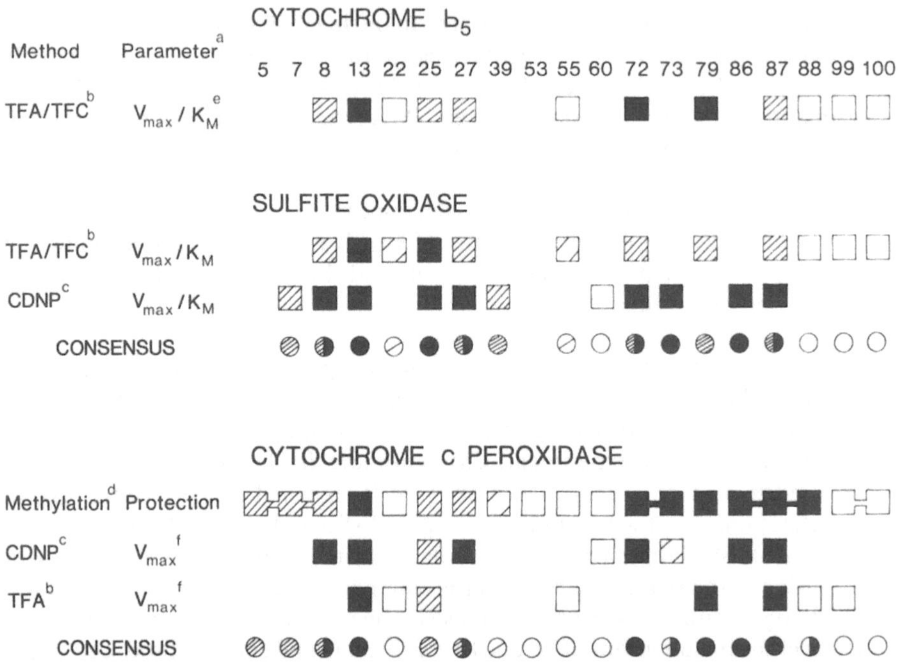

Fig. 2.16. Mapping the interaction surface on cytochrome c for cytochrome b_5, sulfite oxidase and cytochrome c peroxidase. *a* Results were calculated as described in Fig. 2.14. *b* Modifications are defined in Fig. 2.14. Trifluoroacetylation (*TFA*) or trifluoromethylphenylcarbamoylation (*TFC*) (Ng et al. 1977; M. B. Smith et al. 1980; Webb et al. 1980; H. T. Smith and Millet 1980. *c* 4-Carboxy-2,6-dinitrophenylation (Kang et al. 1978; Speck et al. 1981). *d* Methylation using formaldehyde and sodium borohydride (Pettigrew 1978). *e* V_{max}/K_M values were obtained from the y-intercept of Eadie-Hofstee plots. *f* V_{max} values were obtained from the "high affinity phase" of Eadie-Hofstee plots

modification because a single cytochrome c molecule has a small chance of receiving more than one modifying group. On the other hand, differences in shielding could be due to conformational changes induced by binding; for example, see the discussion of cytochrome c peroxidase in Section 2.5.

The criticism has also been made that shielding or protection experiments do not define electron transfer events, but the formation of complexes which may or may not be kinetically competent. However, there is strong agreement between the results of the kinetic and the shielding approaches outlined in Figs. 2.14–2.16.

2.4.2 Cytochrome Oxidase and CoQ Cytochrome c Oxidoreductase

All methods used agree on the importance of lysines 13, 86 and 87 in the interaction of cytochrome c with both oxidase and reductase (Fig. 2.14). There is also agreement on the lack of involvement of lysines 22, 39, 53, 55, 60, 99 and 100. Lysines 72 and 79 may also form part of a common interaction surface but the former is not shielded by cytochrome c_1 and the latter is not shielded by CoQ-cytochrome c oxidoreductase (Rieder and Bosshard 1980). The evidence for the role of other lysines is more ambivalent. Lysine 8 is shielded by both oxidase and reductase but its modification does not greatly affect activity with the former, while lysine 27 is only weakly shielded although its modification considerably inhibits activity with the reductase alone. However, in spite of minor anomalies, the strong impressions given by Fig. 2.14 are (1) that the different methods are supportive rather than contradictory and (2) that the interaction domain on cytochrome c is common to both the oxidase and reductase. This clearly has important implications for the models of cytochrome c action proposed in Fig. 2.2 and appears to preclude the simultaneous binding of cytochrome c to both oxidase and reductase with the cytochrome c acting as an electron-conducting wire. This conclusion will be discussed in Section 2.8.

There remains unexplained data from other studies which seem incompatible with a common site for reductase and oxidase. L. Smith et al. (1973b) and Kuo et al. (1984) found that anti-cytochrome c antibodies differentially inhibited reductase and oxidase activities although Osheroff et al. (1979) could not confirm their findings. However, differential inhibition of the two activities by antibodies may reflect general steric hindrance of access to a binding pocket by a bulky antibody rather than specific blocking of distinct interaction sites. In this respect, it is worth noting that L. Smith et al. (1973b) and Kuo et al. (1984) examined the low affinity phase of the oxidase reaction, a phase in which one cytochrome c molecule already occupies the binding cleft (Sect. 2.5), whereas Osheroff et al. (1979) studied the high affinity phase.

Although different cytochromes c show different patterns of reactivity with oxidase and reductase (Smith et al. 1976), such data are difficult to interpret in view of the complexity of oxidase kinetics (Sect. 2.2.3) and cannot be considered as persuasive evidence for distinct sites. Several chemical modification studies have suggested differential inhibition of reactivities with oxidase and reductase preparations (Margoliash et al. 1973, Myer et al. 1980). However, in many of these cases the modifications perturb the protein structure and thus cannot be used to support a proposal of separate binding sites (Dickerson and Timkovich 1975; Erecinska and Vanderkooi 1978). Only two modifications, the 4-nitrobenzo-2-oxa-1,3-diazole derivative of Lys 13 in horse cytochrome c and the bisphenylglyoxal derivative of Arg 13 in *Candida* cytochrome c did not perturb the protein structure, yet resulted in altered oxidase reactivity but normal reductase reactivity (Margoliash et al. 1973).

Nor is there a unique interpretation for the finding that chemically cross-linked and immobilised cytochrome c remains competent in electron transfer (Erecinska et al. 1975; Erecinska et al. 1980; Sect. 2.5). In these studies considerable cross-linking to phospholipid occurred which would still allow cytochrome c lateral and rotational mobility on the mitochondrial membrane.

Indeed, such lateral mobility is one possible mechanism by which cytochrome c acts as an electron-transferring link betwen the reductase and oxidase (Sect. 2.8). However, if cytochrome c binds to the phospholipid surface with the same cluster of positive charge as is involved in its interaction with its redox partners, the phospholipid surface might be expected to compete with, rather than enhance, electron transfer. A similar problem arises from the work of Petersen and Cox (1980) and Yoshimura et al. (1981) who found that the highly acidic protein, phosvitin, greatly enhanced the rates of reaction of cytochrome c with cationic reagents including cytochrome c itself. Presumably, the enhancement effect of restricting the reaction to two dimensions outweighs the inhibition by the competing negative surface.

In conclusion, there are anomalous results but these do not present a substantial challenge to the present consensus that cytochrome c has a single binding surface for its reductase and oxidase.

2.4.3 The Intermembrane Redox Reactions

Cytochrome b_5. Ng et al. (1977) and M.B. Smith et al. (1980) used the microsomal NADH cytochrome c oxidoreductase system to study the effect of trifluoroacetylation (TFA) and trifluoromethylphenyl-carbamoylation (TFC) on the reaction between cytochrome b_5 and cytochrome c. They used the intercept on the ordinate of the Eadie-Hofstee plot (v/S) as an empirical measure of electron transfer and found that modification of lysines 13, 72, 79, 25, 27, 87 and 8 lowered this figure while modification of lysines 22, 55, 88, 99 and 100 had no effect (Fig. 2.16).

Sulfite Oxidase. The groups of Millet and Margoliash have studied the effect of lysine modification on the reactivity of cytochrome c with sulfite oxidase. Webb et al. (1980) found that modification of lysines 13, 25, 79, 87, 8, 27 and 72 with TFA or TFC groups caused a lowering of the parameter V_{max}/K_M to an extent ranging from 66% for lysine 13 to 23% for lysine 72 (Fig. 2.17) while modification of lysines 22, 55, 88, 99 and 100 had little effect on activity. They proposed that the different degrees of inhibition for the modified lysines reflect their involvement in different electrostatic charge-pair bonds ranging in length from 4.8 Å for lysine 13 to 10 Å for lysine 72. They supported this model by a theoretical analysis of the relationship between V_{max}/K_M and $\sqrt{I}$ which will be considered in the theoretical Section 2.7.

At lower ionic strengths, and with CDNP derivatives of cytochrome c, Speck et al. (1981) found much more severe inhibitions of activity (Fig. 2.17). However, they proposed that this is due to two distinct effects. One, which is common but variable in all the derivatives, is a consequence of the work required to re-orient correctly the dipole of cytochrome c as it approaches the sulfite oxidase (Koppenol et al. 1978; Koppenol and Margoliash 1982). They proposed that for cytochrome molecules modified at lysines 60, 39, 7, 25, 8,

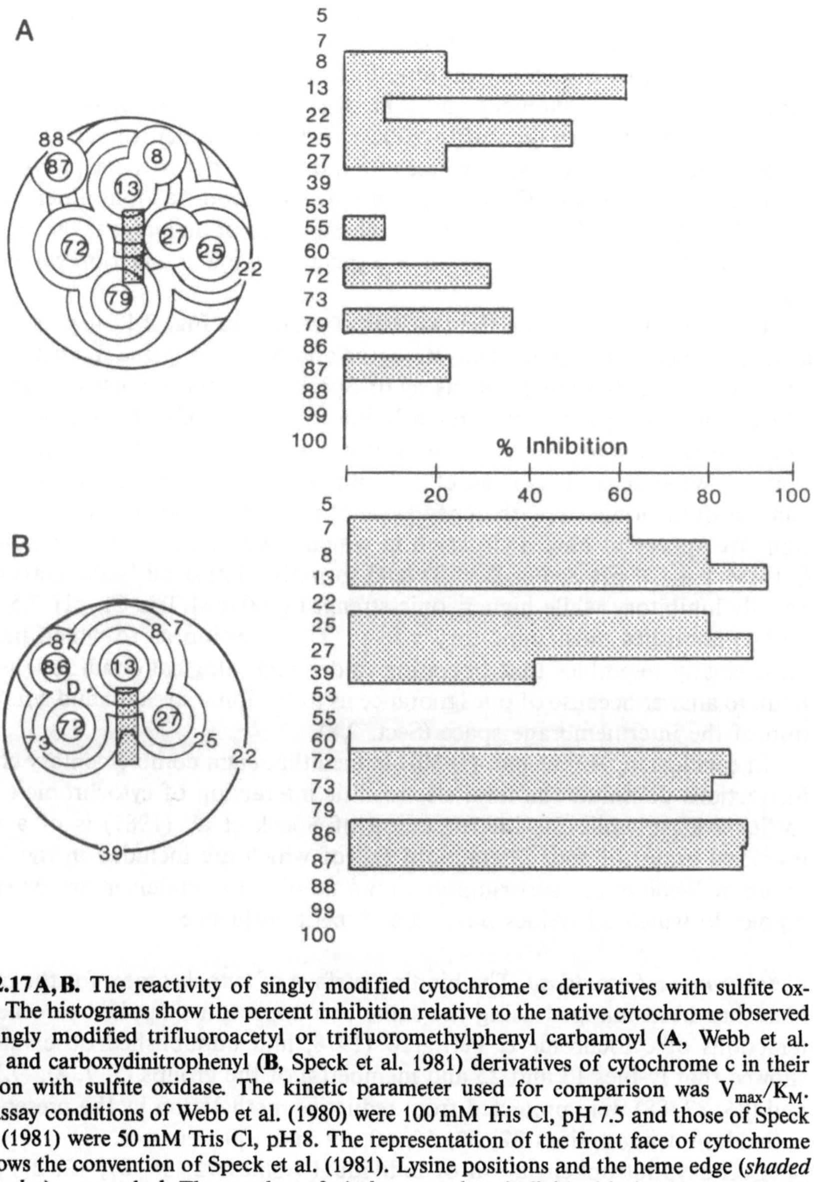

Fig. 2.17 A, B. The reactivity of singly modified cytochrome c derivatives with sulfite oxidase. The histograms show the percent inhibition relative to the native cytochrome observed for singly modified trifluoroacetyl or trifluoromethylphenyl carbamoyl (**A**, Webb et al. 1980) and carboxydinitrophenyl (**B**, Speck et al. 1981) derivatives of cytochrome c in their reaction with sulfite oxidase. The kinetic parameter used for comparison was V_{max}/K_M. The assay conditions of Webb et al. (1980) were 100 mM Tris Cl, pH 7.5 and those of Speck et al. (1981) were 50 mM Tris Cl, pH 8. The representation of the front face of cytochrome c follows the convention of Speck et al. (1981). Lysine positions and the heme edge (*shaded rectangles*) are marked. The number of *circles* around an individual lysine is proportional to the percent inhibition of the TFA or TFC derivative of that lysine (Webb et al. 1980). Radii are multiples of 2.5 Å and each circle represents a 10% inhibition. This is also true for the representation in **B** using the results of Speck et al. (1981). However, in this case the contour frequency is proportional to the percent inhibition of the CDNP derivative of that lysine unaccounted for by the alteration in the dipole moment. The inhibition, solely due to the change in the dipole moment, is calculated as described in Speck et al. (1981) but we direct the reader to the discussion of the theoretical basis for this procedure in Section 2.7

73 and 87, the loss of activity is proportional to this re-orientation energy. However, for the derivatives modified at lysines 13, 27, 72 and 86, Speck et al. proposed that loss of activity on modification cannot be accounted for by a change in the dipole moment and that these lysines are involved in specific charge interactions with the sulfite oxidase. We will consider the validity of this model in Section 2.7 but we note here that it does give rise to some peculiarities in the analysis of results. For example in Fig. 2.17, modification of neighbouring lysines (86 and 87) results in the same high loss of activity yet the analysis concludes that lysine 86 is involved in a specific charge interaction while 87 is not.

The differences between the two sets of results in Fig. 2.17 is presumably a consequence of different ionic strengths and different lysine modifications. The lower ionic strength of the assay of Speck et al. and the use of a reagent which converts a positively charged lysine to a negatively charged derivative would be expected to give rise to more severe inhibitory effects than the conditions of Webb et al. This is an effect anticipated by the latter group in their analysis of the ionic strength dependence of the electrostatic energy of interaction. We appear to have a situation in which lower ionic strength conditions (50 mM Tris-Cl, pH 8; Fig. 2.17B) tend to make almost all lysine derivatives equally inhibitory while higher ionic strength (100 mM Tris Cl, pH 7.5; Fig. 2.17A) abolishes most inhibitory effects. The question as to which pattern most closely resembles that operating under physiological conditions is difficult to answer because of our ignorance as to the ionic strength and composition of the intermembrane space (Sect. 2.8).

In conclusion, Webb et al. (1980) claimed that eight complementary charge interactions dominate the total electrostatic interaction of cytochrome c with sulfite oxidase while the interpretation of Speck et al. (1981) is of a more restricted group of four interactions (all of which are included in the larger group of Webb et al.) superimposed on a model of complementary orienting dipoles in which all lysines have some kinetic influence.

Cytochrome c Peroxidase. The binding surface of cytochrome c for the peroxidase has been mapped using the same two approaches applied to other interactions of cytochrome c. Pettigrew (1978, and unpublished observations) showed that lysines 13 and 25 and member(s) of the groups (5, 7, 8), (72, 73) and (86, 87, 88) were protected from reductive methylation by the presence of peroxidase, while lysines 22, 53, 55, 60, 99 and 100 were not.

Carboxydinitrophenylation of individual lysines 72, 86, 87, 13, 27, 8, 25 and 73 caused a lowering of activity to a degree indicated by the order given but with no effect on K_M. Modification of Lys 72 caused the greatest effect (Kang et al. 1978). The modifications also shifted the optimal ionic strength for the reaction to lower values and the degree of shift correlated both with the lowering of activity under conditions optimal for the native cytochrome and the distance of the modified lysine from the centre of the deduced interaction domain. As with yeast and horse cytochrome c at their respective optimal ionic strength (Table 2.6), 0.3 mol CDNP Lys 72 cytochrome per mol peroxidase was bound at the new optimal ionic strength of 0.01 M after non-equilibrium molecular exclusion chromatography. Modification of the lysines on the front face of cytochrome c thus appears to have two effects on the reaction with peroxidase. Firstly, the binding is weakened by the reduction of 2 in the net positive charge, and consequently the optimal ionic strength shifts to lower values. Secondly, effective binding is established at these lower ionic strengths and thus the decrease in V_{max} that accompanies modification may be due to a subtle alteration in the geometry of the complex rather than to a change in binding.

The TFA studies (H. F. Smith and Millet 1980) are in broad agreement with the CDNP studies (Fig. 2.16).

In conclusion, virtually all the studies described above agree in implicating the front face of cytochrome c as the interaction domain for binding to its redox partners.* This face includes the exposed heme edge (Vol. 2, Chap. 4) and, although it is not proven, it is most probable that this is the site of both electron exit and entry on cytochrome c (Vol. 2, Chap. 8).

2.5 Structural Studies on the Electron Transfer Complexes of Cytochrome c

No direct crystallographic study of an electron transfer complex involving cytochrome c is yet available. The structures of these complexes have been inferred by optimising the alignment of the electrostatic surfaces of the individual crystallographic components. These models have been tested by chemical modification studies which map the area of interaction of cytochrome c (Sect. 2.4) and the redox partner (this section). Chemical cross-linking has been used to define the close approach of surface regions and the distance between redox centres has been measured by fluorescence quenching.

2.5.1 With Cytochrome Oxidase and CoQ Cytochrome c Oxidoreductase

A variety of chemical cross-linking methods have been used to study the binding sites for cytochrome c in cytochrome c-depleted mitochondria and on purified oxidase and reductase. In the case of the purified preparations the nature of the attachment can be defined by the disappearance of certain subunit bands on SDS gel electrophoresis and the appearance of new bands having higher molecular weights and containing heme c. The identification can be confirmed immunologically and in some cases by reversal of the cross-link. Further chemical characterisation of these cross-linked preparations will allow insight into the structure of the electron transfer complexes but at present their

* See appendix note 5

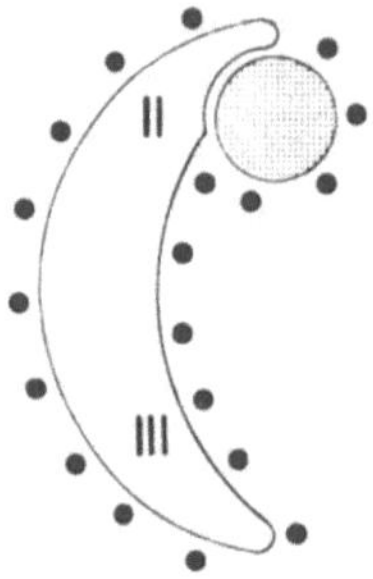

PROTECTION

(a) Methylation

$$R\text{-}NH_2 + H^{14}CHO \longrightarrow R\text{-}N{=}CH_2 \xrightarrow{NaBH_4} R\text{-}N(H)\text{-}CH_3$$

(b) Carbodiimide-promoted amide formation (nucleophile added)

$$R\text{-}COOH + NH_2CH_2\text{-}C({=}O)\text{-}O\text{-}C_2H_5 \xrightarrow{R\text{-}N{=}C{=}N\text{-}R'} R\text{-}C({=}O)\text{-}N(H)\text{-}CH_2\text{-}C({=}O)\text{-}O\text{-}C_2H_5$$

ACTIVITY

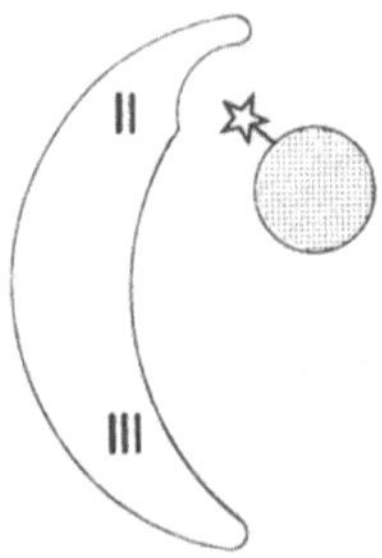

(a) S-ethylthioltrifluoroacetate

$$R\text{-}NH_2 + CF_3\text{-}C({=}O)\text{-}S\text{-}C_2H_5 \longrightarrow R\text{-}N(H)\text{-}C({=}O)\text{-}CF_3 + C_2H_5SH$$

(b) Trifluoromethylphenylisocyanate

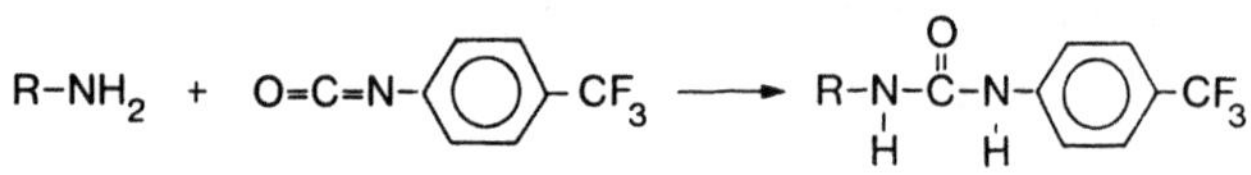

(c) 4-chloro-3,5-dinitrobenzoate

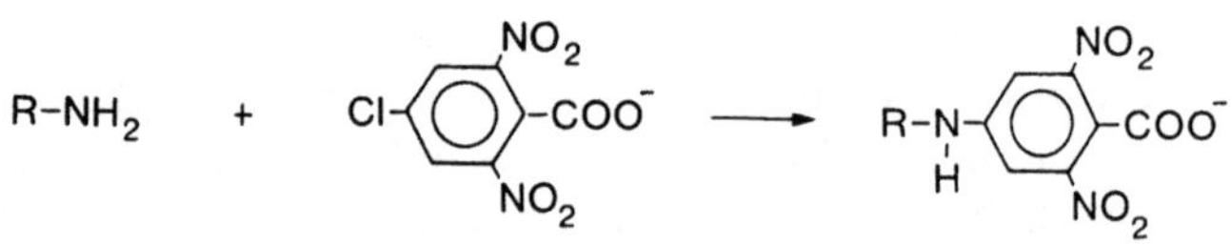

CROSSLINKING

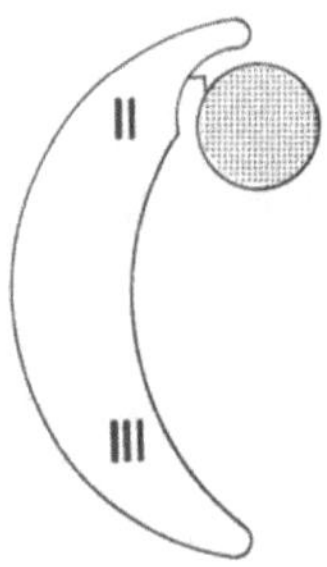

(a) Arylazido-lys 13-cytochrome c

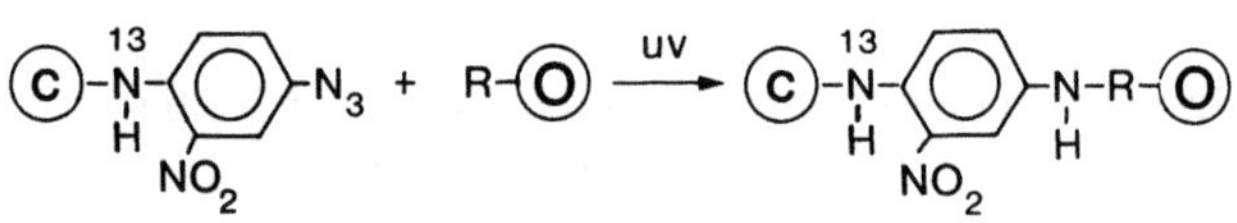

(b) Carbodiimide promoted amide formation

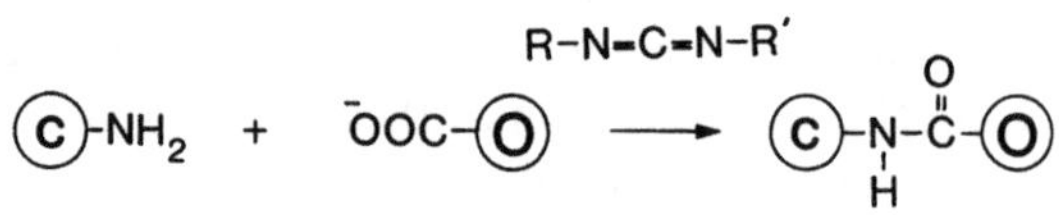

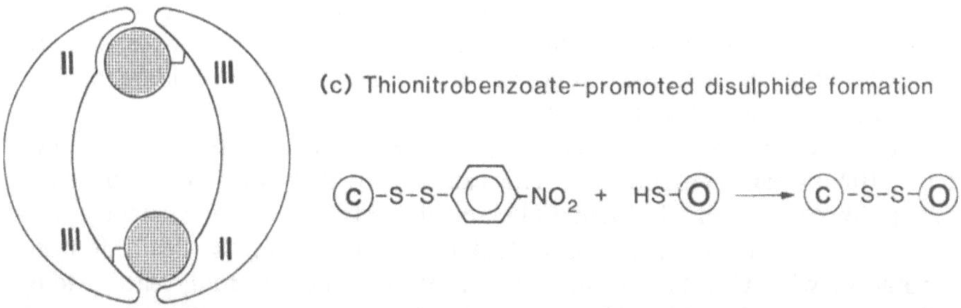

Fig. 2.18. Chemical modifications used to study the interactions of cytochrome c. The diagrams are based on the view of cytochrome oxidase described in Fig. 2.19. Cytochrome c is represented by the *filled circles.* The portion of cytochrome oxidase which protrudes from the C-surface of the membrane is shown as the *curved structure* which, on dimerisation, forms a *cleft* accommodating cytochrome c molecules

main contribution has been in the assessment of the two-site model for cytochrome oxidase discussed in Section 2.8.

One approach has been to prepare arylazido cytochrome c using fluoronitrophenylazide (Bisson et al. 1978a, b, 1980). Certain lysines are preferentially labelled and a purified preparation of arylazido Lys 13 cytochrome c can be obtained (Fig. 2.18). If a complex of this molecule with cytochrome oxidase is irradiated with UV light, the reactive nitrene that is formed allows covalent cross-linking of cytochrome c to subunit II of the oxidase. Under conditions in which only high affinity binding should be present, 0.5 mol cytochrome c were bound per mole of subunit II and a further 0.5 mol to phospholipid tightly associated with the oxidase. The cross-linked complex was inactive both as a cytochrome c oxidase and an ascorbate-TMPD oxidase and this "half-of-sites" effect led Bisson et al. (1980) to postulate a negative interaction between the two high affinity sites present in an oxidase dimer. In contrast, in monomeric oxidase preparations, the loss of activity is proportional to the degree of cross-linking to subunit II (Georgevich et al. 1983).

Tryptic digestion of the purified cytochrome c: subunit II complex followed by amino acid sequence determination of one of the heme c containing peptides gave the pair of sequences (Bisson et al. 1982):

Met – Leu – Val – Ser – Ser – Glu – Asp – Val – Leu – His –*
155

Ile – Phe – Val – Gln –*
10

The former is located near the C-terminus of subunit II (Steffens and Buse 1979) and the latter precedes the heme attachment site in cytochrome c. The poor yields of derivatised amino acids obtained after the marked positions (*) in the sequence determination probably define the positions of the cross-links.

Preliminary results on the binding of arylazido Lys 13 cytochrome c to two-dimensional crystalline arrays of monomeric cytochrome oxidase have indicated that the cytochrome c interacts with the surface of the oxidase involved in dimer formation (Capaldi et al. 1982).

A second approach has been to exploit the presence of the single free cysteine (102) in yeast cytochrome c. This can be activated by a thionitrobenzoate group (Birchmeier et al. 1976; Fuller et al. 1981) or more efficiently by a thiopyridyl group (Moreland and Dockter 1981) resulting in disulfide bond formation with Cys 115 in subunit III of both bovine and yeast oxidase (Malatesta and Capaldi 1982; Fig. 2.18). Cytochrome c oxidase activity is abolished after modification of 1 mol cytochrome c/mol oxidase but ascorbate-TMPD oxidase activity in the presence of added cytochrome c is unaffected (Moreland and Dockter 1981). It is claimed that the covalently bound cytochrome occupies the high affinity site and therefore that the low affinity site has no intrinsic electron transfer activity (Fuller et al. 1981) but these authors appear to ignore the evidence from both kinetic and direct binding studies that bovine oxidase contains two high affinity sites per mole for yeast cytochrome c (Dethmers et al. 1979).

Covalent attachment of TNB Cys 102 yeast cytochrome c to monomeric cytochrome oxidase is much less efficient and results in a complex which, in contrast to that of the dimeric enzyme, retains the ability to oxidise cytochrome c but cannot oxidise ascorbate TMPD (Darley-Usmar et al. 1984). This leads to the proposal, outlined in Fig. 2.19 for the dimeric enzyme, in which the cytochrome c binding sites are situated in a cleft in the monomer: monomer interface. Cytochrome c binds electrostatically to subunit II of one monomer while disulfide bond formation occurs with subunit III of the second monomer. The "half-of-sites" effect seen with the arylazido Lys 13 derivative may derive from steric hindrance within this cleft.

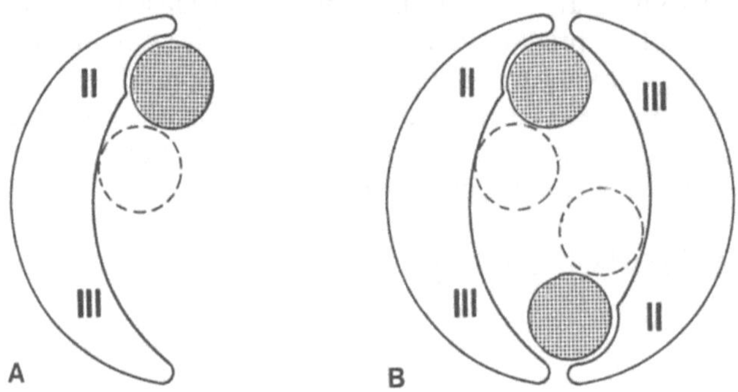

Fig. 2.19 A, B. A model for cytochrome c binding to cytochrome oxidase. The diagrams show transverse sections of the C-surface projetions of the cytochrome oxidase momoner (**A**) and dimer (**B**). In the dimer there is a central cavity in which a maximum of four cytochrome c molecules can bind. The *two filled circles* represent cytochrome c in the high affinity sites associated with subunit *II*. The *broken circles* represent cytochrome c binding with low affinity to an already occupied oxidase

The interaction of cytochrome c with subunit II of cytochrome oxidase has been confirmed by studies with water-soluble carbodiimide (Bisson and Montecucco 1982; Millet et al. 1982). The enzyme is inactivated by carbodiimide in the presence of ^{14}C-glycine ethyl ester as a nucleophile and cytochrome c protects against this. Of four carboxylates in subunit II which become heavily labelled, three – Asp 112, Glu 114 and Glu 198 – are shielded by cytochrome c (Millet et al. 1983). In the absence of nucleophile, the carbodiimide catalyses cross-linking between amines and carboxylates on cytochrome c and subunit II (and also to some extent on "subunit VII") (Fig. 2.18). Bisson and Montecucco (1982) found that the bulkier reagent cyclohexylmorphilinyl, 4-ethylcarbodiimide metho-p-toluenesulfonate gave more selective cross-linking with subunit II. This work was repeated by Kadenbach and Stroh (1984) using a more resolving gel and looking particularly at the small subunits which show tissue specificity (Sect. 2.2.1.2). In addition to subunit II, subunits VIa, VIIa and VIII are protected by cytochrome c in the pig liver but not the heart enzyme and these are the same three subunits which show the largest electrophoretic and immunological differences (Sect. 2.2.1.2). The authors suggested that these small subunits may be tissue-specific modulators of cytochrome c affinity.

Similar experiments have been carried out with CoQ cytochrome c oxidoreductase. In keeping with the binding studies of Weiss and Juchs (1978), the predominant reaction of arylazido Lys 13 cytochrome c in 50 mM Tris Cl is with cytochrome c_1 while, at lower ionic strengths, binding to phospholipid becomes extensive (Broger et al. 1980). Arylazido Lys 13 cytochrome c can also be cross-linked to purified cytochrome c_1 and attaches to a highly acidic region (165–174) of the molecule (Broger et al. 1983). The positively charged reagent, 1-ethyl-3(3-^{14}C-trimethyl amino propyl) carbodiimide (ETC) selectively attacks negative clusters and the product will either re-arrange to give an N-acyl urea or catalyse formation of an isopeptide cross-link. When used with either the CoQ cytochrome c oxidoreductase or purified cytochrome c_1 (King preparation, see Sect. 2.2.2), cross-linking of cytochrome c to both cytochrome c_1 and the "hinge peptide" is observed (Stonehuerner et al. 1985; Gutweniger et al. 1983). Labelling of the highly acidic regions 57–81 (cytochrome c_1) and 1–21 (hinge) is blocked by the presence of cytochrome c.

Cytochrome c modified with methyl 4-azidobenzoimidate cross-linked to cytochrome c_1 in cytochrome c-depleted mitochondria (Erecinska et al. 1980). Remarkably, the mitochondria not only retained succinate cytochrome c oxidoreductase activity but also succinate oxidase activity (although additional cytochrome c stimulated this). This led to the proposal that a cytochrome covalently bound to the reductase may be able to interact with the oxidase. However, it should be pointed out that a portion of the bound cytochrome appears to be bound not to protein but to phospholipid. Also the cross-linking does not preclude a rotation or swinging of cytochrome c between the two enzymes.

2.5.2 With Cytochrome b_5

Salemme (1976), using the established X-ray structures for bovine liver microsomal cytochrome b_5 and tuna cytochrome c, produced an optimised electrostatic "fit" of the two proteins in which four electrostatic interactions dominate the interface. These involve lysine side chains 13, 27, 72 and 79 on cytochrome c and the acid groups of Glu 48, Glu 44, Asp 60 and an exposed heme propionate on cytochrome b_5 (Fig. 2.20). This remarkable model, which preceded any direct experimental information, is consistent with studies on the reactivity of singly substituted lysine derivatives of cytochrome c with cytochrome b_5 (Sect. 2.4.3) and with NMR spectroscopy of the cytochrome c:cytochrome b_5 complex (Eley and Moore 1983).

However, the effects of esterification of the heme propionates of cytochrome b_5 are complex (Mauk et al. 1986). Consistent with the Salemme model, the complex of the modified b_5 with cytochrome c is probably less stable at pH 7 but this situation is reversed at pH 8. Calculations of the electrostatic surface potential reveal that an alternative "side-on" binding mode which is unfavourable for the complex of the native proteins becomes favoured for the esterified b_5 at pH 8 (Mauk et al. 1986). The role of carboxyls in the interaction of cytochrome b_5 with its flavoprotein reductase has been studied by Dailey and Strittmatter (1979). They found that modification of Glu 43, Glu 44, Glu 48, one heme propionate and a further unidentified carboxyl group, by reaction with a water-soluble carbodiimide followed by nucleophilic attack by methylamine or glycine ethyl ester, resulted in a tenfold rise in K_M for NADH cytochrome b_5 reductase. More extensive modification did not produce further changes in K_M.

In view of the correspondence between this group of acidic residues and that proposed by Salemme it would seem highly probable that cytochrome b_5 presents the same face to its reductase and to cytochrome c. This is supported by the properties of the cross-linked complex of cytochrome b_5 with its reductase, promoted by carbodiimide in the absence of added nucleophile (Hackett and Strittmatter 1984). This complex retains a high efficiency of internal electron transfer but only slowly reduces cytochrome c. Salemme had shown that in the optimised electrostatic "fit" of the b_5:c complex, the two hemes were coplanar and their edges approached approx. 8 Å. The reductase:b_5 complex may achieve a similar orientation of flavin and heme.

There is uncertainty as to whether Salemme's model is applicable to other members of the cytochrome b_5 family. Guiard and Lederer (1979b) pointed out that only Glu 44 of the three amino acid carboxylates involved in salt bridges is conserved in sulfite oxidase and cytochrome b_2.* Two studies which sought to determine the distance between the redox centres in the cytochrome c:cytochrome b_2 complex by measuring fluorescence quenching of the horse Zn II cytochrome c derivative gave conflicting results. Vanderkooi et al. (1980) and McLendon et al. (1985) found an Fe-Zn distance of 18 Å in agreement with the Salemme model (Fig. 2.18) but Thomas et al. (1983) proposed that

* See appendix note 6

it was the flavodehydrogenase domain, not the cytochrome b_5 domain of the enzyme, that bound cytochrome c (Sect. 2.3.1).

The X-ray crystal structure of cytochrome b_2 (Xia et al. 1986) shows two of the four hemes orientated in such a way that they are coplanar with their corresponding flavins and at a separation of only 10 Å edge to edge. This involves a tight association of the two domains that precludes a cytochrome c binding of the type seen in the proposed b_5:c model. However, the remaining two heme domains appear disordered and this may reflect a flexibility in the connection with the flavin domains that is also consistent with the susceptibility to proteolysis of this region (Xia et al. 1986). Thus intermolecular electron transfer to cytochrome c may involve an opening of the interface between the heme and flavin domains so that the cytochrome c can bind. This proposal awaits experimental support.

2.5.3 With Cytochrome c Peroxidase

The peroxidase : cytochrome c complex is an attractive subject for cross-linking and detailed structural analysis. A cross-linked complex of M_r 46 K consistent with a 1 : 1 stoichiometry could be isolated after treatment with the cleavable bifunctional reagent dithiobissuccinimidyl propionate (DSP) (Waldmeyer et al. 1980; Pettigrew und Seilman 1982). This complex has residual activity as a cytochrome c peroxidase (10–40%) and can be reduced by ascorbate or cytochrome c_1 much better at high (e.g. 0.2 M) rather than low (e.g. 0.01 M) ionic strengths, features which suggest that the cross-linking may be flexible enough to allow access of molecules to the interface when the non-covalent electrostatic interactions are broken by high salt concentrations. In contrast, the cross-linked complex formed using water-soluble carbodiimide as a catalyst contains "zero-length" cross-links (Fig. 2.18) and has low heme accessibility which is not improved by raising the ionic strength (Waldmeyer and Bosshard 1985).

In the latter case the cross-links were formed between Lys 13 (cytochrome c) and the acidic cluster (33–37), and between Lys 86 (cytochrome c) and the same cluster. A similar cross-linking pattern was found for arylazido Lys 13 cytochrome c (see Sect. 2.5.1) (Bisson and Capaldi 1981) although the peroxidase site was not so narrowly defined. The negative surface on the peroxidase, complementary to the positive binding region of cytochrome c, has been investigated using water-soluble carbodiimide and glycine ethyl ester as a nucleophile (Fig. 2.18). Waldmeyer et al. (1982) found that modification of four to five carboxyl groups lowered the peroxidase activity to 14% and two of these are protected by the presence of cytochrome c. In a similar, but more detailed study, Bechtold and Bosshard (1985) found that the acidic groups 33, 34, 35, 37, 221, 224 and the cluster (290, 291, 294) were strongly shielded by cytochrome c. Unlike the case of cytochrome b_5, no contribution of the heme propionates to binding was observed since their esterification did not affect

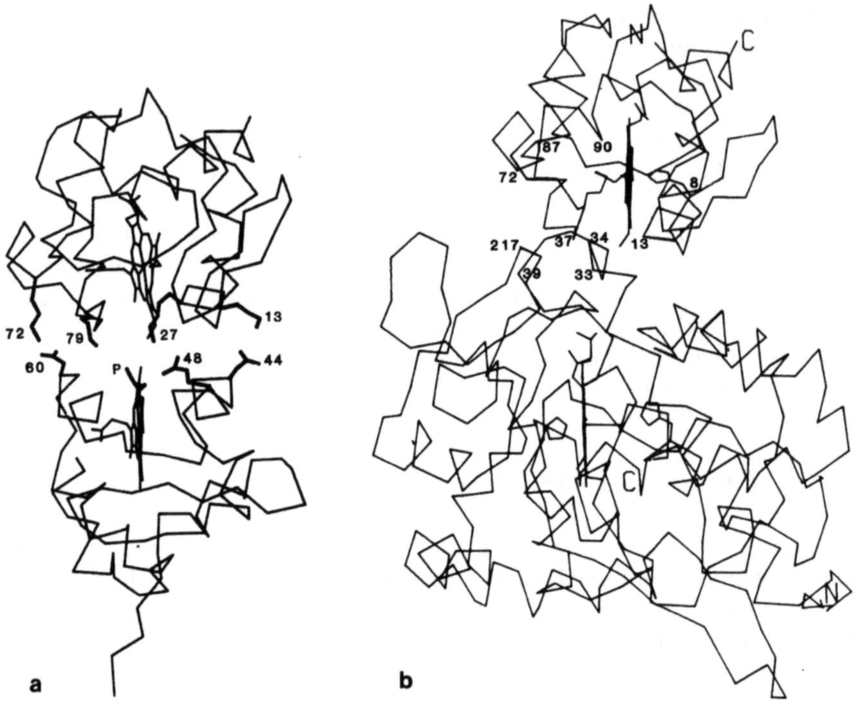

Fig. 2.20 a, b. Proposed intermolecular complexes between **a** cytochrome c (top) and cytochrome b_5 and **b** cytochrome c (top) and cytochrome c peroxidase. **a** The planes of the hemes are approximately parallel, with an interheme edge separation of 8.5 Å and an Fe-Fe separation of 16 Å. The α-carbon backbones are represented by *light lines* and the interacting side chains by *heavy lines.* The side chain orientations have been adjusted to enhance their interactions. (From Mathews 1984 after Salemme 1976). **b** The planes of the hemes are approximately parallel, with an interheme edge separation of 17–18 Å and an Fe-Fe separation of 25–26 Å. (Poulos and Kraut 1980)

complex formation (Mochan 1970). Bosshard et al. (1984) showed that of the six histidines of cytochrome c peroxidase, four were modified by photo-oxidation in the absence of cytochrome c while only three were modified in its presence.

These chemical modification studies and the investigation of the binding surface on cytochrome c for the peroxidase (Sect. 2.4) support the model of optimised electrostatic "fit" of the two proteins using the known X-ray crystal structures (Fig. 2.20) (Poulos and Kraut 1980; Poulos and Finzel 1984; Finzel et al. 1984). The heme of the peroxidase is buried within the protein but it approaches the surface of the molecule at a region of negative charge, proposed to form the cytochrome c binding site. In this site, Asp 33, Asp 34, Asp 37, Asp 217 and Lys 39 are proposed to form charge interactions with the cytochrome c residues Lys 8, Lys 87, Lys 13, Lys 72 and Asp 90 respectively, a proposal compatible with the map of the cytochrome c binding surface (Fig. 2.16) and with the protection of His 181 by cytochrome c. In the orientation defined by the electrostatic interactions, further stabilising forces can be identified. These include hydrogen bond formation between Gln 86 (peroxidase) and Gln 16 (cytochrome c) and Asn 87 (peroxidase) and Gln 12 (cytochrome c). Also a dipole interaction may exist between the two helices 2–18 (cytochrome c) and 85–95 (peroxidase), while Ile 81 (cytochrome c) may facilitate desolvation at the interface of the two molecules by interacting hydrophobically with the peroxidase. The heme-heme distance of the model (16.5 Å) is in good agreement with the value of 14 Å measured by fluorescence quenching (Leonard and Yonetani 1974).

We note as a caution against too uncritical an interpretation of protection experiments that Bosshard et al. (1984) found that cytochrome c binding protected Trp 211 and Trp 223 of peroxidase from photo-oxidation. These two tryptophans are not part of the binding region and, if the model of the interface is to be retained, it is necessary to propose an indirect shielding via a conformational change. A similar rationalisation is required to explain the apparent protection of the C-terminal region (290–294) by cytochrome c (Bechtold and Bosshard 1985).

2.6 Ion Binding to Cytochrome c

Cytochrome c binds a variety of small ions, at an unknown number of sites, with a range of dissociation constants varying from 10^{-2} M to 10^{-6} M (Barlow and Margoliash 1966; Margoliash et al. 1970; Kayushin and Ajipa 1973; Margalit and Schejter 1973a, b, 1974; Stellwagen and Shulman 1973; Nicholls 1974; Morton and Breskvar 1977; Andersson et al. 1979; Osheroff et al. 1980). As a result of this ion binding, cytochrome c is usually associated with small ions under the conditions of most experiments and the bound ions must be regarded as integral components of the protein structure. An indication of their importance is that redox reactions of cytochrome c with reagents as diverse as Fe^{2+} (Taborsky 1979), dithionite (Miller and Cusanovich 1973) and cytochrome oxidase (Fig. 2.11) (Ferguson-Miller et al. 1976; L. Smith et al. 1979b, 1980) are strongly influenced by the presence of phosphate. Indeed, so marked are the ion-binding properties of cytochrome c that it has been argued they have an important biological role (Osheroff et al. 1980) and cytochrome c itself was suggested to be an ion carrier within the inner mitochondrial membrane (Margoliash et al. 1970; Margalit and Schejter 1973b). Such a role is not, however, compatible with our view of cytochrome c as a water-soluble peripheral membrane protein.

The problems in kinetic assays caused by ion binding have led some workers to prepare ion-free cytochrome c by isoelectric focussing or electrolysis at the isoelectric point (L. Smith 1978; Peterman and Morton 1977). Tris-cacodylate buffers are generally used since these are believed not to bind to cytochrome (Barlow and Margoliash 1966; Margalit and Schejter 1973a, b) although this has been contested by Brooks and Nicholls (1982). Many binding studies, which are not physiologically relevant, have been done using small charged redox reagents in order to investigate mechanisms of electron transfer. These will be discussed in Volume 2, Chapter 8.

Binding of Phosphate. There are at least three phosphate binding sites on horse cytochrome c though only two have been well-characterised. One of these is near Lys 87 (called phosphate site I) and has a dissociation constant of 2×10^{-4} M while phosphate site II is close to Lys 25-His 26-Lys 27 with a dissociation constant of $>2\times10^{-3}$ M (Stellwagen and Schulman 1973; Brautigan et al. 1978; Taborsky and McCollum 1979; Osheroff et al. 1980). The binding is observed by NMR for both ferricytochrome c and ferrocytochrome c (Kayushin and Ajipa 1973; Stellwagen and Schulman 1973a) although dissociation constants were not determined for the latter. However, these NMR studies have generally not confirmed the very pronounced differences observed in earlier studies for anions binding to the two redox states (Margoliash et al. 1970; Margalit and Schejter 1973a, b, 1974).

Nucleoside di- and triphosphates bind at both sites in both redox states (Stellwagen and Schulman 1973; Osheroff et al. 1980) and Kayushin and Ajipa (1973) found a $K_D = 6.7\times10^{-3}$ M (although this study did not distinguish between binding sites).

Competition experiments establish that these two phosphate sites do not bind chloride, acetate, borate, cacodylate or carbonate, in spite of the fact that carbonate has been shown to bind close to lysines 72, 73, 86 and 87 (Osheroff et al. 1980). However, phosphate and chloride do compete for a binding site at which the K_D for phosphate is 4.7×10^{-2} M in the ferricytochrome and 1.7×10^{-2} M in the ferrocytochrome while that for chloride is 5.9×10^{-2} M and 10^{-2} M in the two redox states (Andersson et al. 1979). This third phosphate site may be at the heme crevice close to Lys 13.

Cation Binding. Calcium and magnesium ions bind to cytochrome c but the number of binding sites is not known (Margoliash et al. 1970; Margalit and Schejter 1974). Lanthanide ions, which are good probes for calcium sites (Williams 1970), bind at a minimum of three sites. One of these is close to the N-terminus (perhaps Asp 2 and Asp 93), another is close to the C-terminus (perhaps the C-terminus itself) and a third site is close to Met 65 and Tyr 74 (perhaps Glu 66 and Glu 69) (G.R. Moore and G. Williams, unpublished data).

In summary, there are at least three phosphate binding sites on horse cytochrome c, one of moderately high affinity and one competitive with chloride. Cation binding sites are less well characterised although binding of lanthanides such as Gadolinium occurs at a minimum of three positions. An important but unresolved question is the nature and extent of ion binding to cytochrome c under physiological conditions.

2.7 Theoretical Aspects of Complex Formation by Cytochrome c

On the basis of the experimental evidence considered in Sections 2.4 and 2.5, cytochrome c forms specific complexes with its redox partners stabilised by electrostatic linkages. However, there are conflicting views as to the nature of the electrostatic effects leading to complex formation and the appropriate theoretical model for their quantitative description.

It is clear that some mechanism of rate enhancement over that of random collision must exist for cytochrome c. The exposed heme edge of cytochrome c, the proposed site of electron exit and entry, occupies only 0.6% of the protein surface (Stellwagen 1978) and if a similar percentage applies to its physiological redox partners, then a randomly formed complex has only a small probability of being in the correct orientation for electron transfer (Koppenol and Margoliash 1982). This should produce a reaction at least 10^3 times slower than the diffusion-controlled rate. Yet kinetic measurements show that such reactions are close to the diffusion limit.

We can make qualitative suggestions as to mechanisms operating to enhance the rate of fruitful complex formation. Firstly, this rate will be increased by the presence of opposite net charge on the reactants. Secondly, if charge distribution is asymmetric, electrostatic orientation may take place prior to collision. Thirdly, surface diffusion may occur after collision. Finally, the orientation required for electron transfer may be stabilised by specific local interactions. The problem has been that assessment of the relative importance of these effects requires a rigorous quantitative description of the electrostatic energies involved and although considerable progress has been made, uncertainties remain.

2.7.1 Monopole–Monopole Interactions

According to Debye-Hückel theory, for the reaction:

$$A+B \rightleftharpoons AB \rightarrow \text{products} \tag{2.13}$$

the activity coefficient of a participating ion is given by:

$$\ln \gamma_i = \frac{-Z_i e}{2DkT}\left(\frac{K}{1+Ka_i}\right), \tag{2.14}$$

where a_i is the hydrated ion radius, K is a temperature-dependent function of the $\sqrt{I}$ and equals $0.329\sqrt{I}$ at 25 °C, Z_i is the integral charge, e is the charge of the electron, D is the dielectric constant, k is the Boltzman constant and T is the absolute temperature. The dependence of the rate constant on the activity of the reacting ionic species is given by the Bronsted relationship:

$$k_I = k_{I=0}\,\frac{\gamma_A \gamma_B}{\gamma_{AB}} \tag{2.15}$$

and therefore

$$\ln k_I = \ln k_{I=0} - \left(\frac{e^2}{2DkT}\right) K \left(\frac{Z_A^2}{1+Ka_A} + \frac{Z_B^2}{1+Ka_B} - \frac{(Z_A+Z_B)^2}{1+Ka_{AB}}\right) . \quad (2.16)$$

This full Bronsted-Debye-Hückel relationship can be simplified to:

$$\ln k_I = \ln k_{I=0} + (e^2/DkT)\, Z_A Z_B \ , \quad (2.17)$$

if the ionic radii are equal and if $a_i \ll 1$.

An alternative approach (Wherland and Gray 1976) uses Marcus theory to describe the effect on the rate constant of the electrostatic free energy (V) of the activated complex:

$$k_I = k_{I=\infty} e^{-V/RT} \ . \quad (2.18)$$

From Debye-Hückel theory:

$$V = 2.1175 \left[\frac{e^{-Ka_A}}{1+Ka_B} + \frac{e^{-Ka_B}}{1+Ka_A}\right] \frac{Z_A Z_B}{(a_A+a_B)} \quad (2.19)$$

and therefore,

$$\ln k_I = \ln k_{I=\infty} - 3.576 \left[\frac{e^{-Ka_A}}{1+Ka_B} + \frac{e^{-Ka_B}}{1+Ka_A}\right] \frac{Z_A Z_B}{(a_A+a_B)} \ . \quad (2.20)$$

We can offer comments on the applicability of these equations to protein : protein interactions and those of cytochrome c in particular.

1. Equation (2.17), which can be successfully applied to the reactions of small molecules should not be applied to proteins because neither condition leading to the simplification is met. The apparent charge product (Z_A Z_B) that is sometimes derived by graphical analysis is artefactual and cannot be considered to reflect the local charges at the interface.
2. Equations (2.16) and (2.20) were both derived from relationships valid only at low ionic strength, yet have been applied to data collected at $I > 0.01$ M.
3. Equations (2.16) and (2.20) both assume homogeneous spherical charge distributions. This is certainly not true for cytochrome c and has led several authors to examine the kinetic consequences of asymmetric charge distributions as discussed in Section 2.7.2 (Koppenol et al. 1978; Koppenol and Margoliash 1982; van Leeuwwen 1983; Matthew et al. 1983).
4. Loss of positive charge due to specific lysine modification in cytochrome c can lead to either little or no change in the reaction rate, or a dramatic lowering of the rate depending on the particular lysine involved (Sect. 2.4). This would not be expected if simple monopole-monopole interactions were major determinants of productive collision. Similarly, many c-type cytochromes react with the strongly negatively charged redox protein flavodoxin with an ionic strength dependence consistent with a positive-negative interaction. This pattern is observed both for cytochromes with a net positive charge and for those with a net negative charge (Tollin et al. 1984).

2.7.2 Pre-Orientation by Electrostatic Interaction

Pre-orientation of cytochrome c and its reaction partner prior to their making physical contact has been suggested to allow the formation of an optimally reactive complex (Koppenol et al. 1978). Such favourable pre-orientation could occur through the interaction of the dipole moment of cytochrome c with the electrostatic potential field of the redox partner. The dipole moment (μ) is calculated from the distance between the "centres of gravity" of negative and positive charge in the molecule and in horse cytochrome c has a value of 325 Debye (Table 2.7). The positive end of the dipole lies at the top of the heme crevice (Fig. 2.17B), and at an angle of 33° to the heme plane. It lies close to the centre of the interaction domain on the front face of the molecule which was mapped by the methods described in Section 2.4.

Several points can be made regarding the calculation and significance of this dipole. Firstly, incorporation of a dipole term into the monopole Eqs. (2.16) and (2.20) allows a much better description of the ionic strength dependence of rate for redox proteins (van Leeuwwen 1983). An important feature of the dipole term is that it varies with I in contrast to the monopole term which varies with $\sqrt{I}$. Because of this, the dipole term becomes the more important of the two as the ionic strength is raised. Secondly, the calculations do not account for fluctuations in the dipole due to relative movement of the charged groups, the changing protonation state of those groups or the binding

Table 2.7. The dipole moment of horse cytochrome c. From Koppenol and Margoliash (1982)

Protein[a]	θ[b]	Dipole moment (debye)	Approximate points at which the dipole axis crosses the protein surface: Positive end	Negative end
Ferric (His 33 unprotonated)	0	325	Phe 82 amide N	Between Phe 36 (βC) and amide of Asn 103
Ferric (His 33 protonated)	9.4	304	Ile 81 carbonyl O	Phe 36 (βC)
Ferrous (His 33 unprotonated)	3.0	308	Phe 82 (βC)	Asn 103 amide
CDNP-Lys 7-ferric	27.2	428		
CDNP-Lys 8-ferric	37.7	384		
CDNP-Lys 13-ferric	32.7	243		
CDNP-Lys 72-ferric	35.9	187		
CDNP-Lys 73-ferric	41.2	272		

[a] His 33 has a pK_a of 6.4 and is probably mostly protonated under physiological conditions. The calculations for the CDNP derivatives assume this. The calculations also assume that the heme propionates are protonated. The additional negative charge on HP-7 (Vol. 2, Chap. 4) does not greatly affect the calculations. For example, the dipole moment of mouse ferricytochrome c is changed from 340 to 361 D by inclusion of the HP-7 negative charge, and θ changes by 5° only (Koppenol, personal communication).

[b] θ is the angle between the dipole vectors of ferricytochrome c and the other proteins.

of ions to the protein. Thirdly, the dipole moment of 325 Debye is not especially large for a protein in spite of the large number of positively and negatively charged residues on cytochrome c. Many globular proteins such as myoglobin (170 D), ovalbumin (250 D), and horse serum albumin (380 D) have comparable dipoles (although smaller in proportion to size) and a sphere of the same diameter as cytochrome c will generate a dipole of 170 D with a single positive surface charge diametrically opposed to a single negative one. The fourth point concerns the relative strength of the orientation effect at physiological ionic strength. Calculations of the electrostatic potential field (Koppenol and Margoliash 1982) suggest that at zero ionic strength the forces are great enough to allow orientation even at a distance of 40 Å. However, this distance falls dramatically as the ionic strength is raised. For example, Matthew et al. (1983) calculated that pre-orientation of cytochrome c and flavodoxin at I = 0.08 M will only occur when the two proteins are approx. 6 Å apart. Finally, in situ, the environment of cytochrome c contains not only its redox partners but also negatively charged membrane surfaces which may perturb any specific electrostatic orientation of the cytochrome.

The main question is whether pre-orientation by means of the dipole can be a significant contributor to rate enhancement. Experimental evidence relevant to this problem comes from the study of the effects of modification of individual lysines on the reaction rate as discussed in Section 2.4. For the singly modified CDNP lysine derivatives, Koppenol and Margoliash (1982) calculated the perturbed dipole moments which are shifted through an angle θ relative to the native protein. The work required (U) to adjust these dipoles to give the native protein orientation in an electric field (E) is given by:

$$U = E\mu(1-\cos\theta) \ . \tag{2.21}$$

This work represents an increase in the activation energy for the CDNP derivative and therefore:

$$r = k_{deriv.}/k_{native} = \frac{Ae^{-(E_a+U)/kT}}{Ae^{-E_a/kT}} \tag{2.22}$$

or

$$r = e^{-U/kT} \ . \tag{2.23}$$

In Fig. 2.21, the values of ln r are plotted against U/E [calculated from Eq. (2.21)] for different CDNP derivatives of cytochrome c in reaction with sulfite oxidase. Koppenol and Margoliash (1982) came to the following conclusions from this analysis.

1. The loss in activity on modification of certain lysines (13, 27, 72 and 86) is too great to be accounted for by changes in the dipole alone. These lysines are proposed to form the interaction site and be involved in specific salt bridges.

2. Modification of a second group of lysine (60, 39, 7, 25, 8, 73 and 87) causes effects on activity, roughly in proportion to the value of U/E. It is proposed that these effects are due to perturbation of the dipole.

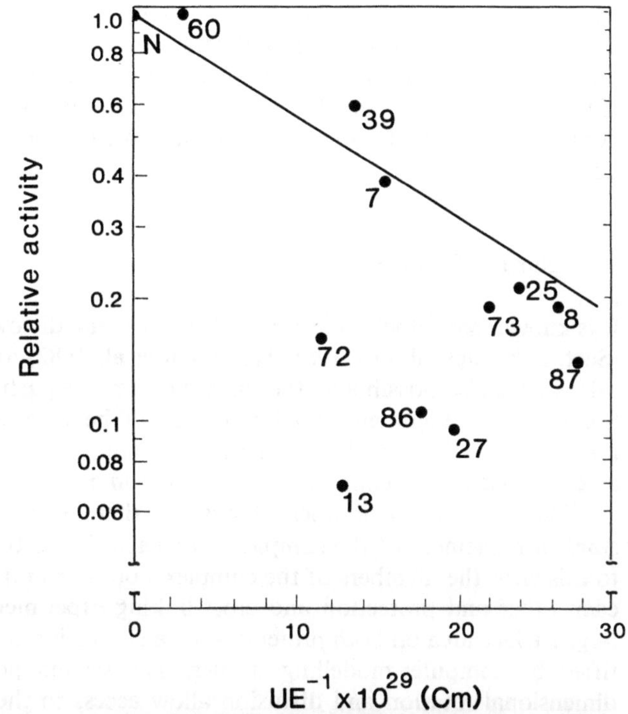

Fig. 2.21. The relative activities of CDNP-cytochromes c with sulfite oxidase. The *numbers* refer to the individual CDNP lysine derivatives. Ln r is calculated from the relative activity of the derivatives according to Eq. (2.22). These values are plotted against *U/E* calculated from $\mu(1-\cos\theta)$ according to Eq. (2.21) and expressed as Coulomb-meters. The *line* is arbitrary. It is proposed that for derivatives that fall on the line, the inhibition is due solely to the amount of work required to turn the dipole through an angle θ to form a productive complex. For those that fall below the line the inhibition cannot be explained solely by the change in the dipole moment and these lysines are proposed to define a specific interaction site. *N* native (Speck et al. 1981)

Their general conclusion is that dipole orientation is a significant contributor to rate enhancement in cytochrome c reactions. However, some critical comments should be offered. Firstly, the line that is drawn, representing the effect on r of the re-orientation work required on the dipole, is not a theoretical line because E is not known. The position of the line is therefore arbitrary and, if drawn for example through the experimental points for the native protein and the CDNP Lys 60 derivative in Fig. 2.21, would have led to the conclusion that no other derivative behaved according to the dipole model. Secondly, values of k_{on} are required for the calculation of r. In only a few cases are these values available (Koppenol and Margoliash 1982). In others, the limiting intercept on the ordinate of Eadie-Hofstee plots is used as an estimate of k_{on}, but while this may be valid in some instances (for example cytochrome b_5, Sect. 2.3), it is certainly not valid in others (for example cytochrome oxidase, where the kinetic plot is biphasic). Finally, in the case of TFA and TFC derivatives of cytochrome c, modification of certain lysines produce no effect on the rate of reaction with any of the several enzymes studies (Smith et al. 1980). These authors concluded that global monopole and dipole effects are not important contributions to rate enhancement, at least under the conditions tested. *

2.7.3 Surface Diffusion

It is known that cytochrome c can diffuse very rapidly over membrane surfaces (Sect. 2.8; Hackenbrock 1981; Hochman et al. 1982) and if two-dimensional diffusion is also possible on the surface of a redox partner, the probability for the productive alignment of redox centres will be increased. Such a mechanism raises the question of whether there is a unique complex for each interaction as envisaged by the computer-modelling studies (e.g. Fig. 2.20).

There are two possibilities. One is that there is not just a single optimally stable arrangement of the complex, competent in electron transfer. According to this view, the members of the complex would be in rapid motion relative to each other and protection and cross-linking experiments would implicate a large surface area on both proteins, only a subsection of which would be identified by computer-modelling studies. The second possibility is that two-dimensional rotation and diffusion allow access to the most stable arrangement of the complex where electron transfer occurs. According to this view, protection, cross-linking and computer-modelling experiments describe this trapped final state while kinetic experiments on modified cytochromes describe the ability of the cytochrome c to achieve the trapped state. This ability may be affected not only by the interactions in the final state but also by the adhesiveness of the initial collision and by the freedom of surface diffusion that occurs after collision.

There is no direct experimental evidence bearing on this question of surface diffusion.

2.7.4 Thermodynamic Measurements of Complex Formation

The association of cytochrome c with cytochrome b_5 is the only interaction of cytochrome c for which detailed thermodynamic information is available (Mauk et al. 1982). An analysis of the temperature dependence of the association equilibrium yielded $\Delta H^0 = 4 \pm 12\ \mathrm{kJ\ mol^{-1}}$ and $\Delta S^0 = 139 \pm 46\ \mathrm{JK^{-1}\ mol^{-1}}$ for $I = 0.001$ M and 25 °C indicating that the stabilisation of the complex is largely entropic in nature.

This entropic stabilisation is due to changes in the solvation of the proteins which outweigh the entropy change due to loss of rotational and translational energy of the individual proteins (Ross and Subramanian 1981). Both hydro-

* See appendix note 3

phobic association and charge neutralisation can contribute to the net entropy increase and their relative importance cannot be decided from the thermodynamic data alone. It seems likely that both make a significant contribution as the front face of cytochrome c has both hydrophobic regions (see Vol. 2, Chap. 4) and several charged residues.

The importance of water exclusion from such an interface is demonstrated by the interaction of cytochrome c with the highly charged bacterial redox protein, flavodoxin. The energy of stabilisation conferred by the interacting electrostatic potential fields can be calculated (Matthew et al. 1983) and is found to increase if water is excluded during complex formation.

2.7.5 The Role of Local Complementary Charge Interactions

The previous sections have tended to discount the role of monopole and dipole interactions in influencing the reactivity of cytochrome c. Even if dipole orientation effects were important they would require, in addition, a mechanism for locking the axially spinning dipole in the correct position for electron transfer. In this section we examine the role of a limited group of specific lysine residues involved in local complementary charge interactions with a redox partner. As shown in Section 2.4, there is good agreement that these lysines are situated at the front face of the molecule. Millet and co-workers have attempted to evaluate the energetic contribution of each of these lysines to the formation of the complex (Stonehuerner et al. 1979; Webb et al. 1980; H. T. Smith et al. 1981).

The contribution is given by:

$$V_i = -RT\ln\frac{(V_M/K_M)_{\text{native}}}{(V_M/K_M)_{\text{derivative (i)}}}, \tag{2.24}$$

where V_i is the interaction energy of a particular lysine (i) on cytochrome c with all charged groups on the reaction partner. Equation (2.24) was applied to each derivative of cytochrome c in which a single lysine had been modified. For each complementary charge pair, the separation distance r_{AB} was calculated from:

$$V_{AB} = 2.1175\left(\frac{e^{Ka_A}}{1+Ka_A}+\frac{e^{Ka_B}}{1+Ka_B}\right)\frac{Z_A Z_B e^{-Kr_{AB}}}{r_{AB}}, \tag{2.25}$$

where a_A and a_B are the effective radii of charged groups (assumed to be 1.7 Å).

In the case of sulfite oxidase in 0.1 M Tris Cl, pH 7.5 (Sect. 2.4.3 and Fig. 2.17), Webb et al. (1980) calculated that there are eight complementary charge interactions with separations of 4.8 to 10 Å. At the ionic strength of 0.1 M Tris Cl, the steep fall in V_{AB} as r_{AB} increases means that interactions separated by greater than 10 Å contribute very little to the total energy of stabilisation (V). However, at lower ionic strengths, longer range interactions become significant and more lysines are implicated (Sect. 2.4.3).

Although offering a coherent model for the role of local complementary charge interactions in complexes of cytochrome c, the approach of Millet and co-workers is open to several criticisms.

1. Equation (2.25) is based on extended Debye-Huckel theory and is suspect when applied to proteins at these relatively high ionic strengths. Cases of its successful application are for small molecules, not proteins, while many other examples of small molecule studies do not obey the equation.

2. The treatment is based on the assumption that lysine modification affects only the electrostatic component of the association step of the reaction. Steric effects can probably be excluded because of the agreement between the results for the bulky TFC modifying group and those for the much smaller TFA group. However, minor conformational changes that accompany the modification could be difficult to identify, yet cause large changes in reactivity. For example, the salt bridge between Lys 13 and Glu 90 is known to be important in maintaining the structural integrity of ferricytochrome c (Osheroff et al. 1980) yet when this salt bridge is lost by chemical modification of Lys 13 the subsequent reactivity change is ascribed solely to the charge change on the lysine. A further possibility, not considered by the model, is that modification may affect the rate constant for electron transfer. This is assumed to be constant because the redox potentials of the derivatives are very similar to that of unmodified cytochrome c but this does not exclude changes in the intrinsic reactivity of the cytochrome (see Vol. 2, Chap. 8).

Cusanovich and co-workers (Tollin et al. 1984) have employed a conceptually similar approach to analyse the ionic strength dependence of electron transfer between flavodoxin and members of the group of structurally homologous class I cytochromes (Chap. 3). Instead of examining the contributions of individual charge pairs, they described the interaction domain as parallel plates of radius ρ with net charges Z_1 (for cytochrome) and Z_2 (for flavodoxin). They propose that the local charge product $Z_1 Z_2$ contributes to the electrostatic interaction energy V_{ii} according to:

$$V_{ii} = a\rho^{-2}D^{-1}Z_1Z_2r_{12} \ , \tag{2.26}$$

where a is a constant, D is the dielectric constant of the interface and r_{12} is the distance between the plates. The analysis requires a number of assumptions making it largely empirical but it does produce reasonably good agreement between the calculated charge of the interaction domain and that expected from inspection of the crystallographic structures of the two molecules when $D_e = 10$ and $r_{12} = 3.5$ Å.

Equation (2.26) implies that as the distance between the charges increases, the electrostatic energy also increases. However, this is contrary to established theory and it arises because, as r_{12} increases ρ also increases, so that r_{12}/ρ^2, and hence V_{ii} decrease. Unfortunately, the derivation and theoretical justification for this equation have not been published.

In conclusion, the role of electrostatic effects in the reaction of cytochrome c with its redox partners is complex and is difficult to analyse quantitatively. Electrostatic interactions may be important in the orientation of the cytochrome either during its approach to its redox partner or during its diffusion over the surface of that partner. Electron transfer probably occurs in a precisely orientated complex or set of complexes stabilised by a small number of complementary charge interactions.

2.8 An Assessment of the Role of Cytochrome c in the Mitochondrion

In this section we draw together different lines of research in an attempt to describe the role of cytochrome c in the mitochondrion.

2.8.1 The Reaction with Cytochrome Oxidase

Under certain ionic strength conditions, the cytochrome oxidase monomer can bind two cytochrome c molecules with different affinity (Sect. 2.2). The higher affinity binding ($K_D \sim 10^{-7}$ M) involves electrostatic interaction of the positive front face of cytochrome c with an acidic region of subunit II (Sects. 2.4 and 2.5). These interaction areas probably enclose the route of electron transfer from heme c to heme a. The low affinity binding ($K_D \sim 10^{-6}$ M) probably also involves the front face of cytochrome c but in interaction with tightly bound phosphatidyl glycerol on the oxidase. This binding is abolished (Vik et al. 1981) or diminished in contribution (Thompson and Ferguson-Miller 1983) if most of the phospholipid is removed from the oxidase.

The binding of two molecules of cytochrome c gives rise to biphasic kinetic plots which reflect the high and low affinity binding (Sect. 2.3). The high affinity kinetic phase is characterised by low turnover, limited by the rate of dissociation of the product, ferricytochrome c. At higher substrate concentrations, the binding of a second molecule facilitates dissociation of the first and gives rise to a low affinity phase of high turnover. Thus the low affinity binding appears to play an effector role in the dissociation of cytochrome c. It is probably not itself competent in electron transfer because covalent attachment of cytochrome c to the high affinity site only, blocks cytochrome c oxidation.

It is not yet certain whether this facilitated dissociation is an intrinsic property of the monomer or arises from negative cooperativity of binding to the two sites of the dimer (Sect. 2.2.3.1). The balance of evidence favours the former and that view is indicated diagrammatically in Fig. 2.19. According to this Fig., two cytochromes c may bind with high and low affinity to each monomer of the dimer. The binding takes place in an interface between the monomers such that binding of yeast (TNB Cys 102) cytochrome c to subunit II of one monomer can lead to covalent cross-link formation to subunit III of the second monomer with loss of cytochrome c oxidase activity proportional to the degree of cross-linking. In contrast, covalent attachment of (arylazido Lys 13) cytochrome c to the dimeric enzyme leads to complete loss of activity after only 50% cross-linking to protein had occurred. This is an apparent "half-of-the-sites" effect which may indicate steric hindrance in the dimer cleft.

2.8.2 The Transfer of Electrons Between Complex III and Cytochrome Oxidase

We are now in the position to assess the three possible models which were proposed for the action of cytochrome c in the mitochondrion (Fig. 2.2). The large body of evidence discussed in Sects. 2.4 and 2.5 suggests that there is one site of electron exit and entry on cytochrome c and that this site is enclosed by an interaction surface recognised by both complex III and cytochrome oxidase. This eliminates the static, electron-conducting "wire" model of Fig. 2.2a.

A requirement of model 2.2c is that the dissociation rate constant for the cytochrome c : oxidase complex must be large enough to support the observed electron transfer rates of $100\,s^{-1}$ or greater (Nicholls and Chance 1974; Ferguson-Miller et al. 1976). Available estimates of k_{off} include the $2.5\,s^{-1}$ for the reduction of the cytochrome c : oxidase complex by cytochrome c_1 (Wilms et al. 1981b) and the turnover numbers of between 5 and $40\,s^{-1}$ for the high affinity phase of the spectroscopic assay which is probably limited by the dissociation of cytochrome c (Errede et al. 1976; Rosevear et al. 1980). Thus the high affinity phase of cytochrome c oxidation, which is expected to be the predominant phase in most mitochondria due to the low cytochrome c content cannot support observed electron transfer rates. This problem has not yet been fully resolved but there are two possible solutions. One, which is considered in the following subsection, proposes that under physiological ionic conditions, dissociation of cytochrome c is not rate-limiting. The other involves restricted mobility of cytochrome c between the oxidase and reductase without dissociation (model 2.2b). The most restricted form of this model is one in which cytochrome c rotates between an oxidase site and a reductase site in a ternary complex of the three proteins (Mochan and Nicholls 1972; Nicholls 1974; Ahmed et al. 1978). Although such a ternary complex has been observed on mixing purified components (Chiang and King 1979) several lines of evidence suggest that such complexes have no physiological role and that cytochrome c acts by diffusing rapidly but locally over the membrane surface.

First, the rotational relaxation time for cytochrome oxidase in membranes is not influenced by the presence of complex III (Kawato et al. 1981). Second, succinate oxidase activity is markedly inhibited by raising the viscosity of the medium while ascorbate-TMPD oxidase activity is unaffected (Swanson et al.

1982). Also, replacement of vesicle phospholipids by dimyristoyl glycerophosphate which has a sharp phase transition, greatly inhibits auinol oxidase activity at lower temperatures in parallel with a progressive immobilisation of spin-labelled cytochrome c (Froud and Ragan 1984). Third, the mitochondrial membrane is unusually fluid and ideally suited to rapid diffusion of cytochrome c associated with phospholipid (Hackenbrock 1981). A local lateral diffusion coefficient of $10^{-8}\ cm^2\ s^{-1}$ for cytochrome c is feasible by analogy with surface diffusion of apolipoprotein on phospholipids. It can be calculated that such a diffusion rate would be more than enough to account for maximal rates of mitochondrial electron transfer (Kawato et al. 1981; Hochman et al. 1985).

However, direct measurement of cytochrome c mobility by the FRAP technique (fluorescence recovery after photobleaching) give values near $10^{-10}\ cm^2\ s^{-1}$ (Hochman et al. 1982, 1985; Vanderkooi et al. 1985) which is too slow to account for electron transfer. This discrepancy has not been resolved but it is possible that cytochrome c is rapidly mobile within a limited domain on the membrane bounded by bc_1 and aa_3 complexes. The "choice" of a diffusible link rather than a solid-state system is almost certainly an evolutionary vestige derived from the requirement for branching in the more complex bacterial respiratory systems (Chap. 3). As we will see in the following section, a few elements of branching remain in some mitochondria and involve enzymes of the intermembrane space.

2.8.3. The Role of Cytochrome c in the Intermembrane Space

From the arguments in the preceding section, it is probable that cytochrome c diffuses rapidly across the membrane surface to interact separately with its reductase and oxidase. In this mode, the heme electron transfer site is probably oriented towards the bilayer, an arrangement not suited for reaction with soluble intermembrane enzyme systems. Therefore, there is a requirement for a second pool of soluble cytochrome c.

Since all the interactions of cytochrome c that have been discussed in this chapter are electrostatic, a knowledge of the ionic conditions that exist in the intermembrane space is a prerequisite to any consideration of how cytochrome c is distributed. Indeed, we might look forward to the day when the distribution of cytochrome c might be calculated by considering the relevant binding constants and the concentrations of participating species. Unfortunately, present knowledge falls far short of this. The intermembrane space is thought to be represented by the sucrose-permeable volume of prepared mitochondria and may therefore share with the cytoplasm a pool of low molecular weight compounds including ions which would amount to an ionic strength of about 0.15 M (Wojtczak and Sottocasa 1972). According to Hackenbrock (1981), the size of the intermembrane space varies from 50% of the mitochondrial volume in state III respiration to 10% in state IV. On the other hand, Ferguson-Miller et al. (1978b) argued that because of repulsion effects by the two charged membrane surfaces, ions will be excluded from the space and the ionic strength will be low. In the extreme, Sjostrand (1978) proposed that there is no intermembrane space and that the inner and outer membranes are attached, a view which, if confirmed, would require modification of the present models of cytochrome c action.

There is no direct evidence bearing on the crucial question of the ionic make-up of the intermembrane space but we can ask whether any free cytochrome c is available in this compartment and this should be a reflection of the ionic strength conditions that pertain there. However, there is a sense in which we dictate the answer we obtain since mitochondria have to be prepared and supported in a particular osmotic and ionic medium which presumably will influence the composition of the intermembrane space. Thus for Ferguson-Miller et al. (1978b) to claim that cytochrome c is all bound in vivo because potentiometric measurements show maximum lowering of redox potential (Dutton et al. 1970), ignores the fact that the mitochondria of these studies were prepared in media of relatively low ionic strength (40 mM MOPS).

It is well known that cytochrome c can be washed from Keilin-Hartree particles with 0.15 M KCl and Matlib and O'Brien (1976) showed that in 0.15 M KCl, much of the cytochrome c in rat liver mitochondria is soluble in the intermembrane space because removal of the outer membrane by mild digitonin treatment released most of the cytochrome c and the adenylate kinase. Sulfite oxidase, which was CN^- insensitive in mitochondria prepared in 0.33 M sucrose, became partially CN^- sensitive in 0.15 M KCl suggesting that free cytochrome c was mediating electron transfer between the sulfite oxidase and cytochrome oxidase. Most remarkable of all, sulfite oxidase in intact liver cells was fully CN^- sensitive. Similar results were obtained by Bernardi and Azzone (1981) who showed a steady-state level of cytochrome c reduction of approximately 15% in 0.125 M KCl with exogenous NADH as donor. Finally, Chance et al. (1972) allowed rat liver mitochondria to take up yeast cytochrome c peroxidase and found that 30% of the cytochrome c could be oxidised in a very fast reaction in the presence of peroxide while the remainder was oxidised at a much slower rate. These authors postulated the existence of two pools of cytochrome c, one available for reaction in the intermembrane space, and the other associated with the inner membrane electron transfer system. The relative proportions of the two pools is defined by the ionic strength but if we assume what seems a reasonable figure for this parameter (e.g. 0.05 – 0.15 M) it is clear that a sizeable fraction of the cytochrome c will not be bound to the membrane. The proportion of free cytochrome will also be dependent on the total amount of cytochrome c present relative to its membrane redox partners. Mitochondria in which redox processes in the intermembrane space appear to play dominant roles (for example those of fungi) do appear to have more cytochrome c (Table 2.1).

2.8.4 The Physiological Relevance of the Biphasic Kinetics

In view of these doubts about the ionic environment of cytochrome c it is perhaps surprising that work on the binding of cytochrome c to the oxidase has been so dominated by experimental conditions of low ionic strength. It is clear that our evaluation of the physiological significance of high and low affinity sites would be completely altered if ionic strengths in vivo were as high as 0.15 M. Thus at low ionic strength, striking kinetic differences were found in the reaction of higher and lower primate cytochromes c with beef oxidase but these were not evident at high ionic strength (Ferguson-Miller et al. 1978b). Similarly, L. Smith et al. (1973a) showed that at high ionic strength no differences in activity were observed for a wide range of eukaryotic cytochromes c. Perhaps most striking of all, as little as 25 mM phosphate or 3 mM ATP totally abolishes the biphasic nature of the Eadie-Hofstee kinetic plots and, therefore, by implication, the high affinity site (Ferguson-Miller et al. 1976).* It is not known what effect high ionic strength would have on the nature of the binding site deduced from modification experiments.

Margoliash and co-workers appear now to be as committed to functional differences in the eukaryotic cytochromes c as a result of evolutionary processes as they were previously to the apparent lack of functional differences (L. Smith et al. 1973a). It is as hazardous to draw evolutionary conclusions from results in vitro which appear to show functional differences in proteins, as from results which show no such differences. In the first case, the differences may be physiologically irrelevant; in the second, physiologically important parameters of function may have been missed.

As a partial analogy we offer the study of the K_M for pyruvate kinase from a variety of animals (Fig. 2.22 from Somero and Low 1976). The conclusion was that at the natural temperature of operation, the K_M for this enzyme

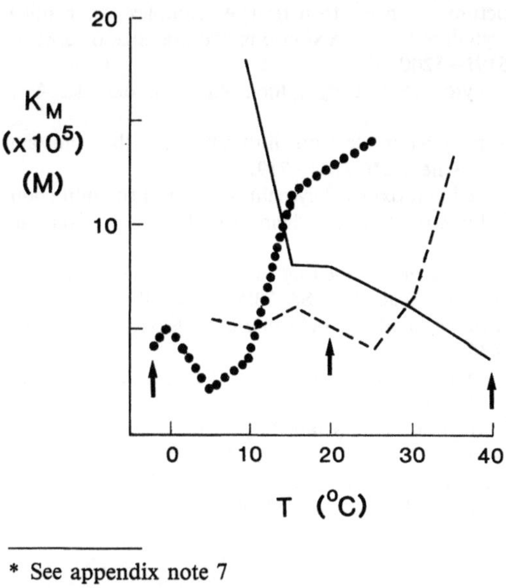

Fig. 2.22. The effect of assay temperature on the apparent K_M for pyruvate kinase from different species of fish. ● ● ● *Trematomus borchgrevinki,* an antarctic teleost living at −1.86 °C throughout the year. – – – *Gillichthys mirabilis,* an eurythermal estuarine fish from southern California living in water between 10 °C and 30 °C. ——— *Cyprinodon macularius,* a warm-adapted eurythermal fish found in ponds in southern Californian deserts at a temperature of approx. 40 °C. The *lines* are experimental determinations of K_M. *Arrows* indicate the average temperature of the environment for each fish

* See appendix note 7

is a remarkably conserved phenomenon. A corresponding comparison for cytochrome c oxidation would be the relationship between K_M and ionic strength for different mitochondria. The problem in this case, however, is that we do not know the ionic strengths at which to make comparisons.

References

Abe K, Kimura S, Kizawa R, Anan FK, Sugita Y (1985) Amino acid sequences of cytochrome b_5 from human, porcine and bovine erythrocytes and comparison with liver microsomal cytochrome b_5. J Biochem 97:1659–1668

Ahmed AJ, Smith HT, Smith MB, Millet FS (1978) The effect of specific lysine modification on the reduction of cytochrome c by succinate cytochrome c reductase. Biochemistry 17:2479–2483

Andersson T, Thulin E, Forsen S (1979) Ion-binding to cytochrome c studied by nuclear magnetic quadrupole relaxation. Biochemistry 18:2487–2493

Antalis TM, Palmer G (1982) Kinetic characterisation of the interaction between cytochrome c oxidase and cytochrome c. J Biol Chem 257:6194–6206

Appleby CA, Morton RK (1959) Purification of cytochrome b_2. Biochem J 71:492–499

Argos P, Mathews FS (1975) The structure of cytochrome b_5 at 2.8 Å resolution. J Biol Chem 250:747–751

Azzi A, Bill K, Broger C (1982) Affinity chromatography purification of cytochrome c binding enzymes. Proc Natl Acad Sci USA 79:2447–2450

Barlow GH, Margoliash E (1966) Electrophoretic behaviour of mammalian-type cytochromes c. J Biol Chem 241:1473–1477

Baudras A (1971) Pure lactate: cytochrome c oxidoreductase of the yeast *Hansenula anomala*. Biochimie 53:929–933

Baudras A, Capeillere-Blandin C, Iwatsubo M, Labeyrie F (1972) The formation of a stable complex between cytochrome b_2 and cytochrome c and study of its role in the overall electron transfer by rapid kinetics. In: Akeson A, Ehrenberg A (eds) Structure and function of oxidation reduction enzymes. Pergamon, New York, pp 273–290

Bechtold R, Bosshard HR (1985) Structure of an electron transfer complex – chemical modification of carboxyl groups of cytochrome c peroxidase in the presence or absence of cytochrome c. J Biol Chem 260:5191–5200

Beetlestone J (1960) The oxidation of cytochrome c by cytochrome c peroxidase. Arch Biochem Biophys 89:35–40

Bernardi P, Azzone GF (1981) Cytochrome c as an electron shuttle between the outer and inner mitochondrial membranes. J Biol Chem 256:7187–7192

Bill K, Azzi A (1984) Interaction of reduced and oxidised cytochrome c with the mitochondrial cytochrome c oxidase and bc_1-complex. Biochem Biophys Res Commun 120:124–130

Birchmeier W, Kohler CE, Schatz G (1976) Affinity labelling of yeast cytochrome oxidase by modified yeast cytochrome c. Proc Natl Acad Sci USA 73:4334–4338

Bisson R, Capaldi RA (1981) Binding of arylazido cytochrome c to yeast cytochrome c peroxidase. J Biol Chem 256:4362–4367

Bisson R, Montecucco C (1982) Different polypeptides of bovine heart cytochrome c oxidase are in contact with cytochrome c. FEBS Lett 150:49–53

Bisson R, Gutweniger H, Azzi A (1978a) Photoaffinity labelling of yeast cytochrome oxidase with arylazido cytochrome c derivatives. FEBS Lett 92:219–222

Bisson R, Azzi A, Gutweniger H, Collona R, Montecucco C, Zanotti A (1978b) Interaction of cytochrome c with cytochrome oxidase. J Biol Chem 253:1874–1880

Bisson R, Montecucco C, Gutweniger H, Azzi A (1979) Hydrophobic labelling of bovine heart cytochrome c oxidase with an azidophosphatidyl choline. Biochem Soc Trans 7:156–160

Bisson R, Jacobs B, Capaldi RA (1980) Binding of arylazido cytochrome c derivatives to beef heart cytochrome c oxidase – cross-linking in the high and low affinity sites. Biochemistry 19:4173–4178

Bisson R, Steffens GCM, Capaldi RA, Buse G (1982) Mapping of the cytochrome c binding site on cytochrome oxidase. FEBS Lett. 144:359–363

Bolli R, Nalecz KA, Azzi A (1985a) The aggregation state of bovine heart oxidase and its kinetics in monomeric and dimeric form. Arch Biochem Biophys 240:102–116

Bolli R, Nalecz KA, Azzi A (1985b) The interconversion between monomeric and dimeric bovine heart cytochrome c oxidase. Biochimie 67:119–128

Bosshard HR, Zurrer M, Schagger H, Jagow G von (1979) Binding of cytochrome c to the cytochrome bc_1 complex (complex III) and its subunits, cytochrome c_1 and b. Biochem Biophys Res Commun 89:250–258

Bosshard HR, Banziger J, Hasler T, Poulos TL (1984) The cytochrome c: cytochrome c peroxidase complex – the role of histidine residues. J Biol Chem 259:5683–5690

Boswell AP, McClune GJ, Moore GR, Williams RJP, Pettigrew GW, Inubishi T, Yonetani T, Harris DE (1980) NMR study of the interaction of cytochrome c with cytochrome c peroxidase. Biochem Soc Trans 8:637–638

Bowyer JR, Edwards CA, Trumpower BL (1981) Involvement of the iron-sulfur protein of the mitochondrial cytochrome bc_1 complex in the oxidant-induced reduction of cytochrome b. FEBS Lett 126:93–97

Brautigan DL, Ferguson-Miller S, Margoliash E (1978) Reaction with 4-chloro-3,5-dinitro benzoate and chromatographic separation of singly substituted cytochrome c derivatives. J Biol Chem 253:130–139

Briggs MM, Capaldi RA (1978) Cross-linking studies on a cytochrome c – cytochrome oxidase complex. Biochem Biophys Res Commun 80:553–559

Broger C, Nalecz MJ, Azzi A (1980) Interaction of cytochrome with the cytochrome bc_1 complex of the mitochondrial respiratory chain. Biochim Biophys Acta 592:519–529

Broger C, Salardi S, Azzi A (1983) Interaction between isolated cytochrome c_1 and cytochrome c. Eur J Biochem 131:349–352

Brooks SPJ, Nicholls P (1982) Anion and ionic strength effects upon the oxidation of cytochrome c by cytochrome oxidase. Biochim Biophys Acta 680:33–43

Brunori M, Wilson MT (1982) Cytochrome oxidase. Trends Biochem Sci 7:295–299

Buse G, Meinecke L, Bruch B (1985) The protein formula of beef heart cytochrome c oxidase. J Inorg Biochem 23:149–153

Capaldi RA, Darley-Usmar V, Fuller S, Millet F (1982) Structural and functional features of the interaction of cytochrome c with Complex III and cytochrome oxidase. FEBS Lett 138:1–7

Capaldi RA, Malatesta F, Darley-Usmar VM (1983) Structure of cytochrome oxidase. Biochim Biophys Acta 736:135–148

Capeillere-Blandin C (1982) Transient kinetics of the one electron transfer reaction between reduced flavocytochrome b_2 and oxidised cytochrome c. Eur J Biochem 128:533–542

Capeillere-Blandin C, Bray RC, Iwatsubo M, Labeyrie F (1975) Flavocytochrome b_2 – kinetic studies by absorbance and EPR spectroscopy of electron distribution among prosthetic groups. Eur J Biochem 54:549–566

Casey RP, Broger C, Azzi A (1981) Structural studies on the cytochrome c oxidase proton pump using a spin-label probe. Biochim Biophys Acta 638:86–93

Caughey WS, Wallace WJ, Volpe JA, Yoshikawa S (1976) Cytochrome oxidase. In: Boyer PD (ed) The enzymes. Academic Press, London New York, pp 299–344

Cerletti N, Schatz G (1979) Cytochrome c oxidase from bakers' yeast – photolabelling of subunits exposed to the lipid bilayer. J Biol Chem 254:7746–7751

Chance B (1967) The reactivity of hemoproteins and cytochromes. Biochem J 103:1–18

Chance B (1974) The function of cytochrome c. Ann N Y Acad Sci 227:613–626

Chance B, Schoener B, DeVault D (1964) Reaction velocity constants for electron transfer and transport reactions. In: King TE, Mason M, Morrison M (eds) Oxidases and related redox systems, vol 2. Wiley, New York London, pp 907–942

Chance B, Erecinska M, Wilson DF, Dutton PL, Lee CP (1972) The function of cytochrome c in mitochondrial membranes. In: Akeson A, Ehrenberg A (eds) Structure and function of oxidation reduction enzymes. Pergamon, UK, pp 263–272

Chiang YL, King TE (1979) Cytochrome c complexes. J Biol Chem 254:1845–1853

Chiang YL, Kaminsky LS, King TE (1976) A complex of cardiac cytochrome c_1 and cytochrome c. J Biol Chem 251:29–36

Cohen HJ, Fridovich I (1971) Hepatic sulfite oxidase. J Biol Chem 246:359–373

Cohen HJ, Betcher-Land S, Kessler DL, Rajagopalan KV (1972) Hepatic sulfite oxidase – congruency in mitochondria of prosthetic groups and activity. J Biol Chem 247: 7759–7766

Corbley MJ, Azzi A (1984) Resolution of bovine heart cytochrome c oxidase into smaller complexes by controlled subunit denaturation. Eur J Biochem 139:535–540

Cumsky MG, Trueblood CE, Poyton RO (1985) Two non-identical forms of subunit V are functional in yeast cytochrome c oxidase. Proc Natl Acad Sci USA 82:2235–2239

Dailey HA, Strittmatter P (1979) Modification and identification of cytochrome b_5 carboxyl groups involved in protein: protein interaction with cytochrome b_5 reductase. J Biol Chem 254:5388–5396

Darley-Usmar VM, Georgevich G, Capaldi RA (1984) Reaction of thionitrobenzoate-modified yeast cytochrome c with monomeric and dimeric forms of beef heart cytochrome c oxidase. FEBS Lett 166:131–135

Deatherage JF, Henderson R, Capaldi RA (1982) Three-dimensional structures of cytochrome c oxidase crystals in negative stain. J Mol Biol 158:487–499

Dethmers JK, Ferguson-Miller S, Margoliash E (1979) Comparison of yeast and beef oxidases. J Biol Chem 254:11973–11981

Dickerson RE, Timkovich R (1975) Cytochromes c. In: Boyer PD (ed) The enzymes, 3rd edn, Vol 11. Academic Press, New York, pp 397–547

Diggens RJ, Ragan CI (1982) Properties of UQ oxidase reconstituted from UQ cytochrome c reductase, cytochrome c and cytochrome oxidase. Biochem J 202:527–534

Dixon HBF (1976) Removal of a bound ligand from a macromolecule by gel-filtration. Biochem J 159:161–162

Dowe RJ, Vitello LB, Erman JE (1984) Sedimentation equilibrium studies on the interaction between cytochrome c and cytochrome c peroxidase. Arch Biochem Biophys 232:566–573

Dutton PL, Wilson DF, Lee CP (1970) Oxidation-reduction potentials of cytochromes in mitochondria. Biochemistry 9:5077–5082

Eigen M, Hammes CG (1963) Elementary steps in enzyme reactions. Adv Enzymol 25:1–38

Eley CGS, Moore GR (1983) ^{1}H-NMR investigation of the interaction between cytochrome c and cytochrome b_5. Biochem J 215:11–21

Erecinska M, Vanderkooi JM (1978) Modification of cytochrome c. Methods Enzymol 53:165–181

Erecinska M, Vanderkooi JM, Wilson DF (1975) Cytochrome c interaction with membranes – a photoaffinity labelled cytochrome. Arch Biochem Biophys 171:108–116

Erecinska M, Davis JS, Wilson DF (1980) Interactions of cytochrome c with mitochondrial membranes. J Biol Chem 255:9653–9658

Erman JE, Vitello LB (1980) The binding of cytochrome c peroxidase and ferricytochrome c. J Biol Chem 255:6224–6227

Errede BJ, Kamen MD (1978) Comparative kinetic studies of cytochromes c in reactions with mitochondrial cytochrome c oxidase and reductase. Biochemistry 17:1015–1027

Errede BJ, Kamen MD (1979) Mechanisms for the steady state kinetics of reaction of cytochrome c with mitochondrial cytochrome oxidase. In: King TE, Orii Y, Chance B, Okunuki K (eds) Cytochrome oxidase. Elsevier, Amsterdam, pp 269–280

Errede BJ, Haight GP, Kamen MD (1976) Oxidation of ferrocytochrome c by mitochondrial cytochrome c oxidase. Proc Natl Acad Sci USA 73:113–117

Falk KE, Jovall PA, Angstrom J (1981) NMR and EPR characterisation of 4-carboxy-2,6-dinitrophenyl lysine cytochromes c. Biochem J 193:1021–1024

Ferguson-Miller S, Brautigan DL, Margoliash E (1976) Correlation of the kinetics of electron transfer activity of various eukaryotic cytochromes c with binding to mitochondrial cytochrome oxidase. J Biol Chem 251:1104–1115

Ferguson-Miller S, Brautigan DL, Margoliash E (1978a) Definition of cytochrome c binding domains by chemical modification. J Biol Chem 253:149–159

Ferguson-Miller S, Brautigan DL, Margoliash E (1978b) The electron transfer function of cytochrome c. In: Dolphin D (ed) The Porphyrins, vol 7. Academic Press, London New York, pp 149–240

Finzel BC, Poulos TL, Kraut J (1984) Crystal structure of yeast cytochrome c peroxidase refined at 1.7 Å resolution. J Biol Chem 259:13027–13036

Frey TG, Chan SH, Schatz G (1978) Structure and orientation of cytochrome c oxidase in crystalline membranes. Studies by electron microscopy and by labelling with subunit specific antibodies. J Biol Chem 253:4389–4395

Froud RJ, Ragan CI (1984) Cytochrome c mediates electron transfer between ubiquinol cytochrome c reductase and cytochrome c oxidase by free diffusion along the surface of the membrane. Biochem J 217:561–571

Fuller SD, Capaldi RA, Henderson R (1979) Structure of cytochrome c oxidase in deoxycholate derived two-dimensional crystals. J Mol Biol 134:305–327

Fuller SD, Darley-Usmar VM, Capaldi RA (1981) Covalent complex between yeast cytochrome c and beef heart oxidase which is active in electron transfer. Biochemistry 20:7046–7053

Fuller SD, Capaldi RA, Henderson R (1982) Preparation of two-dimensional arrays of purified beef heart cytochrome c oxidase. Biochemistry 21:2525–2529

Gelder BF van, Buuren KJH van, Wilms J, Verboom CN (1975) The effect of ionic strength on the oxidation of cytochrome c by cytochrome oxidase. In: Quaglierello E, Papa S, Palmieri F, Slater EC, Siliprandi N (eds) Electron transport chains and oxidative phosphorylation. Elsevier/North Holland Biomedical Press, New York Amsterdam, pp 63–68

Gelder BF van, Veerman E, Wilms J, Dekker HL (1979) The electron accepting site of cytochrome oxidase. In: King TE et al. (eds) Cytochrome oxidase. Elsevier/North Holland Biomedical Press, New York Amsterdam, pp 305–314

Gelder BF van, Veerman E, Wilms J, Dekker HL (1982) The reaction between cytochrome c and cytochrome oxidase. In: King TE, Mason HS, Morrison M (eds) Proc Int Symp Oxidases and related redox systems. Pergamon, New York, pp 1055–1062

Gellerfors P, Nelson BD (1977) Topology of the peptides in free and membrane bound complex III (UQ cytochrome c reductase) as revealed by lactoperoxidase and p-diazonium benzene (^{35}S) sulfonate labelling. Eur J Biochem 80:275–282

Georgevich G, Darley-Usmar VM, Malatesta F, Capaldi RA (1983) Electron transfer in monomeric forms of beef and shark cytochrome oxidase. Biochemistry 22:1317–1322

Gervais M, Risler Y, Corazzin S (1983) Proteolytic cleavage of *Hansenula anomala* flavocytochrome b_2 into its two functional domains. Eur J Biochem 130:253–259

Gibson QH, Greenwood C (1965) Kinetic observations on the near-infra-red band of cytochrome c oxidase. J Biol Chem 240:2694–2698

Gibson QH, Greenwood C, Wharton DC, Palmer G (1965) The reaction of cytochrome oxidase with cytochrome c. J Biol Chem 240:888–894

Greenwood C, Brittain T, Wilson M, Brunori M (1976) Studies on partially reduced mammalian cytochrome oxidase reactions with ferrocytochrome c. Biochem J 157:591–598

Guiard B, Lederer F (1979a) The sequence of b_5-like binding domain of chicken sulphite oxidase. Eur J Biochem 100:441–453

Guiard B, Lederer F (1979b) The cytochrome b_5 fold – structure of a novel protein superfamily. J Mol Biol 135:639–650

Guiard B, Groudinsky O, Lederer F (1974) Homology between bakers yeast cytochrome b_5 and liver microsomal cytochrome b_5. Proc Natl Acad Sci USA 71:2539–2543

Gupta RK, Yonetani T (1973) Nuclear magnetic resonance studies of the interaction of cytochrome c with cytochrome c peroxidase. Biochim Biophys Acta 292:502–508
Gupta RK, Konig SH, Redfield AG (1972) On the electron transfer between cytochrome c molecules as observed by NMR. J Mag Res 7:66–73
Gutweniger H, Bisson T, Montecucco C (1981) Hydrophobic labelling of the yeast cytochrome c oxidase subunits in contact with lipids. Biochim Biophys Acta 635:187–193
Gutweniger HE, Grassi C, Bisson R (1983) Interaction between cytochrome c and ubiquinone cytochrome c reductase – a study with water-soluble carbodiimides. Biochem Biophys Res Commun 116:272–283
Hackenbrock CR (1981) Lateral diffusion and electron transfer in the mitochondrial inner membrane. Trends Biochem Sci 6:151–154
Hackett CS, Strittmatter P (1984) Covalent cross-linking of the active site of vesicle-bound cytochrome b_5 and NADH cytochrome b_5 reductase. J Biol Chem 259:3275–3282
Hakvoort TBM, Sinjorgo KMC, Gelder BF van, Muijsers AO (1985) Separation of enzymatically active bovine cytochrome c oxidase monomers and dimers by high performance liquid chromatography. J Inorg Biochem 23:381–388
Hartzell CR, Hansen RE, Beinert H (1973) Electron carriers of cytochrome oxidase detectable by EPR and their relationship to those traditionally recognised in this enzyme. Proc Natl Acad Sci USA 70:2477–2481
Hatefi Y, Rieske JS (1967) The preparation and properties of DPNH-cytochrome c reductase. Meth Enzymol 10:255–231
Hatefi Y, Haavik AG, Griffiths DE (1962) Studies on the electron transfer system – reduced CoQ cytochrome c reductase. J Biol Chem 237:1681–1685
Hauska G, Hurt E, Gabellini N, Lockau W (1983) Comparative aspects of quinol-cytochrome c/plastocyanin oxidoreductases. Biochim Biophys Acta 726:97–133
Henderson R, Unwin PNT (1975) Three-dimensional model of purple membrane obtained by electron microscopy. Nature (London) 257:28–32
Henderson R, Capaldi RA, Leigh JS (1977) Arrangement of cytochrome oxidase molecules in two dimensional vesicle crystals. J Mol Biol 112:631–648
Hill BC, Greenwood C (1984) Kinetic evidence for the re-definition of electron transfer pathways from cytochrome c to O_2 within cytochrome oxidase. FEBS Lett 166:362–366
Hochman JH, Schindler M, Lee JG, Ferguson-Miller S (1982) Lateral mobility of cytochrome c on intact mitochondrial membranes as determined by fluorescence redistribution after photobleaching. Proc Natl Acad Sci USA 79:6866–6870
Hochman JH, Ferguson-Miller S, Schindler M (1985) Mobility in the mitochondrial electron transport chain. Biochemistry 24:2509–2516
Hoffman BM, Roberts JE, Kang CH, Margoliash E (1981) Electron paramagnetic and electron nuclear double resonance of the hydrogen peroxide compound of cytochrome c peroxidase. J Biol Chem 256:6556–6564
Hovmoller S, Leonard K, Weiss H (1981) Membrane crystals of a subunit complex of cytochrome reductase containing cytochrome b and c_1. FEBS Lett 123:118–122
Hummel JP, Dreyer WJ (1962) Measurement of protein binding phenomena by gel filtration. Biochim Biophys Acta 63:530–532
Huttquist DE, Passon PG (1971) Catalysis of methemoglobin reduction by erythrocyte cytochrome b_5 and cytochrome b_5 reductase. Nature (London) 229:252–254
Ito A (1980) Cytochrome b_5 – like hemoprotein of outer mitochondrial membranes. J Biochem 87:63–71
Jacobs EE, Sanadi DR (1960) The reversible removal of cytochrome c from mitochondria. J Biol Chem 235:531–534
Jacobs EE, Andrews EC, Crane FL (1964) A reaction scheme for cytochrome oxidase suggested by its characteristic reactions with high potential electron donors. In: King TE, Mason HS, Morrison M (eds) Oxidases and related redox systems. Wiley, London New York, pp 784–802
Jacq C, Lederer F (1972) Sur les deux formes moléculaires du cytochrome b_2 de *Saccharomyces cerevisiae.* Eur J Biochem 25:41–48

Jacq C, Lederer F (1974) Cytochrome b_2 from bakers yeast – a double headed enzyme. Eur J Biochem 41:311–320

Jarausch J, Kadenbach B (1985a) Structure of the cytochrome c oxidase complex from rat liver – topographical orientation of polypeptides in the membrane as studied by proteolytic digestion and immunoblotting. Eur J Biochem 146:219–225

Jarausch J, Kadenbach B (1985b) Structure of the cytochrome c oxidase complex of rat liver – studies on nearest neighbour relationships of polypeptides with cross-linking agents. Eur J Biochem 146:211–217

Johnson JL, Rajagopalan KV (1977) Tryptic cleavage of rat liver sulfite oxidase. Isolation and characterisation of molybdenum and heme domains. J Biol Chem 252:2017–2025

Johnson JL, Rajagopalan KV (1980) The oxidation of sulfite in animal systems. In: Elliott K, Whelan J (eds) Sulfur in biology (CIBA foundation symposium). Exerpta Medica, Amsterdam, pp 119–133

Kadenbach B, Stroh A (1984) Different reactivity of carboxylic groups of cytochrome oxidase polypeptides from pig liver and heart. FEBS Lett 173:374–380

Kadenbach B, Ungibauer M, Jarausch J, Buge U, Kuhn-Nentwig L (1983) The complexity of respiratory complexes. Trends Biochem Sci 8:398–400

Kang CH, Ferguson-Miller S, Margoliash E (1977) Steady state kinetics and binding of eukaryotic cytochromes c with yeast cytochrome c peroxidase. J Biol Chem 252:919–926

Kang CH, Brautigan DL, Osheroff N, Margoliash E (1978) Definition of cytochrome c binding domains by chemical modifications. J Biol Chem 253:6502–6510

Kang DS, Erman JE (1982) The cytochrome c peroxidase catalysed oxidation of ferrocytochrome c by hydrogen peroxide. J Biol Chem 257:12775–12779

Kawato S, Sigel E, Carafoli E, Cherry RJ (1981) Rotation of cytochrome oxidase in phospholipid vesicles. J Biol Chem 256:7518–7527

Kayushin LP, Ajipa YI (1973) Cytochrome c nucleotide complexes and the role of unpaired electrons in coupling processes. Ann NY Acad Sci 222:255–265

Keller RM, Groudinsky O, Wuthrich K (1973) Proton magnetic resonance of cytochrome b_2 core-structural similarities with cytochrome b_5. Biochim Biophys Acta 328:233–238

Kessler DL, Rajagopalan KV (1974) Hepatic sulfite oxidase – effects of anions on interaction with cytochrome c. Biochim Biophys Acta 370:389–398

Kim CH, King TE (1981) The indispensibility of a mitochondrial 15 K protein for the formation of the cytochrome c_1 : cytochrome c complex. Biochem Biophys Res Commun 101:607–614

Kim CH, King TE (1983) A mitochondrial protein is essential for the formation of the cytochrome c_1 : cytochrome c complex. J Biol Chem 258:13543–13551

Kimura S, Abe K, Sugita Y (1984) Differences in C-terminal amino acid sequence between erythrocyte and liver cytochrome b_5 isolated from pig and human. FEBS Lett 169:143–146

King TE (1978) Cytochrome c_1 from mammalian heart. Meth Enzymol 53:181–191

King TE (1982) Cardiac cytochrome c_1. Adv Enzymol 54:267–360

Konig BW, Schilder LTM, Tervoort MJ, Gelder BF van (1980) The isolation and purification of cytochrome c_1 from bovine heart. Biochim Biophys Acta 621:283–295

Konig BW, Wilms J, Gelder BF van (1981) The reaction between cytochrome c_1 and cytochrome c. Biochim Biophys Acta 636:9–16

Koppenol WH, Margoliash E (1982) The asymmetric distribution of charges on the surface of cytochrome c. J Biol Chem 257:4426–4437

Koppenol WH, Vroonland CAJ, Braams R (1978) The electric potential field around cytochrome c and the effect of ionic strength on reaction rates. Biochim Biophys Acta 503:499–508

Kuhn-Nentwig L, Kadenbach B (1985a) Isolation and properties of cytochrome c oxidase from rat liver and quantification of immunological differences between isozymes from various rat tissues with subunit specific antisera. Eur J Biochem 149:147–158

Kuhn-Nentwig L, Kadenbach B (1985b) Orientation of rat liver cytochrome c oxidase subunits investigated with subunit specific antisera. Eur J Biochem 153:101–104

Kuo LM, Davies HC, Smith L (1984) Effects of monoclonal antibodies to bovine and *Paracoccus denitrificans* cytochromes c on reactions with oxidase, reductase and peroxidase. Biochim Biophys Acta 766:472–482

Labeyrie F, Groudinsky O, Jacqot-Armand Y, Naslin L (1966) Propriétés d'un hoyau cytochromique b_2 résultant d'une protéolyse de la L-lactate: cytochrome c oxydoreductase de la levure. Biochim Biophys Acta 128:492–503

Labeyrie F, Franco A di, Iwatsubo M, Baudras A (1967) Fluorimetric and spectrophotometric study of heme binding of the apoprotein from a cytochrome b_2 derivative. Biochemistry 6:1791–1797

Lederer F, Ghrir R, Guiard B, Cortail S, Ito A (1983) Two homologous cytochromes b_5 in a single cell. Eur J Biochem 132:95–102

Leeuwen JM van (1983) The ionic strength dependence of a reaction between two large proteins with a dipole moment. Biochim Biophys Acta 743:408–421

Leonard JJ, Yonetani T (1974) Interaction of cytochrome c peroxidase with cytochrome c. Biochemistry 13:1465–1468

Leonard K, Wingfield P, Arad T, Weiss H (1981) Three-dimensional structure of ubiquinone cytochrome c reductase from *Neurospora* mitochondria determined by electron microscopy of membrane crystals. J Mol Biol 149:259–274

Li Y, DeVries S, Leonard K, Weiss H (1981) Topography of the iron sulphur subunit in mitochondrial UQ cytochrome c reductase. FEBS Lett 135:277–280

Loo S, Erman JE (1975) Kinetic study of reaction between cytochrome c peroxidase and H_2O_2-dependence on pH and ionic strength. Biochemistry 14:3467–3470

Loon APGM van, DeGroot RJ, DeHaan M, Dekker A, Grivell LA (1984) The DNA sequence of the nuclear gene coding for the 17000 subunit VI of yeast ubiquinol cytochrome c reductase. EMBO 3:1039–1043

Ludwig B (1980) Haem aa_3 type cytochrome c oxidase from bacteria. Biochim Biophys Acta 594:177–189

Ludwig B, Schatz G (1980) A two subunit cytochrome c oxidase (aa_3) from *Paracoccus denitrificans.* Proc Natl Acad Sci USA 77:196–200

Ludwig B, Downer NW, Capaldi RA (1979) Labelling of cytochrome c oxidase with (^{35}S) diazobenzene sulfonate. Orientation of the electron transfer complex in the inner mitochondrial membrane. Biochemistry 18:1401–1407

Macleod RM, Farkas W, Fridovich I, Handler P (1961) Purification and properties of hepatic sulfite oxidase. J Biol Chem 236:1841–1846

Madden TD, Cullis PR (1985) Preparation of reconstituted cytochrome oxidase vesicles with defined transmembrane protein orientations employing a cytochrome c affinity column. Biochim Biophys Acta 808:219–224

Malatesta F, Capaldi RA (1982) Localisation of Cys-115 in subunit III of beef heart cytochrome oxidase to the C-side of the mitochondrial inner membrane. Biochem Biophys Res Commun 109:1180–1185

Malmström BG (1979) Cytochrome oxidase, structure and catalytic activity. Biochim Biophys Acta 549:281–303

Margalit R, Schejter A (1973a) The effects of temperature, pH and electrostatic media on the standard redox potential of cytochrome c. Eur J Biochem 32:492–499

Margalit R, Schejter A (1973b) Ion-binding linked to oxidation state of cytochrome c. Eur J Biochem 32:500–505

Margalit R, Schejter A (1974) Cation-binding to horse heart ferrocytochrome c. Eur J Biochem 46:387–391

Margoliash E, Barlow GH, Byers V (1970) Differential binding properties of cytochrome c: possible relevance for mitochondrial ion transport. Nature (London) 228:723–726

Margoliash E, Ferguson-Miller S, Tulloss J, Kang CH, Feinberg BA, Brautigan DL, Morrison M (1973) Separate intramolecular pathways for reduction and oxidation of cytochrome c in electron transport chain reactions. Proc Natl Acad Sci USA 70:3245–3249

Mathews FS (1984) The structure, function and evolution of cytochromes. Progr Biophys Mol Biol 45:1–56

Matlib MA, O'Brien PJ (1976) Properties of rat liver mitochondria with intermembrane cytochrome c. Arch Biochem Biophys 173:27–33

Matthew JB, Weber PC, Salemme FR, Richards FM (1983) Electrostatic orientation during electron transfer between flavodoxin and cytochrome c. Nature (London) 301:169–171

Mauk MR, Reid LS, Mauk AG (1982) Spectrophotometric analysis of the interaction between cytochrome b_5 and cytochrome c. Biochemistry 21:1843–1846

Mauk MR, Mauk G, Weber PC, Matthew JB (1986) Role of the cytochrome b_5 heme propionates in the interaction between cytochrome b_5 and cytochrome c. Biochemistry 25:7085–7091

McLendon GL, Winkler JR, Nocera DG, Mauk MR, Mauk AG, Gray HB (1985) Quenching of zinc-substituted cytochrome c excited states by cytochrome b_5. J Am Chem Soc 107:739–740

Merle P, Kadenbach B (1982) Kinetic and structural differences between cytochrome c oxidase from beef liver and heart. Eur J Biochem 125:239–244

Michel B, Bosshard HR (1984) Spectroscopic analysis of the interaction between cytochrome c and cytochrome aa_3. J Biol Chem 259:10085–10091

Miller WG, Cusanovich MA (1973) The reaction of horse cytochrome c with anionic reductants. Biophys Str Mech 1:97–111

Millet F, Darley-Usmar V, Capaldi RA (1982) Cytochrome c is cross-linked to subunit II of cytochrome c oxidase by a water soluble carbodiimide. Biochemistry 21:3857–3862

Millet F, DeJong C, Paulson L, Capaldi RA (1983) Identification of specific carboxylate groups on cytochrome c oxidase that are involved in binding to cytochrome c. Biochemistry 22:546–552

Minnaert K (1961) The kinetics of cytochrome oxidase. Biochim Biophys Acta 50:23–34

Mitchell P (1980) Proton motive cytochrome system of mitochondria. Ann N Y Acad Sci 341:564–581

Mochan E (1970) The nature of complex formation between cytochrome c and cytochrome c peroxidase. Biochim Biophys Acta 216:80–95

Mochan E, Nicholls P (1971) Complex formation between cytochrome c and cytochrome c peroxidase. Biochem J 121:69–82

Mochan E, Nicholls P (1972) Cytochrome c reactivity in its complexes with mammalian cytochrome c oxidase and yeast peroxidase. Biochim Biophys Acta 267:309–319

Moreland RB, Dockter ME (1981) Interaction of the back of yeast iso-l-cytochrome c with yeast cytochrome oxidase. Biochem Biophys Res Commun 99:339–346

Morton RA, Breskvar K (1977) Ion binding to lysine modified derivatives of cytochrome c. Can J Biochem 55:146–151

Morton RK, Sturtevant JM (1964) Kinetic investigations of yeast L-lactate dehydrogenase (cytochrome b_2). J Biol Chem 239:1614–1624

Myer P, Thallum KK, Pande J, Verma BC (1980) Selectivity of oxidase and reductase activity of horse heart cytochrome c. Biochem Biophys Res Commun 94:1106–1112

Nalecz KA, Bolli R, Azzi A (1983) Preparation of monomeric cytochrome c oxidase: its kinetics differ from those of the dimeric enzyme. Biochem Biophys Res Commun 114:822–828

Ng S, Smith MB, Smith HT, Millet F (1977) Effect of modification of individual cytochrome c lysines on reaction with cytochrome b_5. Biochemistry 16:4975–4978

Nicholls P (1964a) Observations on the oxidation of cytochrome c. Arch Biochem Biophys 106:25–48

Nicholls P (1964b) Cytochrome oxidase in situ: concepts and questions. In: King TE, Mason HS, Morrison M (eds) Oxidases and related redox systems, vol 2. Wiley & Sons, New York, pp 764–783

Nicholls P (1974) Cytochrome c binding to enzymes and membranes. Biochim Biophys Acta 346:271–310

Nicholls P (1976) Catalytic activity of cytochromes c and c_1 in mitochondria and submitochondrial particles. Biochim Biophys Acta 430:30–45

Nicholls P, Chance B (1974) Cytochrome c oxidase. In: Hayaishi O (ed) Molecular Mechanisms of oxygen activation. Academic Press, London New York, pp 479–534

Nicholls P, Mochan E (1971) Complex formation between cytochrome c and cytochrome c peroxidase – kinetic studies. Biochem J 121:55–67

Nicholls P, Mochan E, Kimelberg HK (1969) Complex formation by cytochrome c: a clue to the structure and polarity of the inner mitochondrial membrane. FEBS Lett 3:242–246

Ohnishi T, Kawaguchi K, Hagihara B (1966) Preparation and some properties of yeast mitochondria. J Biol Chem 241:1797–1806

Osheroff N, Borden D, Koppenol WH, Margoliash E (1979) The evolutionary control of cytochrome c function. In: King TE, Orii Y, Chance B, Okunuki K (eds) Cytochrome oxidase. Elsevier, Amsterdam, pp 385–394

Osheroff N, Brautigan DL, Margoliash E (1980) Definition of enzymic domains on cytochrome c. J Biol Chem 255:8245–8251

Osheroff N, Speck SH, Margoliash E, Veerman EC, Wilms J, Konig BW, Muisjers AO (1983) The reaction of primate cytochrome c with cytochrome oxidase. J Biol Chem 258:5731–5738

Oshino N, Chance B (1975) Interaction of sulfite oxidase with the mitochondrial respiratory chain. Arch Biochem Biophys 170:514–528

Peterman BF, Morton RA (1977) The oxidation of ferrocytochrome c in non-binding buffer. Can J Biochem 55:796–803

Petersen LC (1978) Cytochrome c:cytochrome aa_3 complex formation at low ionic strength studied by aqueous two phase partition. FEBS Lett 94:105–108

Petersen LC, Cox RP (1980) The effect of complex formation with polyanions on the redox properties of cytochrome c. Biochem J 192:687–693

Pettigrew GW (1978) Mapping an electron transfer site on cytochrome c. FEBS Lett 86:14–18

Pettigrew GW, Seilman S (1982) Properties of a cross-linked complex between cytochrome c and cytochrome c peroxidase. Biochem J 201:9–18

Poulos TL, Finzel BC (1984) Heme enzyme structure and function. In: Hearn MW (ed) Peptide and protein reviews, vol 4. Dekker, New York, pp 115–141

Poulos TL, Kraut J (1980) A hypothetical model of the cytochrome c peroxidase: cytochrome c electron transfer complex. J Biol Chem 255:10322–10330

Poulos TL, Freer ST, Alden RA, Edwards SL, Skoglund U, Takio K, Eriksson B, Xuong NH, Yonetani T, Kraut J (1980) Crystal structure of cytochrome c peroxidase. J Biol Chem 255:575–580

Power SD, Lochrie MA, Sevarino KA, Patterson TE, Poyton RO (1984a) The nuclear coded subunits of yeast cytochrome c oxidase. J Biol Chem 259:6564–6570

Power SD, Lochrie MA, Patterson TE, Poyton RO (1984b) The nuclear coded subunits of yeast cytochrome c oxidase. J Biol Chem 259:6571–6574

Power SD, Lochrie MA, Poyton RO (1984c) The nuclear coded subunits of yeast cytochrome c oxidase – identification of homologous subunits in yeast, bovine heart and *Neurospora crassa* cytochrome c oxidase. J Biol Chem 259:6575–6578

Prats M (1977) Complexes entre les L(+) lactate cytochrome c oxydoréductase extrait des levures *Saccharomyces cereviseae* on *Hansenula anomala* et le cytochrome c de coeur de cheval. Biochimie 59:621–626

Prochaska L, Bisson R, Capaldi RA (1980) Structure of the cytochrome c oxidase complex: labelling by hydrophilic and hydrophobic protein modifying reagents. Biochemistry 19:3174–3180

Reynafarge B, Alexandre A, Davies P, Lehninger AL (1982) Proton translocation stoichiometry of cytochrome oxidase. Proc Natl Acad Sci USA 79:7218–7222

Rich PR (1984) Electron and proton transfers through quinones and cytochrome bc_1 complexes. Biochim Biophys Acta 768:53–79

Rieder R, Bosshard HR (1978) The cytochrome c oxidase binding site on cytochrome c. Differential chemical modification of lysine residues in free and oxidase-bound cytochrome c. J Biol Chem 253:6045–6053

Rieder R, Bosshard HR (1980) Comparison of the binding sites for cytochrome c oxidase, cytochrome bc_1 and cytochrome c_1. J Biol Chem 255:4732–4739

Rieske JS (1967) Preparation and properties of reduced Coenzyme Q-cytochrome c reductase. Meth Enzymol 10:239–245

Rieske JS, Zaugg WS, Hansen RE (1964) Distribution of iron and of the compound giving an electron paramagnetic resonance signal at g = 1.9 in a subfraction of complex III. J Biol Chem 239:3023–3030

Risler JL, Groudinsky O (1973) Magnetic circular dichroism studies of cytochrome c and cytochrome b_2. Eur J Biochem 35:201–205

Roberts H, Hess B (1977) Kinetics of cytochrome c oxidase from yeast. Biochim Biophys Acta 462:215–234

Robinson NC, Capaldi RA (1977) Interaction of detergents with cytochrome c oxidase. Biochemistry 16:375–381

Robinson NC, Talbert L (1980) Isolation of bovine cytochrome c_1 as a single non-denatured subunit using gel filtration or high pressure liquid chromatography in deoxycholate. Biochem Biophys Res Commun 95:90–96

Rosevear P, Akeem T van, Baxter J, Ferguson-Miller S (1980) Alkyl glycoside detergents – a simpler synthesis and their effects on kinetic and physical properties of cytochrome c oxidase. Biochemistry 19:4108–4115

Ross EM, Schatz G (1976) Purification of cytochrome c_1 from yeast. J Biol Chem 251:1991–1996

Ross EM, Schatz G (1978) Purification and subunit composition of cytochrome c_1 from bakers yeast. Meth Enzymol LIII:222–229

Ross PD, Subramanian S (1981) Thermodynamics of protein association reactions. Biochemistry 20:3096–3102

Salemme FR (1976) An hypothetical structure for an intermolecular electron transfer complex of cytochrome c and cytochrome b_5. J Mol Biol 102:563–568

Saraste M (1983) How complex is a respiratory complex. Trends Biochem Sci 8:139–142

Saraste M, Pentilla T, Wikstrom M (1981) Quaternary structure of bovine cytochrome oxidase. Eur J Biochem 115:261–268

Scatchard G (1949) The attractions of proteins for small molecules and ions. Ann N Y Acad Sci 51:660–672

Sebald W, Neupert W, Weiss H (1979) Preparation of *Neurospora crassa* mitochondria. Meth Enzymol LV:144–148

Senior AE (1983) Secondary and tertiary structure of membrane proteins involved in proton translocation. Biochim Biophys Acta 726:81–95

Shimomura Y, Mishikimi M, Ozawa T (1985) Novel purification of cytochrome c_1 from mitochondrial complex III. J Biol Chem 260:15075–15080

Sjostrand FS (1978) The structure of mitochondrial membranes. A new concept. J Ultrastruct Res 64:217–245

Smith HT, Millet F (1980) ^{19}F NMR study of the interaction between cytochrome c and cytochrome c peroxidase. Biochim Biophys Acta 626:64–72

Smith HT, Staudenmeyer N, Millet F (1977) Use of lysine specific modifications to locate the reaction site of cytochrome c with cytochrome oxidase. Biochemistry 16:4971–4978

Smith HT, Ahmed AJ, Millet F (1981) Electrostatic interaction of cytochrome c with cytochrome c_1 and cytochrome oxidase. J Biol Chem 256:4984–4990

Smith L (1978) Purification of bacterial cytochromes c by isoelectric focusing. Meth Enzymol 53:229–231

Smith L, Conrad H (1956) The kinetics of oxidation of cytochrome c by cytochrome oxidase. Arch Biochem Biophys 63:403–413

Smith L, Nava ME, Margoliash E (1973 a) In: King TE, Mason HS, Morrison M (eds) Oxidases and related redox systems, vol. 2. Univ Park Press, Baltimore, pp 629–638

Smith L, Davies HC, Reichlin M, Margoliash E (1973 b) Separate oxidase and reductase reaction sites on cytochrome c demonstrated with purified site-specific antibodies. J Biol Chem 248:237–243

Smith L, Davies H, Nava M (1974) Oxidation and reduction of soluble cytochrome c by membrane-bound oxidase and reductase systems. J Biol Chem 249:2904–2910

Smith L, Davies HC, Nava ME (1976) Evidence for binding sites on cytochrome c for oxidase and reductases from studies of different cytochromes c of known structure. Biochemistry 15:5827–5831

Smith L, Davies HC, Nava ME (1979a) Kinetics of reaction of cytochrome c with cytochrome oxidase. In: King TE et al. (eds) Cytochrome oxidase. Elsevier, Amsterdam, pp 293–304

Smith L, Davies HC, Nava NE (1979b) Studies on the kinetics of oxidation of cytochrome c by cytochrome oxidase. Comparison of the spectrophotometric and polarographic methods. Biochemistry 18:3140–3146

Smith L, Davies HC, Nava NE (1980) Effect of ATP and ADP on the oxidation of cytochrome c by cytochrome c oxidase. Biochemistry 19:1613–1617

Smith MB, Stonehuerner J, Ahmed AJ, Staudenmeyer N, Millet F (1980) Use of specific trifluoroacetylation of lysine residues in cytochrome c to study the reaction with cytochrome b_5, cytochrome c_1 and cytochrome oxidase. Biochim Biophys Acta 592:303–313

Smith RJ, Capaldi RA (1977) Nearest neighbour relationships of the polypeptides in ubiquinone cytochrome c reductase. Biochemistry 16:2629–2633

Somero GN, Low PS (1976) Temperature – a shaping force in protein evolution. Biochem Soc Symp 41:33–42

Speck SH, Margoliash E (1984) Characterisation of the interaction of cytochrome c and mitochondrial ubiquinol cytochrome c reductase. J Biol Chem 259:1064–1072

Speck SH, Ferguson-Miller S, Osheroff N, Margoliash E (1979) Definition of cytochrome c binding domains by chemical modification – kinetics of reaction with beef mitochondrial reductase and functional organisation of the respiratory chain. Proc Natl Acad Sci USA 76:155–159

Speck SH, Koppenol WH, Dethmers JK, Osheroff N, Margoliash E (1981) Definition of cytochrome c binding domains by chemical modification. J Biol Chem 256:7394–7400

Speck SH, Neu CA, Swanson MS, Margoliash E (1983) Role of phospholipid in the low affinity reactions between cytochrome c and cytochrome c peroxidase. FEBS Lett 164:379–382

Speck SH, Dye D, Margoliash E (1984) Single catalytic site model for the oxidation of ferrocytochrome c by mitochondrial cytochrome c oxidase. Proc Natl Acad Sci USA 81:347–351

Staudenmeyer N, Ng S, Smith MB, Millet F (1977) Effect of specific trifluoroacetylation of individual cytochrome c lysines on the reaction with cytochrome oxidase. Biochemistry 16:600–604

Steffens GJ, Buse G (1979) Primary structure and function of subunit II of cytochrome c oxidase. Hoppe-Seylers's Z Physiol Chem 360:613–619

Stellwagen E (1978) Heme exposure as the determinant of oxidation reduction potentials of heme proteins. Nature (London) 275:73–74

Stellwagen E, Schulman RG (1973) NMR study of exchangeable protons in ferrocytochrome c. J Mol Biol 75:683–695

Stonehuerner J, Williams JB, Millet F (1979) Interaction between cytochrome c and cytochrome b_5. Biochemistry 18:5422–5427

Stonehuerner J, O'Brien P, Geren L, Millet F, Steidl J, Yu L, Yu CA (1985) Identification of the binding site on cytochrome c_1 for cytochrome c. J Biol Chem 260:5392–5398

Strittmatter P (1964) A simple micro stopped flow apparatus. In: Chance B, Eisenhardt RH, Gibson QH, Lunberg KK (eds) Rapid mixing and sampling techniques in biochemistry. Academic Press, London New York, pp 71–84

Strittmatter P, Spatz L, Corcoran D, Rogers MJ, Setlow B, Redline R (1974) Purification and properties of rat liver microsomal stearyl coenzyme A desaturase. Proc Natl Acad Sci USA 71:4565–4569

Swanson M, Speck S, Koppenol WH, Margoliash E (1982) Cytochrome c mobility in the eukaryotic electron transport chain. In: Chien Ho (ed) Electron transport and oxygen utilisation. Elsevier, Amsterdam, pp 51–56

Taborsky G (1979) Interaction of ferrous cytochrome c, ferrous ions and phosphate. J Biol Chem 254:5246–5251

Taborsky G, McCollum K (1979) Phosphate binding by cytochrome c. J Biol Chem 254:7069–7075

Thomas MA, Gervais M, Favaudon V, Valat P (1983) Study of the *Hansenula anomala* yeast flavocytochrome b_2-cytochrome c complex. Eur J Biochem 135:577–581

Thompson DA, Ferguson-Miller S (1983) Lipid and subunit III-depleted cytochrome c oxidase purified by horse cytochrome c – affinity chromatography in lauryl maltoside. Biochemistry 22:3178–3187

Thompson DA, Suarez-Villafane M, Ferguson-Miller S (1982) The active form of cytochrome c oxidase. Biophys J 37:285–292

Tollin G, Cheddar G, Watkins JA, Meyer TE, Cusanovich MA (1984) Electron transfer between flavodoxin semiquinone and c-type cytochromes: correlations between electrostatically corrected rate constants, redox potentials and surface topologies. Biochemistry 23:6345–6349

Trumpower BL, Katki A (1975) Controlled digestion with trypsin as a structural probe for the N-terminal peptide of soluble and membraneous cytochrome c_1. Biochemistry 14:3635–3642

Vanderkooi J, Erecinska M, Chance B (1973a) Use of fluorescent probes in the study of cytochrome c interaction with the mitochondrial membrane. Arch Biochem Biophys 154:219–229

Vanderkooi J, Erecinska M, Chance B (1973b) Comparative study of the interaction of cytochrome c with the mitochondrial membrane. Arch Biochem Biophys 157:531–540

Vanderkooi JM, Glatz P, Casadel J, Woodrow GV (1980) Cytochrome c interaction with yeast cytochrome b_2. Eur J Biochem 110:189–196

Vanderkooi JM, Maniara G, Erecinska M (1985) Mobility of fluorescent derivatives of cytochrome c in mitochondria. J Cell Biol 100:435–441

Veerman ECI, Wilms J, Casteleijn G, Gelder BF van (1980) The pre-steady-state reaction of ferrocytochrome c with cytochrome c: aa_3 complex. Biochim Biophys Acta 590:117–127

Veerman ECI, Leeuwen JW van, Buuren KJH van, Gelder BF van (1982) Reaction of cytochrome aa_3 with porphyrin cytochrome c as studied by pulse radiolysis. Biochim Biophys Acta 680:134–141

Veerman ECI, Wilms J, Dekker HL, Muisjers AO, Buuren KJH van, Gelder BF van (1983) The pre-steady-state reaction of chemically modified cytochromes c with cytochrome oxidase. J Biol Chem 258:5739–5745

Vik SB, Georgevich G, Capaldi RA (1981) Diphosphatidyl glycerol is required for optimal activity of beef heart cytochrome c oxidase. Proc Natl Acad Sci USA 78:1456–1460

Wada K, Okunuki K (1969) Studies on chemically modified cytochrome c – the trinitrophenylated cytochrome c. J Biochem 66:249–262

Wakabayashi S, Matsubara H, Kim CH, King TE (1982a) Structural studies of bovine heart cytochrome c_1. J Biol Chem 257:9335–9344

Wakabayashi S, Takeda H, Matsubara H, Kim CH, King TE (1982b) Identity of a haem-not-containing protein of bovine heart cytochrome c_1 preparation with the protein mediating c_1-c complex formation. J Biochem 91:2077–2085

Waldmeyer B, Bosshard HR (1985) Structure of an electron transfer complex – covalent cross-linking of cytochrome c peroxidase and cytochrome c. J Biol Chem 260:5184–5190

Waldmeyer B, Bechtold R, Zurrer M, Bosshard HR (1980) Cross-linking of cytochrome c to peroxidase: covalent complex catalyses oxidation of cytochrome c by H_2O_2. FEBS Lett 119:349–351

Waldmeyer B, Bechtold R, Bosshard HR, Poulos TL (1982) The cytochrome c peroxidase: cytochrome c electron transfer complex. Experimental support of a hypothetical model. J Biol Chem 257:6073–6076

Webb M, Stonehuerner J, Millet F (1980) The use of specific lysine modifications to locate the reaction site on cytochrome c with sulfite oxidase. Biochim Biophys Acta 593:290–298

Weiss H (1976) Subunit composition and biogenesis of mitochondrial cytochrome b. Biochim Biophys Acta 456:291–313

Weiss H, Juchs B (1978) Isolation of a multiprotein complex containing cytochrome b and c_1 from *Neurospora crassa* mitochondria by affinity chromatography. Eur J Biochem 88:17–28

Weiss H, Kolb HJ (1979) Isolation of mitochondrial succinate UQ reductase, cytochrome c reductase and cytochrome c oxidase from *Neurospora crassa* mitochondria by affinity chromatography. Eur J Biochem 99:139–149

Weiss H, Wingfield P (1979) Enzymology of ubiquinone-utilising electron transfer complexes in non-ionic detergent. Eur J Biochem 99:151–160

Wherland S, Gray HB (1976) Metalloprotein electron transfer reactions – analysis of reactivity of horse cytochrome c with inorganic complexes. Proc Natl Acad Sci USA 73:2950–2954

Wikström M (1984) Pumping of protons from the mitochondrial matrix by cytochrome c oxidase. Nature (London) 208:558–560

Wikström M, Krab K (1979) Proton-pumping cytochrome c oxidase. Biochim Biophys Acta 549:177–222

Wikström M, Krab K, Saraste M (1981) Proton-translocating cytochrome complexes. Annu Rev Biochem 50:623–655

Williams JN (1968) Molecular proportion of the fixed cytochrome components of the respiratory chain of Keilin Hartree particles and beef heart mitochondria. Biochim Biophys Acta 113:175–178

Williams JN, Thorp SL (1970) Influence of degradative procedures, salts, respiratory inhibitors and gramicidin on the binding of a cytochrome c by liver mitochondria. Arch Biochem Biophys 141:622–631

Williams RJP (1970) The Biochemistry of Sodium, Potassium, Magnesium and Calcium. Q Rev 24:331–365

Wilms J, Dekker HL, Boelens R, Gelder BF van (1981a) The effect of pH and ionic strength on the pre-steady-state reaction of cytochrome c and cytochrome aa_3. Biochim Biophys Acta 637:168–176

Wilms J, Veerman ECI, Konig BW, Dekker HL, Gelder BF van (1981b) Ionic strength effects on cytochrome aa_3 kinetics. Biochim Biophys Acta 635:13–24

Wilson MT, Greenwood C, Brunori M, Antonini E (1975) Kinetic studies on the reaction between cytochrome c oxidase and ferrocytochrome c. Biochem J 147:145–153

Wingfield P, Arad T, Leonard K, Weiss H (1979) Membrane crystals of Ubiquinone: cytochrome c reductase from *Neurospora* mitochondria. Nature (London) 280:696–697

Winter DB, Bruyninckx WJ, Foulke FG, Grinich NP, Mason HS (1980) Location of heme a on subunits I and II and copper on subunit II of cytochrome oxidase. J Biol Chem 255:11408–11414

Wojtzcak L, Sottocasa GL (1972) Impermeability of the outer mitochondrial membrane to cytochrome c. J Membr Biol 7:313–324

Xia Z, Shamala N, Bethge PH, Lim LW, Bellamy HD, Lederer F, Mathews FS (1986) 3-dimensional structure of flavocytochrome b_2 from bakers yeast at 3 Å resolution. In press

Yonetani T (1976) Cytochrome c peroxidase. In: Boyer PD (ed) The enzymes, vol 13. Academic Press, London New York, pp 345–361

Yonetani T, Ray GS (1965) Studies on cytochrome c peroxidase. Purification and some properties. J Biol Chem 240:4503–4514

Yonetani T, Ray GS (1966) Studies on cytochrome c peroxidase. J Biol Chem 241:700–706

Yonetani T, Schleyer H, Ehrenberg A (1966) Studies on cytochrome c peroxidase: epr absorption of the enzyme and complex ES in dissolved and crystalline forms. J Biol Chem 241:3240–3243

Yoshimura T, Matsubara A, Aki K (1977) Formation of a complex between yeast L-lactate dehydrogenase (cytochrome b_2) and cytochrome c. Ultracentrifugal and gel chromatographic analysis. Biochim Biophys Acta 492:331–339

Yoshimura T, Sogabe T, Aki K (1981) Electron transfer between horse and *Candida* cytochromes c in the bound and free states. Biochim Biophys Acta 636:129–135

Yu CA, Yu L, King TE (1972) Purification of cytochrome c_1 from cardiac muscle. J Biol Chem 247:1012–1019

Yu CA, Yu L, King TE (1973) Kinetics of electron transfer between cardiac cytochrome c_1 and cytochrome c. J Biol Chem 248:528–533

Yu L, Yu C, King TE (1977) Subunit structure of the reconstitutively active cytochrome bc_1 complex. Biochim Biophys Acta 495:232–247

Yun L, Leonard K, Weiss H (1981) Membrane and water-soluble cytochrome c_1 from *Neurospora* mitochondria. Eur J Biochem 116:199–205

Zhang YZ, Georgevich G, Capaldi PA (1984) Topology of beef heart cytochrome c oxidase from studies on reconstituted membranes. Biochemistry 23:5616–5621

Chapter 3 The Function of Bacterial and Photosynthetic Cytochromes c

3.1 Introduction

This chapter seeks to place our extensive knowledge of the structure and properties of bacterial cytochromes c within the context of the energy-conserving electron transport systems of which they are a part. The role of cytochrome c in mitochondrial electron transport has been considered in detail in Chapter 2 but for comparative purposes in this chapter we will treat both the mitochondrion and the chloroplast as specialised endosymbiotic bacteria. Justification for this derives from the strong similarities between certain aerobic bacteria and mitochondria, on the one hand, and between cyanobacteria and the chloroplast, on the other (Vol. 2, Chap. 6).

Our treatment will emphasise the role of cytochromes c in these systems rather than the diversity of properties of isolated cytochromes: a view from the bacterium rather than a view from the protein. Thus although the present proliferation of named bacterial cytochromes baffle the observer, the underlying functional themes of bacterial respiration and photosynthesis are few and simple. A very large group of bacterial cytochromes c (the class I cytochromes defined in Chap. 1) probably act as diffusible links between cytochrome c reductase systems and terminal reactions (Fig. 3.1) while other c-type cytochromes are involved as part of certain of these reductases and terminal enzymes.

This general role of class I cytochromes c is illustrated in Fig. 3.2a and specific examples appear in the subsequent diagrams. The diagrams illustrate the central importance of cytochrome c in the modular construction of the oxidising end of electron transport systems which allows flexibility in the use of electron donors and acceptors. Also emphasised is the periplasmic location of cytochrome c (Wood 1983), a crucial feature of electrogenic transfer to a terminal oxidant site at the cytoplasmic side of the membrane. This location for c-type cytochromes has not been demonstrated in every case; indeed, in some cases the location is disputed; but in all instances, where clear experimental data is available, a wholly periplasmic location is confirmed.

As to be expected from the functional position of cytochromes c in electron transfer systems, their redox potentials generally occupy the positive end of the biological redox potential scale (Fig. 3.3) (the exceptions are cytochrome c_3, involved in reduction of the relatively poor oxidants associated with sulfur).

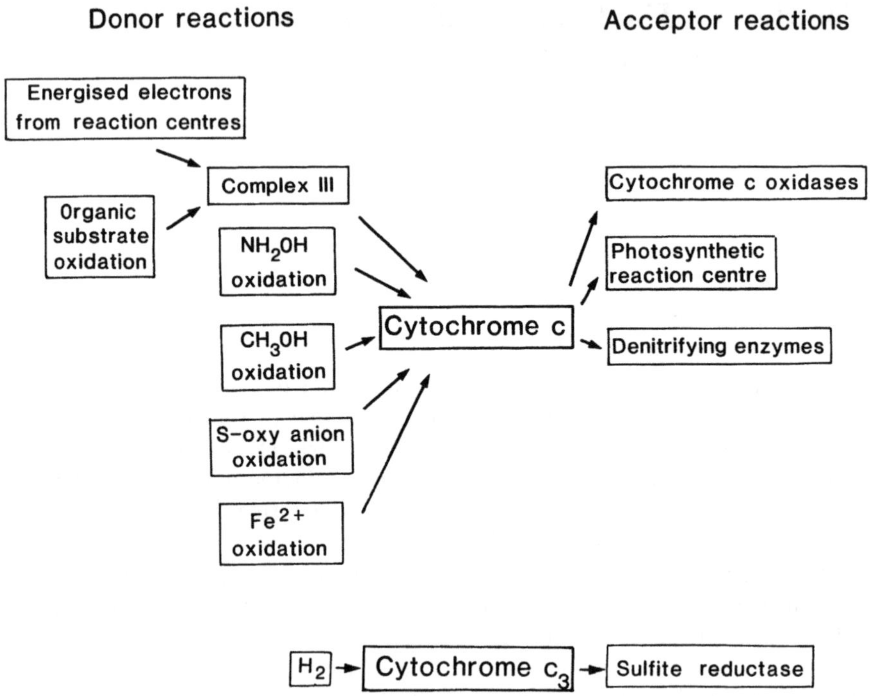

Fig. 3.1. Terminal reactions in bacterial electron transport. Fortunately for those interested in unravelling the respiratory functions of bacteria, these pathways do not all occur in the one organism! However, several may occur together or, at least, an individual species may have the potential to display several different respiratory modes. A second c-type cytochrome, cytochrome c_1 or f, is a part of the integral membrane complex III and c-type cytochromes can occur in association with cytochrome c oxidases, photosynthetic reaction centres, denitrifying enzymes and substrate oxidising enzymes

Only the redox centres associated with the terminal enzymes themselves are more oxidising, while flavins and quinones and most b-type cytochromes and iron-sulfur proteins have more negative redox potentials. Here we will refer to other types of redox centres only insofar as they bear on our understanding of cytochrome c function but in the chapters on redox potential (Vol. 2, Chap. 7) and electron transfer (Vol. 2, Chap. 8) they will form an important part of the comparative discussion.

At present we often know much more about the structure and properties of a particular protein than its function. Definitive proof that a cytochrome is involved in an electron transfer process has been difficult to obtain, particularly when multiple terminal oxidases are present as is strikingly evident in denitrification (Sect. 3.5). Several lines of evidence may implicate a cytochrome in a respiratory or photosynthetic process. Firstly, the process and the cytochrome may have a coincident biological distribution. Secondly, the

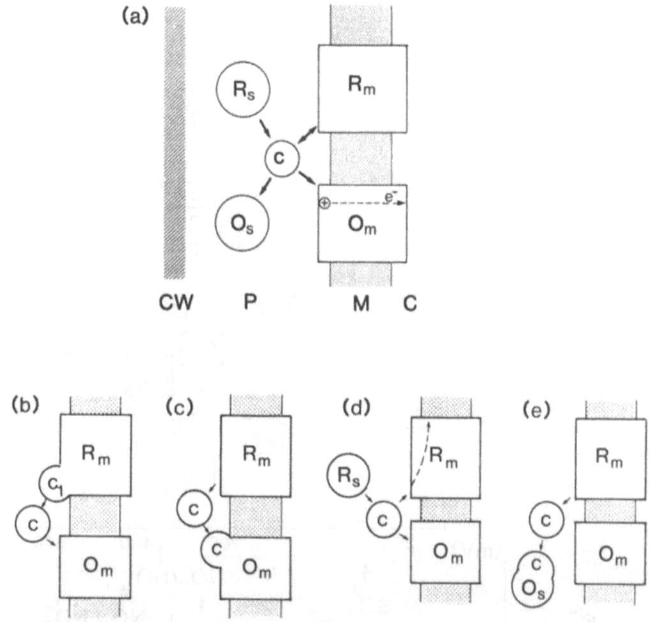

Fig. 3.2a–e. The role of cytochromes c in bacterial electron transport. **a** The respiratory electron transport chain is located in the membrane (*M*) and only the quinol-cytochrome c oxidoreductase (R_m) and the terminal cytochrome c oxidase (O_m) are shown. In photosynthetic electron transport O_m would represent the reaction centre of the photosystem. Electron transport through the terminal enzyme generates a membrane potential, positive outside. In the case of the cytochrome c oxidases, the location of the O_2 reduction site has not been firmly established. This diagram and later figures indicate a cytoplasmic (C)-side site with direct H^+ uptake and transmembrane electron flow. The alternative view that O_2 reduction occurs at the P-side with direct electron transfer and transmembrane H^+ movement is energetically indistinguishable. Electron transport through both R_m and O_m may in addition be accompanied by proton extrusion. Cytochrome c may also interact with soluble reductases (R_s) and soluble oxidases (O_s) situated in the periplasmic space (*P*) and bounded by the cell wall (*CW*). **b–e** Particular forms of the general scheme of **a**. In **b** and **c**, the soluble cytochrome c acts as a diffusible link between membrane reductase and terminal enzyme. In **b**, cytochrome c_1 (or its photosynthetic counterpart, cytochrome f) is a membrane-bound subunit of quinol cytochrome c oxidoreductase and may be a common feature of electron transport systems (Wood 1980b). In **c**, a membrane-bound cytochrome c is attached to the terminal enzyme, an arrangement found in certain aerobic respiratory systems and some photosynthetic systems. The flow of electrons in many chemolithotrophic bacteria is represented in **d**. The chemiosmotic energy made available in the cytochrome c-mediated electron transport from the chemolithotrophic donor to the terminal oxidase is used in part to drive reversed electron transport from the donor to NAD(P), again mediated by cytochrome c (*broken line*). In **e**, electrons flow from a membrane reductase (R_m) to a soluble terminal enzyme which may contain a c-type cytochrome as part of its structure. This arrangement is seen in some of the terminal reduction reactions of denitrification and in the removal of H_2O_2 by peroxidase. Both nitrite reductase (cytochrome cd_1) and cytochrome c peroxidase contain a c-type cytochrome which mediates electron transfer to the enzymic site

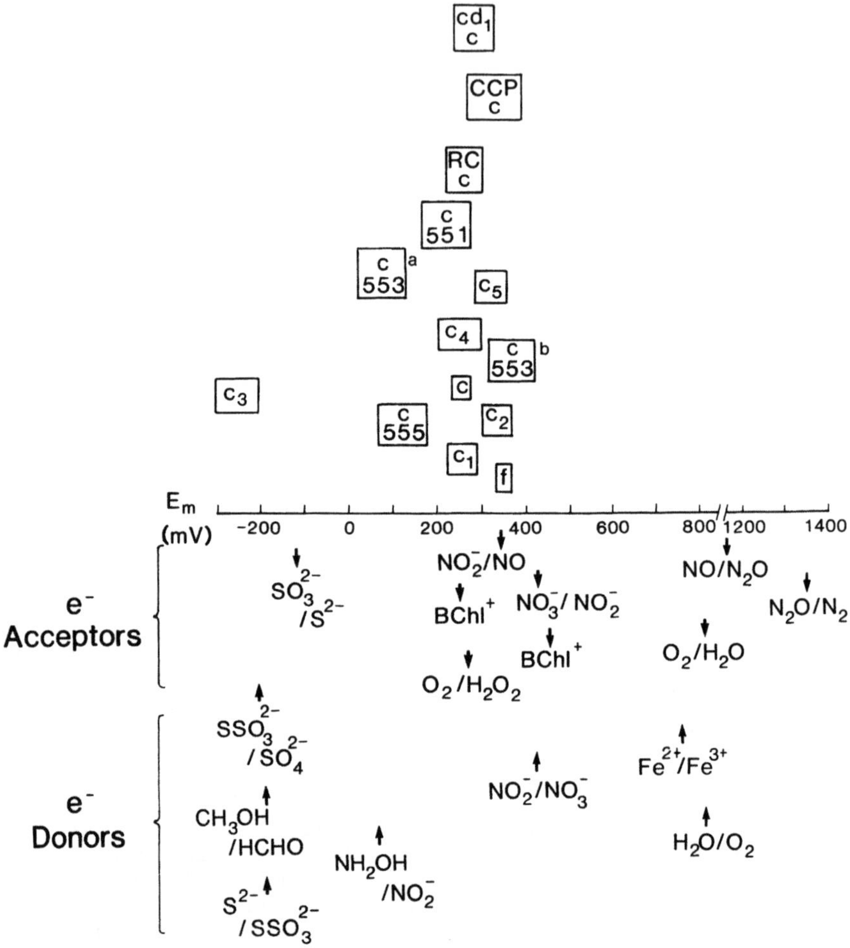

Fig. 3.3. The redox potentials of c-type cytochromes and terminal reactions of electron transport. Each c-type cytochrome is enclosed in a box, the middle of which corresponds to the median midpoint potential of that cytochrome group. cd_1-c, CCP-c and RC-c denote the high potential hemes c associated with *Pseudomonas* nitrite reductase, *Pseudomonas* cytochrome c peroxidase and photosynthetic reaction centres respectively. c-551, *Pseudomonas* cytochrome c-551; c-553[a], *Desulfovibrio* cytochrome c-553; c-553[b], algal cytochrome c-553; c-555, *Chlorobiaceae* cytochromes c-555. The value of 250 mV for $BChl^+$ is for the reaction centre of the *Chlorobiaceae;* those of the *Rhodospirillaceae* and the *Chromatiaceae* have midpoint potentials in the range 400–500 mV. The SO_3^{2-}/S^{2-} midpoint potential of −116 mV is for the 6 e^- process and represents the average of three component 2 e^- reactions which vary widely in midpoint potential (Table 3.16). Individual midpoint potentials appear in the text in appropriate sections

cytochrome may be induced in the presence of an electron donor or terminal acceptor. Thirdly, the reactivity with an oxidase or reductase may be sufficient to support observed rates of electron transfer in vivo. Ideally such reactivity should be measured in the intact state but this may be difficult if several c-type cytochromes are present with similar spectroscopic and potentiometric properties. The use of specific antisera to selectively block reactions is a useful approach that has seldom been employed. Finally, genetic loss should lead to predictable metabolic consequences.

In very few cases has such evidence been obtained. Indeed, doubt remains as to whether the relative lack of specificity of electron transfer between soluble redox systems in vitro is a reflection of the in vivo state. If this were the case, a description of unique routes of electron transfer would not be justified.

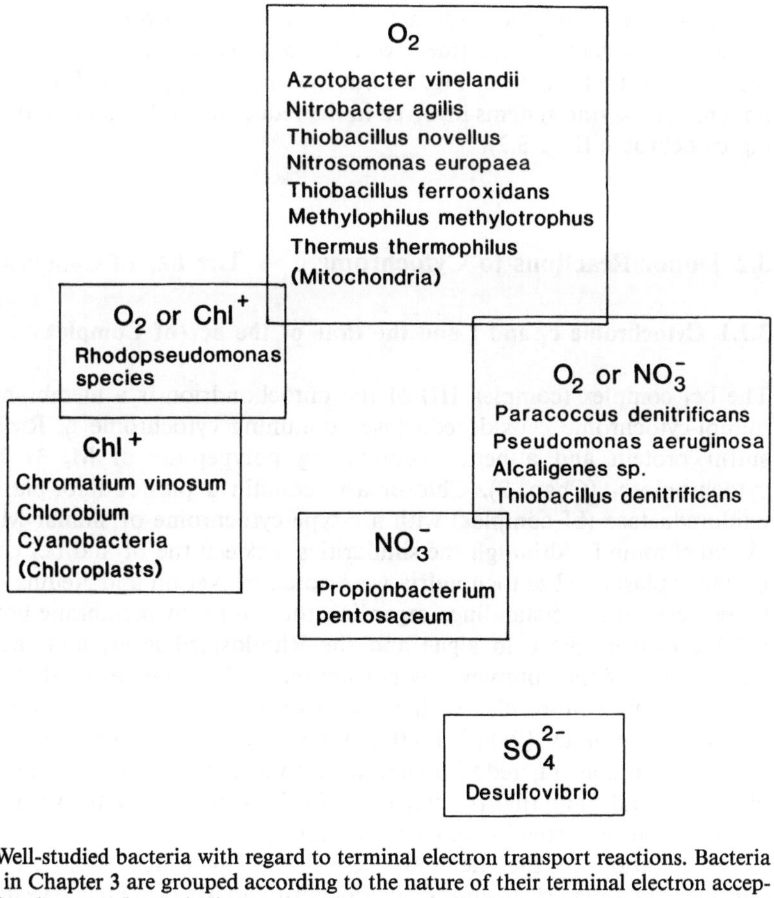

Fig. 3.4. Well-studied bacteria with regard to terminal electron transport reactions. Bacteria discussed in Chapter 3 are grouped according to the nature of their terminal electron acceptor. Overlap between *boxes* indicates that there are similarities in the electron transport systems that may imply evolutionary relatedness associated with loss or gain of a particular terminal function

The major division of this chapter is, on the basis of Fig. 3.1, into processes reducing and processes oxidising cytochrome c. We have done this to emphasise the focal position of cytochrome c at which respiratory and photosynthetic pathways converge and then diverge to terminal acceptors. Many bacteria are bioenergetically versatile and can adapt to different electron donors or acceptors. Such bacteria and their constituent cytochromes will appear more than once (Fig. 3.4).

The disadvantage of this approach is that a complete pathway of respiration or photosynthesis is not dealt with as a single process. This disadvantage is especially pronounced in the case of photosynthetic electron transport where the electron flow can be cyclic but we have applied the division in this case also in order to achieve a unified approach.

Thus this chapter deals with the organotrophic, photosynthetic and chemolithotrophic processes of cytochrome c reduction followed by the aerobic, denitrifying and photosynthetic processes of cytochrome c oxidation. A major emphasis in this treatment will be not only the nature and function of the linking cytochrome c of Fig. 3.1 but the role of c-type cytochromes as components of enzyme systems involved in the reduction and oxidation of the linking cytochrome (Fig. 3.2).

3.2 Donor Reactions to Cytochrome c – The bc_1/bf Complex

3.2.1 Cytochrome c_1 and f and the Role of the bc_1/bf Complex

The bc_1 complex (complex III) of the mitochondrion is a membrane-bound quinol-cytochrome c oxidoreductase containing cytochrome b, Rieske iron-sulfur protein and a heme c-containing polypeptide of M_r 31 K called cytochrome c_1 (Chap. 2). Chloroplasts contain a plastoquinol-plastocyanin oxidoreductase (bf complex) with a c-type cytochrome of similar M_r known as cytochrome f. Although the similarities between the bf and bc_1 complexes of chloroplasts and mitochondria were noted by Nelson and Neumann (1972) there were misunderstandings regarding the nature of membrane-bound and soluble cytochrome c in algae and the Rhodospirillaceae, and the general significance of the complex was not appreciated. Wood resolved these problems by demonstrating clearly that both algae (Wood 1977) and *Rhodopseudomonas sphaeroides* (Wood 1980a, b) contained a membrane-bound cytochrome with spectra, redox potential and molecular weight similar to cytochrome c_1 and f and that the function of this cytochrome was to reduce either plastocyanin or a small soluble cytochrome c.

Purification of the membrane complexes of which these c-type cytochromes are a part (Gabellini et al. 1982; Yu and Yu 1982) has revealed essentially similar compositions and we will refer to the complex generally as bc_1/bf. Hauska et al. (1983) suggested that the bc_1/bf complex may be a feature of all systems having a third site of energy conservation associated with a cytochrome c.

3.2.2 Composition and Mode of Action of the bc_1/bf Complex

Solubilisation of photosynthetic membranes from many plant and bacterial sources using cholate and octylglucoside yields complexes which retain some quinol-cytochrome c oxidoreductase activity (Hauska et al. 1983). Comparison of the subunit composition of the purified complexes with that of the mitochondrial complex III (Table 3.1) indicates a simpler structure lacking the "core" subunits of the mitochondrial enzyme.

Most authors agree on the presence of 2 mol heme b, 1 mol heme c, 1 mol iron-sulfur centre and some bound quinone or plastoquinone per mol complex (Hauska et al. 1983). Heme c can be detected in SDS electrophoretic gels on the basis of peroxidase activity (Thomas et al. 1976; Goodhew et al. 1986) or porphyrin fluorescence (Katan 1976; Wood 1981) and with gentle pretreatment, the subunit attachment of heme b may also persist (Hauska et al. 1983; Guikema and Sherman 1980). However, artefactual migration of heme b to another subunit is difficult to exclude (Goodhew et al. 1986) and purified cytochromes b do not always correspond to heme b-containing subunits identified from SDS gels (Phillips and Gray 1983; Clark and Hind 1983; Gabellini and Hauska 1983). Thus designation of heme b-containing subunits in Table 3.1 should be regarded as tentative.

The probable topography of the redox centres with respect to the membrane is illustrated in Fig. 3.5. Cytochrome c_1 and f are inaccessible to electron acceptors in closed membrane vesicles (Wood and Bendall 1976; Muscatello and Carafoli 1969; Prince et al. 1975) and are therefore proposed to be situated on the side of the membrane opposite the ATP synthase, whether this be the interior of the thylakoid, the cytoplasmic surface of the mitochondrial inner membrane or the periplasmic surface of the chromatophore. Both cytochrome c_1 and f, and the iron-sulfur protein are anchored to the membrane rather than embedded in it by analogy to the well-studied *Neurospora* system (Chap. 2).

A dominant, but still controversial theory of electron transfer and energy conservation in the mitochondrial bc_1 complex is the Q-cycle of Mitchell (1975) whereby the oxidation of quinol by the Rieske centre leads to the production of the semi-quinone reductant of the b-type cytochromes. It is by this process that the phenomenon of oxidant reduction of the b-type cytochromes can be explained. The bc_1 and bf complexes of Table 3.1 share this feature (Hauska et al. 1983; Zhu et al. 1984). They also resemble the mitochondrial complex in having potentiometrically distinct cytochromes b and in their suceptibility to inhibition by 5-n-undecyl-6-hydroxy-4,7-dioxobenzothiazole (UHDBT) which acts at the Rieske iron-sulfur centre. Antimycin A, the classic inhibitor of the mitochondrial bc_1 complex, is less effective with the bf com-

Table 3.1. The composition and properties of bc_1/bf complexes

	Beef heart	*P. denitrificans*	*Rps. sphaeroides*	*A. variabilis*	Spinach
Polypeptides M_r (kD)	49[a] core 47 core 31 b 30 c_1 25 Fe-S 12 10 5 a	[b] 68 c_1	 48 b[c] 30 c 24 (Fe-S) 12	 31 f[d] 22.5 b_6 16.5	 33/34 f[e,f] 23.5 b 20 (Fe-S) 17.5
b-Types	b_K, 562 nm, +93 mV[g] b_T 556 nm, −32 mV	b_K, 562 nm, +40 mV[h] b_T, 565 nm, −60 mV	b, 560 nm {+50 mV[i], −60 mV}	b_6, 563 nm, ND[d]	b_6 563 nm {−50mV[k], −80 mV}
c-Types	c_1, 553 nm, +232 mV[h]	c_1, 553 nm, +190 mV[h]	c_1, 552 nm, +285 mV[j]	f, 557 nm, ND[d]	f, 554 nm, +340 mV[k]
e^- Donor	Ubiquinol	Ubiquinol	Ubiquinol	Plastoquinol	Plastoquinol
e^- Acceptor	Cyt.c	Cyt.c-550	Cyt.c_2	Cyt.c-553 or plastocyanin	Plastocyanin
Relative sensitivity to antimycin (A) and DBMIB (D)	A>D	A>D	A>D	D>A	D>A

Polypeptide compositions are from [a]Nelson and Gellerfors (1976) (beef heart mitochondria), [c]Yu and Yu (1982) and Yu et al. (1984) (*Rps. sphaeroides*), [d]Krinner et al. (1982) (*Anabaena variabilis*) and [e]Hurt and Hauska (1981) (spinach). No complex has yet been isolated from *Para. denitrificans* but the cytochrome c_1 has been purified by Ludwig et al. (1983)[b].

[f] If the sample is heated prior to SDS gel electrophoresis, the cytochrome f migrates with M_r 38 K, a phenomenon also seen with cytochrome f from *Anacystis nidulans* (Guikema and Sherman 1980). Clark and Hind (1983) found subunits of 37 K and 34 K for cytochrome f, 22 K for cytochrome b plus polypeptides of 19 K and 17 K in the spinach bf complex. They claimed that full appearance of the 37-K band requires heating in SDS and suggested that its absence in the preparation of Hurt and Hauska (1981) may be due to proteolysis. The cytochrome bf complex of pea contains 37 K (cytochrome f), 34 K, 19.5 K and 15 K polypeptides.

[g] Nelson and Gellerfors (1976).

[h] Cox et al. (1978); John and Whatley (1975).

[i] The cytochromes b of the purified complex from *Rps. sphaeroides* are potentiometrically but not spectrally distinguishable (Gabellini et al. 1982). In the intact membrane they can be distinguished spectrally and correspond to cytochrome b_K and b_T of the mitochondrion (Bowyer and Crofts 1981).

[j] Wood (1980a); Gabellini et al. (1982).

[k] Hauska et al. (1983).

DBMIB, dibromomethylisopropylbenzoquinone.

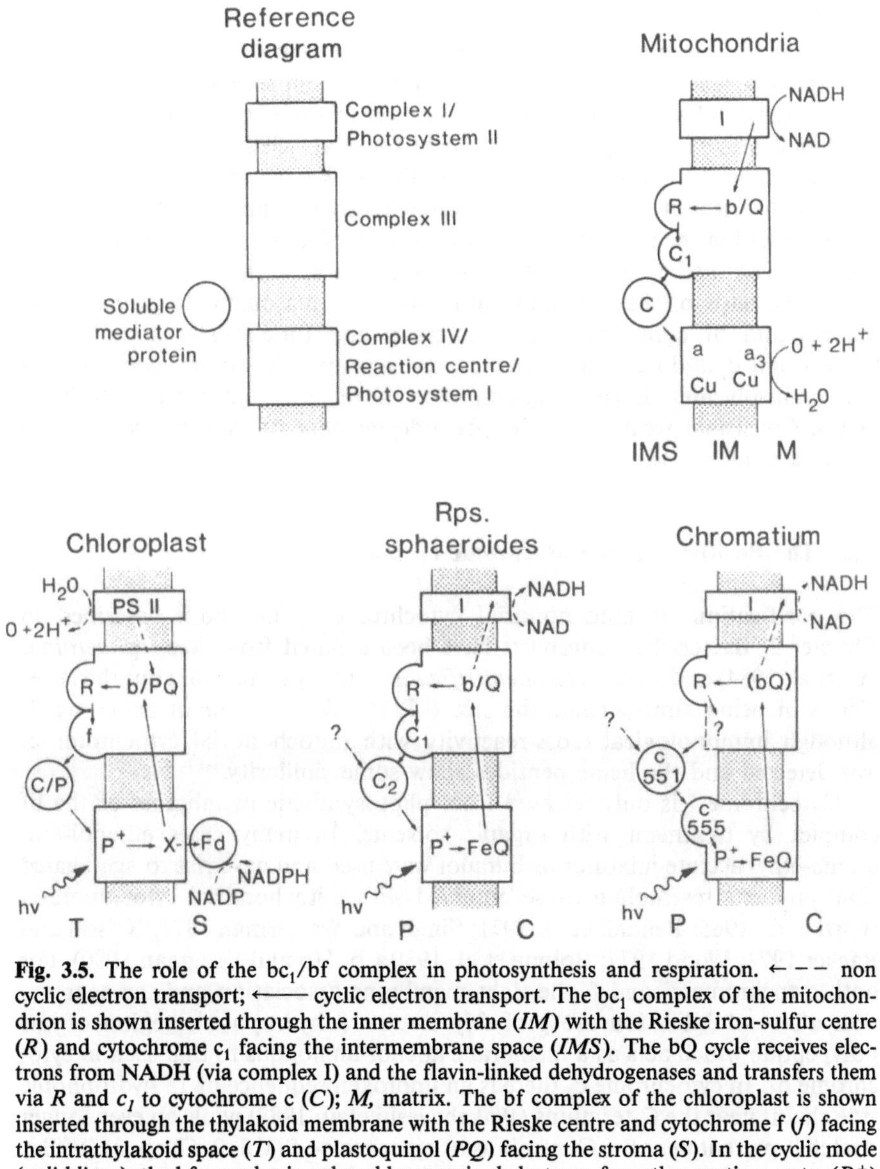

Fig. 3.5. The role of the bc_1/bf complex in photosynthesis and respiration. ← – – non cyclic electron transport; ←—— cyclic electron transport. The bc_1 complex of the mitochondrion is shown inserted through the inner membrane (*IM*) with the Rieske iron-sulfur centre (*R*) and cytochrome c_1 facing the intermembrane space (*IMS*). The bQ cycle receives electrons from NADH (via complex I) and the flavin-linked dehydrogenases and transfers them via *R* and c_1 to cytochrome c (*C*); *M*, matrix. The bf complex of the chloroplast is shown inserted through the thylakoid membrane with the Rieske centre and cytochrome f (*f*) facing the intrathylakoid space (*T*) and plastoquinol (*PQ*) facing the stroma (*S*). In the cyclic mode (*solid lines*), the bf complex is reduced by energised electrons from the reaction centre (P^+) via poorly characterised intermediate carriers (*X*). The electrons are returned to P^+ via a soluble cytochrome c or plastocyanin (C/P). In the non-cyclic mode (*broken lines*), the bf complex is reduced by photosystem II (*PSII*) and supplies electrons to photosystem *I* ($P^+ \rightarrow X$) which then reduce NADP via ferredoxin (*Fd*). The bc_1 complex of *Rps. sphaeroides* resembles that of the mitochondrion in composition but functions like the bf complex of the chloroplast in cyclic photosynthesis. The bQ cycle receives electrons from the reaction centre P^+ via an iron-quinone intermediate carrier (*FeQ*). Non-cyclic photosynthetic electron transport probably occurs by the periplasmic (*P*) reduction of cytochrome c_2 by e^- donors such as thiosulfate. A similar role is suggested for the complex from *Chromatium* but much less information is available. These poorly characterised pathways of cytochrome c reduction are denoted by *?*

plex while dibromomethylisopropylbenzoquinone (DBMIB) has the converse pattern of potency. Both the mitochondrial bc_1 complex and the chloroplast bf complex can be reduced by flash activation of the purified photosynthetic reaction centre of *Rps. sphaeroides* (Packham et al. 1980; Prince et al. 1982). This technique allows the very fast generation of known amounts of reducing equivalents without mixing and at poised redox potentials and has been a powerful tool in the examination of the complex electron transfer process that occurs within the bc_1/bf complex (Bowyer et al. 1981; Zhu et al. 1983, 1984).

On the basis of these striking similarities it is reasonable to propose that the bc_1 and bf complexes are structurally and functionally homologous. Cytochrome c_1 and f are considered to have a simple electron-transferring role with no involvement in the coupled proton movements that occur in the complexes. Consistent with this is the pH independence of their redox potential (Hauska et al. 1983).

3.2.3 The Purification of Cytochrome c_1 and f

The purification of mitochondrial cytochrome c_1 has been described in Chapter 2. Bacterial counterparts have been isolated from *Rps. sphaeroides* (Yu et al. 1984) and *Paracoccus denitrificans* (Ludwig et al. 1983) but the latter differs in being almost twice the size (M_r 60–68 K; 19 nmol heme mg^{-1}) although immunological cross-reactivity with mitochondrial cytochrome c_1 was detected and the heme peptides show some similarity.*

Cytochrome f is only released from photosynthetic membranes or the bf complex by treatment with organic solvents. In many cases ethanol-ammonia-ethyl acetate mixtures or butanol were used and gave rise to aggregated final products resembling those obtained with mitochondrial cytochrome c_1 (Forti et al. 1965; Bendall et al. 1971; Singh and Wasserman 1971, Nelson and Racker 1972; Wood 1977; Bohme et al. 1980a, b; Ho and Krogman 1980). For both cytochrome c_1 and f, the tight membrane association and the aggregation after solubilisation are probably due to a hydrophobic region of the polypeptide which acts as a membrane anchor analogous to that seen in cytochrome b_5. In cytochrome c_1 there is an unbroken sequence of 15 hydrophobic side chains near the C-terminus (Wakabayashi et al. 1982) while an even longer stretch is present near the C-terminus of cytochrome f (Vol. 2, Chap. 3) (Willey et al. 1984; Fig. 3.6). For the latter, Willey et al. (1984) proposed that the hydrophobic sequence forms an α-helix across the membrane, a recurring motif in transmembrane proteins. They showed that the small peptide which remains on the stromal side of the thylakoid membrane is susceptible to protease attack in intact thylakoids while the bulk of the protein is only digested if the membranes are disrupted.

It is not known whether such a model might also be applied to cytochrome c_1. We should note however that the α-helix requires 27 residues to cross a 40-Å lipid bilayer and the hydrophobic stretch of cytochrome c_1 would seem

* See appendix note 8

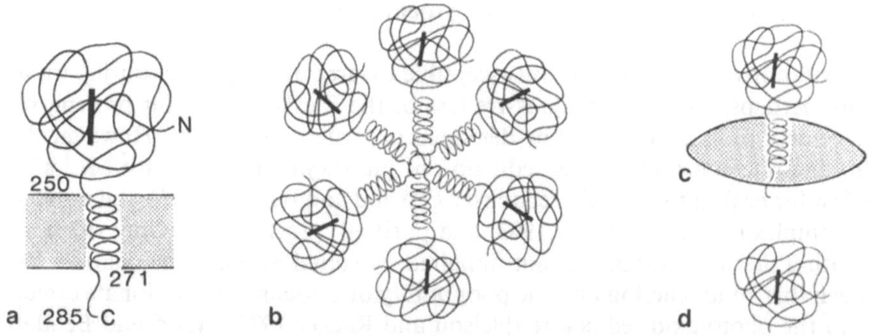

Fig. 3.6a–d. Structural features of cytochrome f. **a** Cytochrome f is depicted as a globular monoheme protein with a hydrophobic α-helical tail near the C-terminus (Willey et al. 1984). It is proposed that the α-helix is inserted through the membrane (residues 250–271) so that the bulk of the molecule lies on one side of the membrane while 14 residues at the C-terminus (*C*) lie on the other. **b** Both cytochrome f and cytochrome c_1 aggregate when solubilised from the membrane and this is represented by a hypothetical interaction of the hydrophobic tail regions to form, in this case, a hexamer. **c** Triton maintains both cytochrome f and c_1 in micellar solution shown here by the interaction of the hydrophobic tail with the Triton micelle. **d** Planned or unintentional proteolysis may yield a form of both cytochrome c_1 and cytochrome f which lacks the hydrophobic tail region and is soluble and monomeric in the absence of detergent

to be too short. In this connection it may be of importance to note (Chap. 4) that cytochrome f is encoded in the chloroplast genome (Gray 1980) and insertion in the membrane is from the stromal side resulting in a transmembrane protein facing the intrathylakoid space (Willey et al. 1984). In contrast, cytochrome c_1 is nuclear-encoded and insertion into the mitochondrial inner membrane is from the same side as its final position so that the C-terminal anchor may not completely cross the membrane. An intriguing parallel to this situation exists for cytochrome b_5 for which synthesis and final location are on the same side of the membrane for the microsomal form but on opposite sides for the form in the mitochondrial outer membrane (Chap. 2). The positions of the anchor sequences in the membrane in these cases are not yet known.

Cytochrome f can be released by butan-2-one from chloroplast membranes of the Cruciferae in a monomeric state (Matsuzaki et al. 1975; Tanaka et al. 1978; Gray 1978) and the algal cytochrome f appears to be monomeric (Bohme et al. 1980a, b; Ho and Krogman 1980). However, Willey et al. (1984) have suggested that the monomeric state of these preparations is a consequence of proteolytic cleavage of the membrane anchor (Fig. 3.6), a proteolysis that can also occur in cytochrome c_1 to release a monomeric fragment of M_r 24 K (Yun et al. 1981).

3.2.4 Electron Transfer Activity of Cytochrome f and c_1

Preparations of bf appear to be susceptible to large activity losses during isolation, perhaps due to the progressive loss of the Rieske iron-sulfur protein, replacement of essential lipids by detergent or structural damage. However, the limiting rate constant for the reduction of plastocyanin ($3-5\times 10^7\ M^{-1}\ s^{-1}$) is fast for broken thylakoid preparations (Wood and Bendall 1976), the isolated bf complex (Clark and Hind 1983) and purified cytochrome f (Gray 1978) indicating that this reaction is not impaired by solubilisation. The contrast between this rapid reaction and the poor ability of cytochrome f to donate directly to the photo-oxidised centre (Nelson and Racker 1972; Wood and Bendall 1975) is consistent with the conclusion that

cyt.f → plastocyanin → reaction centre

is a series arrangement (Wood 1974).

The pattern of specificity of the cytochromes f shows an interesting evolutionary trend. At one extreme, cytochrome f from parsley reacts best with higher plant plastocyanin, moderately well with other acidic proteins such as algal cytochrome c-553 and *Pseudomonas* cytochrome c-551 but poorly with basic cytochromes such as horse cytochrome c (Wood 1974; Wood and Bendall 1976). At the other, the bf complex of the cyanobacterium *Anabaena variabilis* shows good reactivity with basic proteins including cytochrome c-553 and plastocyanin from the same source, and horse cytochrome c but poorly with acidic acceptors such as *Euglena* cytochrome c-552 (Krinner et al. 1982). This pattern of specificity is also found for the reaction centre of *A. variabilis* and has led to the suggestion, discussed in Section 3.6, that the electrostatic interactions have evolved from an original preference for a basic mediator between f and the reaction centre to a preference for an acidic mediator.

Both plastocyanin and cytochrome f from higher plants have isoelectric points below 7 so that the ionic strength dependence of the electron transfer rate at pH 7 must reflect a local opposite-charge interaction of the proteins (Niwa et al. 1980). Chemical conversion of negatively to positively charged residues in plastocyanin or from positive to negative in cytochrome f results in a decrease in the rate constant (Takabe et al. 1984; Takenaka and Takabe 1984). This is consistent with a negative surface on plastocyanin interacting with a positive surface on cytochrome f.

Only mitochondrial cytochromes c and certain cytochromes c_2 show substantial reactivity with the mitochondrial bc_1 complex (assayed in conjunction with complex I as NADH cytochrome c oxidoreductase; Errede & Kamen 1978). For mitochondrial cytochrome c this interaction has been shown to involve a group of lysines on the front face of the molecule (Chap. 2). In view of the considerable sequence homology between the cytochrome c_2 and the mitochondrial cytochromes c (Vol. 2, Chap. 3) we might expect that conservation of these same lysines govern the reactivity of the cytochromes c_2. However, consideration of the sequence evidence suggests that this is not so. Thus there is no strong correlation between positive charge density on the front face of cytochrome c_2 and reactivity with the reductase. Although the most reactive cytochromes c_2 (from *Rps. sphaeroides*, 102% and *Rps. capsulata*, 129%) have a high charge density, cytochrome c_2 from *Rhodospirillum photometricum* has a high charge density but poor reactivity (14%) while cytochrome c_2 from *Rhodomicrobium vannielli* has few front surface lysines but good reactivity (56%) (Table 3.13). Also there is no pattern of conserved lysines on the front face of cytochrome c_2. No one residue is conserved in all cases. Thus the structural basis for the reactivity of the cytochromes c_2 is not evident from the pattern of front face lysines.

3.2.5 Distribution

The bc_1/bf complex is established as the quinol-cytochrome c (or plastocyanin) oxidoreductase in mitochondria, chloroplasts, cyanobacteria and *Rps. sphaeroides*. Its presence and role in other organisms is less certain. For example, in *Pa. denitrificans*, the bc_1 complex appears to be associated with the cytochrome aa_3 and copurifies with it as a quinol oxidase activity after solubilisation of the membrane (Berry and Trumpower 1985). This differs from the mitochondrial case where quinol oxidase activity can only be observed by adding back cytochrome c to membranes or the solubilised complexes. It implies that in *Pa. denitrificans*, there is a fixed relationship between complex III and IV which allows electron transfer without the need for a soluble mediator cytochrome. In the case of *Rsp. rubrum*, Matsuda et al. (1984) have purified a "photoreaction unit" which catalyses antimycin A-sensitive electron transport but does not contain cytochrome b or c. In *Rps. capsulata*, cyclic electron transport is not sensitive to antimycin A (Evans and Crofts 1974b) and there is no potentiometric evidence for a distinct membrane-bound c-type cytochrome (Evans and Crofts 1974a; Prince and Dutton 1977a). However, Bowyer et al. (1981) found similar photo-oxidation spectra for the c-type cytochromes of *Rps. sphaeroides* and *capsulata* (see Sect. 3.6) and *Rps. capsulata* cytochrome c_2 is the most reactive of its class with the mitochondrial bc_1 complex (Errede and Kamen 1978). Thus a thorough search for the bc_1 complex in *Rps. capsulata* would seem advisable. This may be frustrated by the different stabilities of the complex from different sources. The Rieske iron-sulfur centre is readily lost from both the bc_1 and bf complexes (Hauska et al. 1983).

Short of isolation of the complex, some of the structural and functional features may provide acceptable indicators of its presence. Thus cytochrome b reduction in *Chromatium* is inhibited by quinone analogues but enhanced by antimycin A (Bowyer and Crofts 1980). The distinctive EPR signal of the Rieske iron-sulfur centre can be detected in membranes of *Chromatium* and *Chlorobium* species (Trumpower 1981) and in *Rps. viridis* (Wynn et al. 1985). Analogues to cytochrome c_1 in terms of molecular weight have been identified in membranes of *Pseudomonas aeruginosa* (Wood and Willey 1980), *Rps. viridis* (Wynn et al. 1985) and *Nitrosomonas europaea* (Miller and Wood 1982). Of particular interest is the role of the bc_1/bf complex in the chemolithotrophs where it may be involved in reversed electron transport to allow synthesis of NADH. Tanaka et al. (1983) have purified a nitrite-oxidising system from *Nitrobacter agilis* which contains a heme c component of M_r 29 K (Sect. 3.3.4).

3.3 Donor Reactions to Cytochrome c – Chemolithotrophy

3.3.1 The Oxidation of Ammonia

Nitrification, according to the equation:

$$NH_4^+ + 1.5 O_2 \rightarrow NO_2^- + 2H^+ + H_2O$$

is an important part of the natural nitrogen cycle and is of economic relevance in the treatment of sewage effluent and the use of ammonia-based fertilisers. The aerobic oxidation of NH_3 is the sole source of energy for autotrophic nitrifiers such as *Nitrosomonas*. Although the enzymology is poorly understood, the key intermediates are believed to be hydroxylamine (NH_2OH) and nitroxyl (NOH) (Lees 1960; Suzuki 1974). C-type cytochromes play a central role in electron transport outlined in Fig. 3.8 and are also a constituent of the enzyme hydroxylamine oxidase. Their properties are summarised in Table 3.2.

The Oxidation of NH_3 to Hydroxylamine. Present evidence suggests that the initial oxidation of NH_3 to NH_2OH is carried out by a mixed function oxidase. The overall stoichiometry of 1.5 mol oxygen taken up per mol NH_3 oxidised to nitrite is consistent with an oxygenation step (Drozd 1980) and Hollocher et al. (1981) have shown that the O in hydroxylamine is derived from molecular oxygen. The oxygenase requires an electron donor to reduce the second oxygen atom and the lag phase observed in NH_3 oxidation by whole cells, which can be abolished by hydroxylamine, is interpreted as due to the build-up of this donor by further oxidation of hydroxylamine (Hooper 1969). NH_3 oxidase activity can be reconstituted using a membrane fraction and purified cytochrome c-554 (Suzuki and Kwok 1981) and Tsang and Suzuki (1982) found that reduced cytochrome c-554 is partly re-oxidised in a rapid reaction on addition of NH_3. Thus this cytochrome is proposed to be the donor to the NH_3 mono-oxygenase and must be re-reduced by further oxidation of the NH_2OH produced. In this respect the process resembles that catalysed by methane mono-oxygenase in *Methylosinus trichosporium* (Sect. 3.3.2).

The oxygenase has not been characterised. An unusual pigment believed to be a hemoprotein was isolated by Rees and Nason (1965) and Erickson and

Table 3.2. C-type cytochromes of *Nitrosomonas*

Cytochrome	Mol. wt.	Hemes	Spectra[a]	pI	E_m	Electron transfer properties	References
c-554	22–28000	2*	ox 407 red 421 524 554 (as) α/β = 1.7	10–10.7	+ 20	Catalyses reduction of c-552 by NH_2OH oxidase. Reconstitutes NH_3 oxidase activity in membrane preparations	Yamanaka and Shinra 1974; Miller and Wood 1982
c-552	10000	1	ox 410 red 416 523 552 (as) α/β = 1.7	3.7	+255	Reacts well with *P. aeruginosa* cyt. cd_1 and poorly with bovine cyt.oxidase	Tronson et al. 1973; Yamanaka and Shinra 1974; Miller and Wood 1982
c-550	32–36000		550	4.2	+140	Binds CO	Miller and Wood 1982
c-552_{CO}	16000		552		– 50	Binds CO	Miller and Wood 1983a
NH_2OH oxidase	175–225000 (but species of 11 K, 60 K and 125 K can be obtained in SDS)	10–20 heme c, novel pigment absorbing at 460 nm may also be present	ox 408 red 418 460 524 553 (559 sh)	4.7–5.3	**	Only partly reduced by hydroxylamine, donates best to horse cyt.c, poorly to *Nitrosomonas* c-554, not at all to *Nitrosomonas* c-552	Ritchie and Nicholas 1974; Yamanaka et al. 1979; Hooper and Terry 1977; Hooper et al. 1978; Miller and Wood 1982

Tronson et al. (1973) also found a high potential cytochrome c-553 of M_r 50000. The same cytochrome appears to copurify with the c-550 of Miller and Wood (1982, 1983a) and has a midpoint redox potential of 450 mV. Miller and Wood (1982) found traces of NH_2OH oxidase and cytochrome c-554 in the membrane fraction and proposed that cytochrome c_1 with molecular weight 24000 and E_m 170 mV was also present.

[a] sh, Shoulder; as, asymmetric.

* See appendix note 9.

** See appendix note 10.

Hooper (1972). In its red-shifted Soret peak and CO binding ability it resembled cytochromes a, d and P450 but it contained no stable heme, extractable into acid acetone. Although an oxygenase function has been proposed based on these similarities to microsomal P450 (Suzuki 1974), it has not been demonstrated and the relationship between this species and the component of hydroxylamine oxidase which absorbs at 460 nm is unknown.

Oxidation of Hydroxylamine. Hydroxylamine oxidase has a complex subunit structure. Variable amounts of 125, 195 and 225 K species were observed on SDS gel electrophoresis accompanied by a band of 11 K (Terry and Hooper 1981). On removal of heme c, all high molecular weight forms gave rise to a band of 63 K. Terry and Hooper suggested that the molecule may be a hexamer with three subunits of molecular weight 63 K, each containing six heme c, and three of 11 K each containing one heme c. The stability in SDS conferred by the hemes is interesting and merits further investigation.

The spectrum of hydroxylamine oxidase is complex (Fig. 3.7). Although the shoulder at 559 nm in the spectrum of reduced enzyme suggests the presence of a b-type cytochrome, no protoheme IX could be extracted into acid acetone (Yamanaka et al. 1979a) and the red-shifted shoulder presumably reflects a heme c component in an unusual environment. The origin of the 460-nm peak remains obscure. The similarity in spectral position to that of heme d_1 of cytochrome cd_1 does not extend to appropriate α-peaks between 600 and

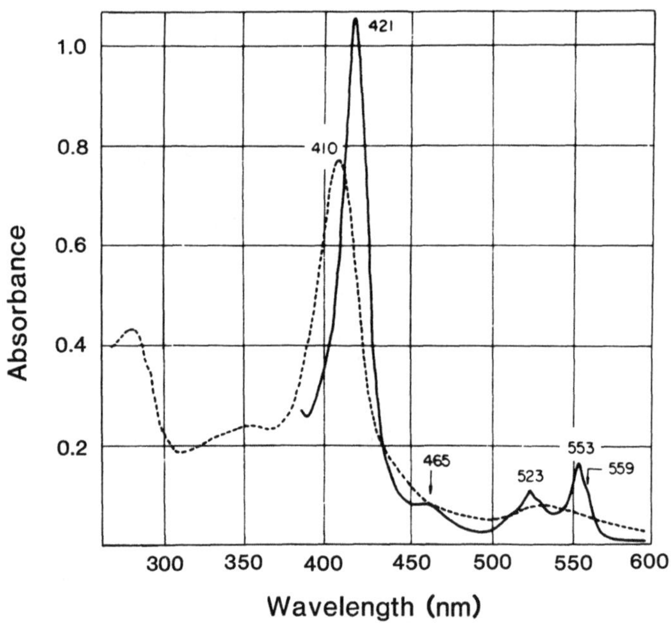

Fig. 3.7. The spectrum of hydroxylamine oxidase from *Nitrosomonas.* – – – Oxidised; ——— reduced with dithionite (Yamanaka and Okunuki 1974)

700-nm and no evidence for a distinctive pyridine ferrohemochrome was found, either in the whole protein or in an acid acetone extract (Ritchie and Nicholas 1974; Yamanaka et al. 1979a). Non-heme cofactors should also be considered. The role of the P460 species in enzymic activity is also uncertain. No natural reductant has been found, although hydroxylamine does partly reduce the c-type hemes (Hooper et al. 1978). However, Hooper and Terry (1977) found that addition of hydrogen peroxide resulted in loss of the 460-nm band in parallel with enzyme activity and substrate protected against this effect. Thus it remains possible that the P460 species is closely involved in the dehydrogenation site on the enzyme.

The product of hydroxylamine oxidation may be nitroxyl (NOH) which is further oxidised to nitrite. The electron economy of the whole sytem is such that if the organism is to derive energy from the process, one of the two 2 e^- steps must involve regeneration of the reductant for further NH_3 oxidation and the other must transfer electrons to an energy-conserving electron transport system terminating in oxygen. This consideration appears to preclude a second mixed-function oxidase reaction at the nitroxyl stage. Purified hydroxylamine oxidase catalyses NH_2OH oxidation to nitrite if phenazine methosulfate is present but it is not known whether the oxidation of NOH is non-enzymic and due to the action of the dye. With cytochrome c as acceptor the predominant product was N_2O (thought to be formed from NOH decomposition) (Rees 1968; Hooper and Nason 1965). In the partially purified state, hydroxylamine oxidase appears to be associated with a Cu-containing nitrite reductase resembling that found in *Alcaligenes* sp. (Sect. 3.5) (Hooper 1968; Ritchie and Nicholas 1974; Miller and Wood 1983b). This system catalysed the production of N_2O from hydroxylamine and nitrite. It seems plausible that this nitrite reductase acts as the physiological NOH oxidase and it would be useful to study its enzymic properties and its interaction with hydroxylamine oxidase and the cytochromes.

Electron Economy and Energy Conservation. In summary, the model of Fig. 3.8 offers an adequate description of the ammonia oxidation process and explanation of the reactivities of individual components. The role of cytochrome c-554 is central to the scheme and it is surprising that this protein is only 10% as reactive with hydroxylamine oxidase as is horse cytochrome c! The topology of the reactions relative to the cell membrane is not known but, by analogy with other bacterial systems, we might expect the c-type cytochromes and therefore the hydroxylamine oxidase to be periplasmic as in the general diagram of Fig. 3.2d and the specific cases of Fig. 3.10. This would allow establishment of a proton gradient, positive outside, across the membrane. If such a process allowed synthesis of a single ATP per 2 e^- transferred, then *Nitrosomonas* would have to oxidise 3 NH_3 to allow synthesis by reversed electron transfer of 1 NADH molecule for biosynthetic reductions. It is this demanding stoichiometry which presumably explains the slow growth of cells in spite of high NH_3 oxidation.

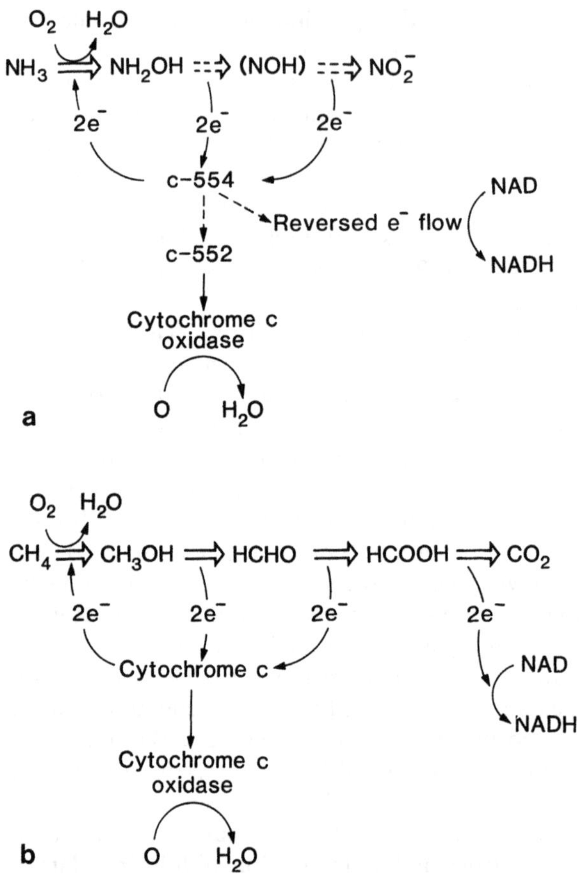

Fig. 3.8 a, b. The electron economy of NH_3 and CH_4 oxidations. **a** NH_3 oxidation by *Nitrosomonas europaea. Broken arrows* indicate uncertainty in the nature of pathways or intermediates. Of four electrons released on oxidation of a molecule of hydroxylamine, two must be used to replace the hydroxylamine via the mixed function oxidase and two remain to generate an electrochemical gradient via the terminal oxidase reaction and to carry out reversed e^- transport to form NADH.* **b** CH_4 oxidation by *Methylosinus trichosporium.* Of the six electrons released on oxidation of a molecule of methanol, two are used to replace the methanol via the action of mixed function oxidase and two are used to reduce NAD. Two remain to generate an electrochemical gradient via the terminal oxidase reaction. The donor to the mixed function oxidase may be NADH in some cases

3.3.2 The Oxidation of Methane and Methanol

The major fraction of atmospheric methane is biologically produced but only a small part of the biological output is actually released into the atmosphere. This is due to the action of the methanotrophs – prolific, usually obligate

* See appendix note 11

bacteria – surviving on the oxidation of methane by oxygen. The methylotrophs, many of which are facultative organisms, cannot oxidise methane but carry out the rest of the pathway of oxidation of single carbon to carbon dioxide. This pathway bears some resemblance to that of oxidation of NH_3 to NO_2^- and again c-type cytochromes play an important role in coupling the oxidation processes to the electron transport sytem (Fig. 3.8b).

As we saw with NH_3 oxidation by *Nitrosomonas*, methane oxidation by the methanotrophic bacteria involves a mixed-function oxidase reaction best studied in *Methylococcus capsulatus* (Bath) and *Methylosinus trichosporium* (Colby and Dalton 1978; Tonge et al. 1977). These enzymes catalyse the incorporation of ^{18}O during oxidation of methane but appear to differ in the nature of the electron donor for the second oxygen atom. This donor is NADH in the case of the *Mc. capsulatus* (Bath) enzyme but that from *Ms. trichosporium* contains a c-type cytochrome which may act to mediate the recycling of electrons between the methanol dehydrogenase and the methane mono-oxygenase (Fig. 3.8b).

Oxidation of Methanol. In whole cell studies, methanol oxidation is insensitive to antimycin A (van Versefeld and Stouthammer 1978; Cross and Anthony 1980b) and methanol causes reduction of cytochrome c but not b (Bamforth and Quayle 1978). A mutant of *Pseudomonas* AMl which lacks cytochrome c cannot oxidise methanol but can use other substrates (Anthony 1975; Widdowson and Anthony 1975). Thus methanol is unusual as a carbon substrate in that it donates to the electron transport chain at the level of cytochrome c.

Methanol oxidation is catalysed by the enzyme methanol dehydrogenase which contains two molecules of pyrroloquinoline quinone having an O-quinone element as the redox centre (Duine and Frank 1980). This novel cofactor is naturally ubiquitous also being involved in glucose dehydrogenase of *Acinetobacter*, alcohol dehydrogenases of non-methylotrophs (Groen et al. 1984, 1986) and mammalian serum amine oxidase (Ferguson 1984). Methanol dehydrogenase isolated under anaerobic conditions from *Hyphomicrobium* x contains a cytochrome c which is reduced on addition of methanol (Duine et al. 1979). In contrast, the enzyme isolated under aerobic conditions has an altered spectrum and reduces cytochrome c without the addition of methanol (Anthony 1975; O'Keefe and Anthony 1980a). Mechanisms have been proposed to explain this "autoreduction" phenomenon (O'Keefe and Anthony 1980b; Beardmore-Gray et al. 1982) but it seems probable that it is the anaerobic form of the enzyme that is the physiological state. This is consistent with the dramatic loss of methanol oxidase activity on cell-breakage under aerobic conditions (Cross and Anthony 1980a; Duine et al. 1979).*

Electron Economy. With two molecules of pyrroloquinoline quinone, methanol dehydrogenase has the electron capacity to oxidise methanol directly to formate and this may occur physiologically (Sperl et al. 1974). In view of the very low redox potential for formaldehyde oxidation (E_m −450 mV) this does

* See appendix note 12

seem wasteful of energy and in *Mc. capsulatus* (Bath), there is an NAD-linked formaldehyde dehydrogenase (Stirling and Dalton 1978). Oxidation of formate (E_m −420 mV) always appears to be coupled to reduction of NAD^+ (Johnson and Quayle 1964).

Figure 3.8 summarises the electron economy of methane oxidation in comparison with that of oxidation of NH_3. A relationship between the two is suggested by the oxidation of NH_3 to NO_2^- by methanotrophs and the inhibition of methane oxidation by NH_3 (O'Neill and Wilkinson 1977). Both processes involve a recycling of electrons to an oxygenase via a cytochrome c but they differ in that NADH can only be produced in *Nitrosomonas* by reversed electron transport while formate dehydrogenase allows net synthesis of NADH during methane oxidation. In the case of *Mc. capsulatus* (Bath) where the mono-oxygenase appears to require NADH as donor, a second NAD-linked enzyme, involved in oxidation of formaldehyde, allows net synthesis of NADH by the pathway.

The properties of the soluble c-type cytochromes found in methanotrophs and methylotrophs are summarised in Table 3.3. In some organisms (e.g. *Pseudomonas* AMl and *Methylophilus methylotrophus*), at least two soluble c-type cytochromes are present, a small basic monoheme protein and a larger acidic monoheme protein. Although the latter is more rapidly reduced by the aerobic form of methanol dehydrogenase, neither have been tested with the active anaerobically-prepared enzyme and there is no strong evidence for functional discrimination. All the cytochrome c of *Pseudomonas* AMl is available for reduction both by NADH-producing substrates and by methanol (Anthony 1975).

The amino acid sequence of the small cytochrome from *Mph. methylotrophus* has been determined (R. P. Ambler, personal communication, Vol. 2, Chap. 3) and is clearly related to *Pseudomonas* cytochrome c-551. On the basis of amino acid composition (Beardmore-Gray et al. 1982) this will also apply to the similar cytochrome of *Pseudomonas* AMl. However, the classification of the cytochromes from the methanotrophs is less certain. The amino acid sequence of *Mc. capsulatus* cytochrome c-555 is unusual in lacking tryptophan and bears no obvious relationship to established sequence groups. The large cytochromes of *Pseudomonas* AMl and *Mph. methylotrophus* are not related to the smaller cytochromes by proteolysis or dimerisation (Beardmore-Gray et al. 1982) and the combination of the relativeley large size and monoheme nature distinguish them from other known cytochrome c groups.

A peculiarity of the soluble cytochromes c of Table 3.3 is their ability to bind CO to a greater or lesser degree. This property is not a preparative artefact because it can be detected in whole cells and it was originally believed to reflect an oxidase function for these proteins (Tonge et al. 1975). However, the binding of CO is very slow in comparison to binding to the cytochrome aa_3 (Widdowson and Anthony 1975) and in several cases the ability to bind CO does not reflect a tendency to autoxidation (O'Keefe and Anthony 1980a; Cross and Anthony 1980a). Thus these cytochromes probably do not contribute signifi-

Table 3.3. C-type cytochromes from methylotrophs and methanotrophs

Source	Mode of growth	α-peak	pI	Mol. wt.	E_m	Hemes	%CO binding	Reactivity with MDH[g]	Reference
Group 1: small									
Pseudomonas AMl	Facultative methylotroph	550.5[a]	8.8 (c_H)[i]	11000	294	1	36	Yes[c]	O'Keefe and Anthony 1980a, b
Methylophilus methylotrophus	Obligate methylotroph	551[a]	8.9 (c_H)	8500	373[b]	1	7	Yes[c]	Cross and Anthony 1980a
Methylomonas J[d]		551		12500	240	1	50	Yes[c]	Ohta and Tobari 1981
Methylosinus trichosporium	Obligate methanotroph	550[e]		13000	310		74	May be donor to oxygenase	Tonge et al. 1975 Tonge et al. 1977
Pseudomonas extorquens	Facultative methylotroph			13000	295		High		Higgins et al. 1976
Group 2: large									
Pseudomonas AMl		549	4.2 (c_L)	21000	256	1	72	Yes[c,f]	O'Keefe and Anthony 1980a, b
Methylophilus methylotrophus[h]		550	4.2 (c_L)	19000	310[b]	1	60[b]	Yes[c,f]	Cross and Anthony 1980a
Methylomonas J		551		16000	290	1	Low	Yes[c]	Ohta and Tobari 1981

A comparison of partial digestion products of the two cytochromes of *M. methylotrophus* demonstrated that the smaller protein was not a proteolytic digestion product of the larger (Beardmore-Gray et al. 1982).

[a] These cytochromes have a free cysteine.

[b] The same potentiometric species were found in the membrane fraction.

[c] The reduction by methanol dehydrogenase occurs independently of added methanol.

[d] An azurin was observed in addition to the two cytochromes.

[e] Similar cytochromes called cytochromes c_{CO} have been found in *M. capsulatus* and *M. organophilum.*

[f] Although both large and small cytochromes present in *Pseudomonas* AMl and *M. methylotrophus* react with CH_3OH dehydrogenase, the large cytochrome reacts faster.

[g] MDH, methanol dehydrogenase.

[h] This organism contains a second acidic cytochrome of pI 4.6, M_r 16.8 K, E_m 336 mV, described by Cross and Anthony (1980a) as c_{LM}.

[i] The *Pseudomonas* AMl and *Methylophilus* cytochromes are also called c_H and c_L depending on whether they have a high or low pI.

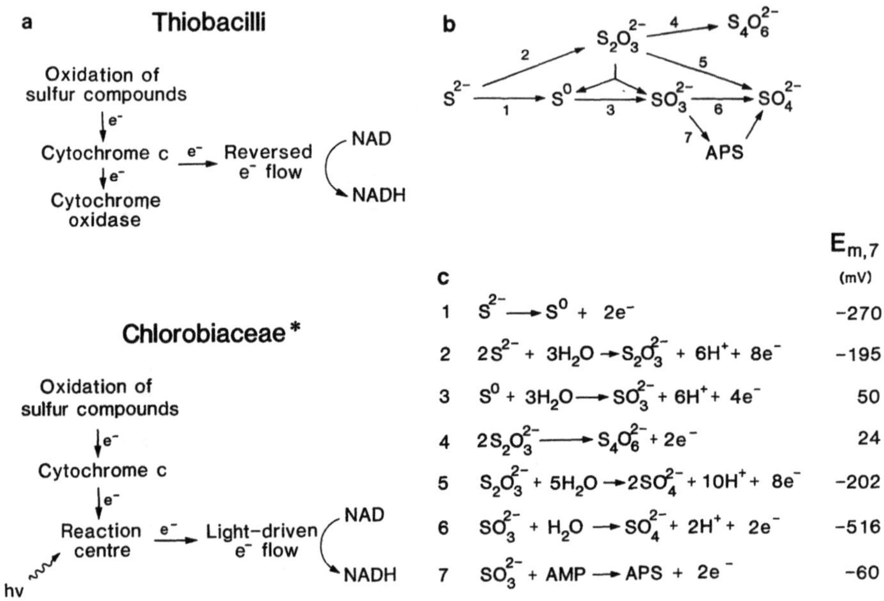

Fig. 3.9 a–c. The electron economy and thermodynamics of the oxidation of sulfur compounds. * Oxidation of sulfur compounds by the *Chromatiaceae* and *Rhodospirillaceae* also probably occurs at the level of cytochrome c (but see text). $E_{m,7}$ values are from Thauer et al. (1977)

cantly to physiological oxygen reduction (Wood 1984) and CO binding may be coincidental. We may note however that methanotrophs can oxidise CO at rates comparable to those for methane (Ferenci 1974) and the cytochrome c of *Ms. trichosporium* that is associated with methane mono-oxygenase binds CO (Tonge et al. 1975).

Energy Conservation. Methanol dehydrogenase is periplasmic in *Paracoccus denitrificans* (Alefounder and Ferguson 1981) and if the proton uptake site on the cytochrome oxidase faces the cytoplasm, this topographical arrangement would lead to the establishment of a proton gradient ($2H^+/O$) without transmembrane movement of protons (Fig. 3.10).

The nature of the cytochrome oxidase present is dependent on the growth conditions. In *Pa. denitrificans,* cytochrome aa_3 is induced by methanol and is associated with the appearance of a third site of energy conservation. Methanol oxidation at this site allows acidification of 3.5 H^+/O indicating that additional proton pumping takes place (van Versefeld and Stouthamer 1978). In *Mph. methylotrophus,* cytochrome aa_3 was observed only under conditions of methanol-limited growth. In methanol excess and oxygen limitation it was replaced by cytochrome o. The poorer growth yield under methanol excess suggests that there is less energy conservation associated with cytochrome o and this may be a response to allow rapid removal of toxic levels of methanol. The cytochrome o from *Mph. methylotrophus* has been isolated (Carver and Jones 1983; Froud and Anthony 1984) and found to be of the *co* type (Sect. 3.4). The c-type component has been claimed to be the large acidic soluble cytochrome already characterised (Table 3.3; Carver and Jones 1983) but this is disputed (Froud and Anthony 1984). The large acidic cytochrome is however an effective electron donor to the cytochrome *co*.

3.3.3 The Oxidation of Sulfur Compounds

The ability to use sulfide, sulfur and the oxides of sulfur is widespread. The *Thiobacilli* are usually obligate chemolithotrophs which obtain energy from the aerobic oxidation of sulfur compounds (exceptions include *Thiobacillus novellus* which can grow heterotrophically and *T. denitrificans* which can grow anaerobically on nitrate). Since oxidation of sulfur compounds appears to occur at the level of cytochrome c, much of the oxidative energy must be used to drive reversed electron transport to form NAD(P)H (Fig. 3.9) as we have seen for other chemolithotrophic donors.

The phototrophic bacteria generally oxidise sulfur compounds anaerobically using the electrons to reduce $NAD(P)^+$. Again the electrons are supplied at the level of cytochrome c and there must be an input of energy to achieve reduction of $NAD(P)^+$. In the green sulfur bacteria (Chlorobiaceae), this is achieved by the light-driven expulsion of an electron from the photosynthetic reaction centre to a primary acceptor with E_m −540 mV. That electron is then replaced from cytochrome c (Fig. 3.9). In the purple bacteria (Chromatiaceae and Rhodospirillaceae), the primary acceptor is not reducing enough to allow synthesis of NAD(P)H and further energy input derived from cyclic photosynthetic electron transport is required (Truper and Fischer 1982; Fig. 3.5). Some of the purple bacteria can grow in the dark in the presence of oxygen like the *Thiobacilli* and an evolutionary relationship between the phototrophic bacteria, sulfate-reducing bacteria and the *Thiobacilli* has been suggested (Truper and Rogers 1971).

The pathways of sulfur utilisation are complex (Fig. 3.9) and no one organism catalyses all the possible variations. Indeed, there are strains of *Chlorobium limicola* for example that utilise thiosulfate and others that cannot. This militates against a unified treatment of the subject. The discussion that follows deals with the main features of sulfur utilisation with emphasis on the central role of the c-type cytochromes.

The Oxidation of Sulfide (S^{2-}). Sulfide is commonly used by both *Thiobacilli* and phototrophic bacteria. Because it will slowly reduce many cytochromes c there has been some doubt about whether enzymic oxidation of sulfide actually takes place. However, certain cytochromes do appear to have a rapid sulfide-cytochrome c oxidoreductase activity (Table 3.4).

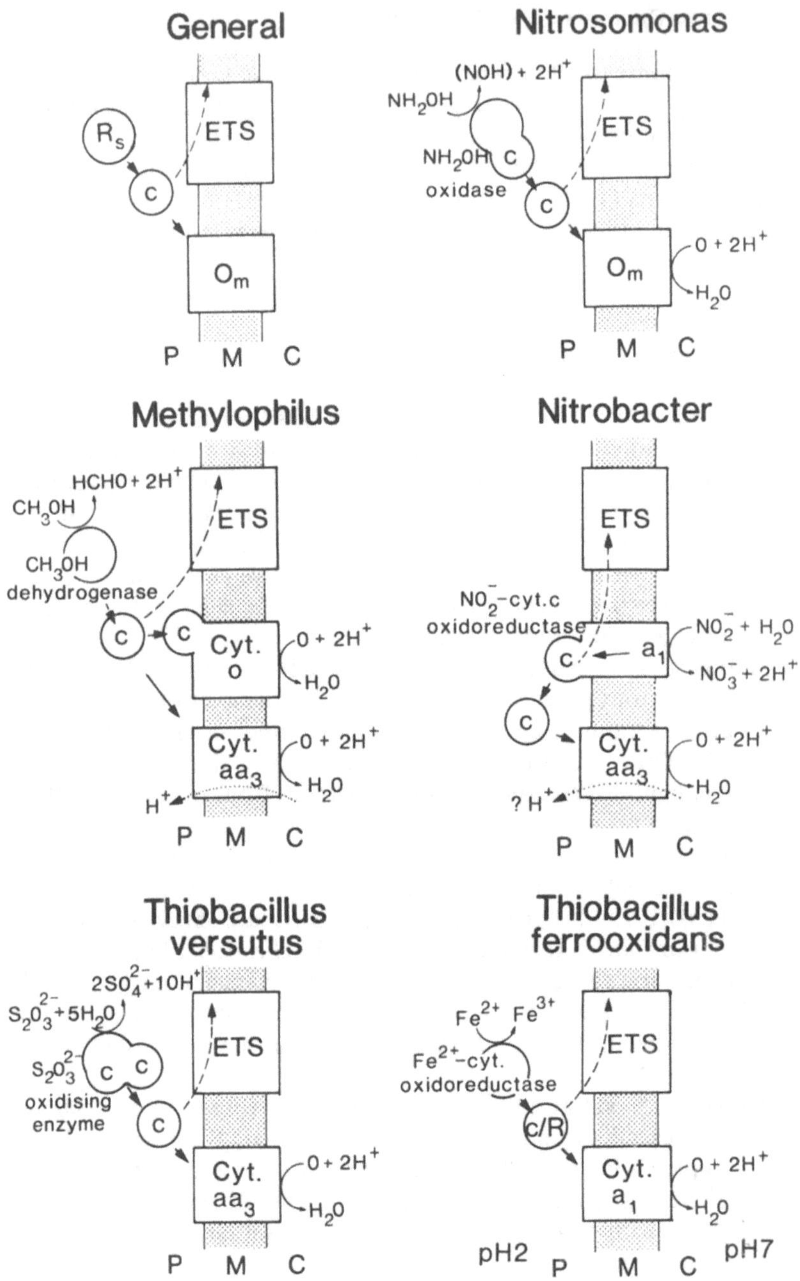

Fig. 3.10. The topography of chemolithotrophy. The diagram is constructed after the pattern in Figs. 3.2 and 3.5. Not all the individual elements have been directly demonstrated. The electron transport system (*ETS*) probably resembles complex III of the mitochondrion but is poorly characterised. – – → Entry to reversed electron transport; · · · · → proton translocation. *R* rusticyanin. Note that although the oxygen reaction in *Thiobacillus ferrooxidans* is shown to occur at pH 7 on the cytoplasmic side of the membrane, this does not affect the overall thermodynamic calculation of energy available in oxidation of Fe^{2+} which is defined by potentials operating in the bulk phase (Ingledew 1982)

The flavocytochrome of *Chlorobium limicola* f. thiosulfatophilum stimulates by a factor of ten the reduction of cytochrome c-555 by sulfide (Kusai and Yamanaka 1973a) and a similar effect is seen with the *Chromatium* flavocytochrome c (Fukumori and Yamanaka 1979). Both flavocytochromes bind SO_3^{2-} and SSO_3^{2-} at the flavin group, an effect which may have regulatory significance in achieving the complete oxidation of sulfide (Meyer and Bartsch 1976). A study of the fast kinetics of reduction of the flavocytochromes by lumiflavin indicates that rapid formation of the semi-quinone anion of the internal flavin is followed by a fast intramolecular electron transfer to the heme group(s) (Tollin et al. 1982). Direct reduction of the heme groups by the lumiflavin does not occur in the holoenzyme but occurs readily in the isolated heme subunit (Meyer et al. 1985) suggesting that the heme groups are concealed by the presence of the flavin subunit. The *Chromatium* flavocytochrome forms a tight complex with 2 mol horse cytochrome c at ionic strengths at which the sulfide-cytochrome c oxidoreductase activity is maximal (Gray and Knaff 1982). A group of lysines on the "front face" of horse cytochrome c are involved in the binding (Bosshard et al. 1986).

However, the distribution of flavocytochrome c is not consistent with an essential role in the oxidation of S^{2-}. Thus, in *Ectothiorhodospora abdelmalekii,* a small monoheme cytochrome c (Table 3.4) catalyses the oxidation of S^{2-} (Then and Truper 1983) and flavocytochrome c is also absent in *Thiocapsa pfennigii* (Meyer et al. 1973) and the *Thiobacilli.* Meyer and Bartsch (1976) suggested that the distribution of flavocytochrome c may correlate with the ability to oxidase *both* SSO_3^{2-} and S^{2-}. Indeed, the product of S^{2-} oxidation might be SSO_3^{2-} (although several studies indicate the simple formation of S^0). However, the non-thiosulfate-utilising strain 6330 of *Chlorobium limicola* does have flavocytochrome c (Steinmetz and Fischer 1981).

The Oxidation of Sulfur. In some organisms sulfur appears to be produced directly from sulfide while in others it is a product of dismutation of SSO_3^{2-}. Very little is known of its oxidation. Both the *Thiobacilli* and the photosynthetic bacteria (apart from the Chlorobiaceae) contain large amounts of the siroheme-containing sulfite reductase analogous to that found in *Desulfovibrio* (Peck et al. 1965) which, in reverse mode, may act to oxidise sulfide and sulfur to sulfite (Truper and Fischer 1982). The electron acceptor is not known.

The Oxidation of Thiosulfate. Thiosulfate-utilising enzymes are widespread. Thiosulfate may be cleaved by rhodanese to sulfur and sulfite, may undergo reductive dismutation to sulfide and sulfite, may be oxidised to tetrathionate or may be oxidised directly to sulfate with no release of intermediate forms.

The last is found in *Thiobacillus versutus* f. A2 and the enzyme contains both the cytochrome c-551 and c-552.5 appearing in Table 3.4 and two colourless subunits (A and B) (Lu and Kelly 1984). The whole complex catalyses the complete oxidation of thiosulfate to sulfate and it has not been possible

Table 3.4. C-type cytochromes involved in the oxidation of sulfur compounds

Cytochrome	Organism	Spectra	M_r	Hemes	pI	E_m	Electron transfer	References
1. Sulfide oxidation								
Flavo-cytochrome c[a]	*Chlorobium limicola* f. thiosulfatophilum[b]	ox 410 480 450 red 417 523 553	50 K_{SED} → 11 K → 47 K	1 (1 Flavin)	6.7	+98	S^{2-}-cyt.c oxidoreductase	Bartsch et al. (1968), Meyer and Kamen (1982), Gray and Knaff (1982), Kusai and Yamanaka (1973), Fukumori and Yamanaka (1979)
	Chromatium vinosum[b]	ox 410 480 450 red 417 523 552	72 K_{SED}[c] → 20 K → 47 K	2 (1 Flavin)	5.1	+20 ~ +20	S^{2-}-cyt.c oxidoreductase. S^0 is the product	
Cytochrome c-550[d]	*Thiocapsa roseopersicina*		34 K_{SEP}				$S^{2-} \rightarrow S^0$	Fischer and Truper (1977)
Cytochrome c-551	*Ectothiorhodospora abdelmalekii*	ox red 417 529 551	9.5 K_{SEP}	1	3.5[e]	−7	$S^{2-} \rightarrow S^0$	Then and Truper (1983)
2. Thiosulfate oxidation								
Cytochrome c-552.5	*Thiobacillus versutus* f. A2	ox 415 red 418 523 552.5	56 K_{SEP} 29 K_{SDS}	2−3	4.8	+220 (65%)[j] −215 (35%)	Part of thiosulfate oxidising enzyme	Lu and Kelly (1984), Lu et al. (1984)
Cytochrome c-551	*Thiobacillus versutus* f. A2	ox 410 red 418 522 551	300 K_{SEP} 43 K_{SDS}	4−5	5.2	+240 (45%) −115 (55%)	Part of thiosulfate oxidising enzyme	Lu and Kelly (1984), Lu et al. (1984)
Cytochrome c-551	*Chlorobium limicola*[f] f. thiosulfatophilum	ox 410 red 416 521 551	45 K_{SED}	2	6	+135	Part of thiosulfate oxidising enzyme	Meyer et al. (1968), Kusai and Yamanaka (1973b)
3. Sulfite oxidation								
Cytochrome c-551	*Thiobacillus novellus*	ox 411 red 416 522 551	23 K_{SDS}	ND	5.2	+260	Part of sulfite cytochrome c oxidoreductase	Yamanaka et al. (1971) and (1981a)

4. Mediator function								
Cytochrome c-550[g]	*Thiobacillus novellus*	ox 410 red 415 520 550	13.3 K_{AA}	1	7.5	+276	Donor to cyt.aa_3	Yamanaka et al. (1971)
Cytochrome c-555[h]	*Chlorobium limicola* f. thiosulfatophilum	ox 413 red 419 523 555	10 K_{AA}	1	10.5	+145	Donor to reaction centre	Meyer et al. (1968)
Cytochrome c-551[i]	*Chromatium vinosum*		15 K_{SEP}	ND		+260	Donor to reaction centre	Gray and Knaff (1982), van Grondelle et al. (1977), Gray et al. (1983)

M_r determined by SED = sedimentation equilibrium, SEP = molecular exclusion chromatography, SDS = SDS gel electrophoresis, AA = amino acid analysis.

[a] Flavocytochrome c with similar properties is also found in *Chl. vibrioforme* (f. thiosulfatophilum) (Steinmetz and Fischer 1982) and *Chl. limicola* strain 6330 (non-thiosulfate-utilising) (Steinmetz and Fischer 1981). A rather different flavocytochrome c was isolated as a p-cresol dehydrogenase from *Ps. putida* (Hopper and Taylor 1977). It contains two subunits of M_r 56 K, one containing a heme with E_m +250 mV, higher than those of sulfur-metabolising organisms, and one containing a novel O-tyrosyl-linked FAD (Hopper 1983; McIntyre et al. 1981). The natural electron acceptor is unknown.*

[b] Both *Chromatium* and *Chlorobium* flavocytochromes c have 8-α-cysteinyl FAD (Kenney and Singer 1977; Kenney et al. 1977).

[c] The protein dimerises at low ionic strengths (Gray and Knaff 1982).

[d] Fischer and Truper (1977) suggested that cytochrome c-552 (550) of *Thiocapsa pfennigii* (Meyer et al. 1973), which is a diheme, 30 K protein, may be a relative of this cytochrome.

[e] Characteristically low pI for an extreme halophile.

[f] *Chlorobium vibrioforme* (f. thiosulfatophilum) contains a cytochrome c-551 of M_r 32 K which may be involved in thiosulfate oxidation (Steinmetz and Fischer 1982).

[g] Basic small monoheme cytochromes c which probably play the same role have been isolated from *Thiobacillus thiooxidans* (Takakuwa 1975), and *Thio. versutus* (f. A2) (Lu and Kelly 1984; Lu et al. 1984). The latter organism also contains a small acidic cytochrome c-550.

[h] Relatives of cytochrome c-555 are found in *Chl. vibrioforme* f. thiosulfatophilum (Steinmetz and Fischer 1982); *Chlorobium limicola* 6330 (Steinmetz and Fischer 1981); *Chlorobium vibrioforme* (non-thiosulfate strain) (Steinmetz et al. 1983); *Chloropseudomonas* 2 K (now known to be a consortium; see p. 192) (Shioi et al. 1972).

[i] Gray et al. (1983) described this cytochrome as c-550 and also found a low potential c-551.

[j] % values are the contributions of individual redox components to a redox titration.

* See appendix note 13.

to assign partial reactions to particular subunits. Thiosulfate oxidase activity can be reconstituted from the individual subunits, horse cytochrome c and *Thiobacillus* cytochrome oxidase (Lu and Kelly 1984). The small basic cytochrome c-550 is probably the natural mediator.

In the Chlorobiaceae, a cytochrome c-551 (Table 3.4) is found only in those strains utilising thiosulfate (Truper and Fischer 1982). It accepts electrons from a colourless, thiosulfate-oxidising enzyme and can reduce cytochrome c-555 (Kusai and Yamanaka 1973b).

Oxidation of Sulfite. Sulfite is commonly utilised by the *Thiobacilli* but rarely by the phototrophic bacteria: indeed, some of the latter can tolerate only trace sulfite in the medium. Two paths of sulfite oxidation are possible: one direct (sulfite-cytochrome c oxidoreductase) and the other involving adenosine phosphosulfate as an intermediate (APS reductase). Both enzymes are widely distributed and their relative importance in sulfite oxidation is uncertain.

Sulfite-cytochrome c oxidoreductase has been isolated from *Thiobacillus novellus* and found to contain a cytochrome c-551 as an essential component (Table 3.4; Yamanaka et al. 1981a). Reconstitution of sulfite oxidase activity was achieved using purified *Thiobacillus* cytochrome aa_3 and yeast cytochrome c. The small basic cytochrome c-550 (Table 3.4) is presumably the natural mediator.

APS reductase was named for its role in sulfate *reduction* in the *Desulfovibrio* (Sect. 3.7). Operating in reverse, it releases the high energy compound adenosine phosphosulfate which can be used to synthesise ATP at the substrate level. However, because of the cosubstrate requirements for AMP, APS reductase must be a cytoplasmic enzyme and therefore unlike the periplasmic chemolithotrophic processes that we have so far considered. Wood (1983) has proposed the general rule that all cytochromes c are situated either free in the periplasmic space or attached to the periplasmic side of the plasma membrane. Thus we would expect neither APS reductase itself nor its electron acceptor would contain cytochrome c. Surprisingly the enzyme from *Thiocapsa roseopercicina* does appear to contain heme c (Truper and Rogers 1971) and constitutes an unresolved anomaly in Wood's thesis.

Energy Conservation. The thiosulfate-oxidising enzyme from *Thiobacillus versutus* is periplasmic (Lu 1986) and the widespread involvement of soluble c-type cytochromes makes it probable that other oxidations of sulfur compounds also occur in the periplasm. Such oxidations coupled to a cytoplasmic reduction of oxygen will establish a proton gradient across the membrane (Fig. 3.10). Some of the compounds involved (for example SO_3^{2-}) are quite strong reductants and their oxidation at the level of cytochrome c does seem wasteful of available energy. If the APS reductase route is used, however, substrate level phosphorylation can occur.

Measurements of the magnitude of the proton gradient are still in their infancy but suggest values of 2 for H^+/O using SSO_3^{2-}, S^{2-} and SO_3^{2-} as substrates. If however SO_3^{2-} is oxidised via APS reductase it does not give rise to a proton gradient and the process is insensitive to uncouplers (Drozd 1974, 1977).

3.3.4 The Oxidation of Nitrite

Nitrobacter agilis grows autotrophically by aerobic oxidation of nitrite ($E_{m,7}$ 430 mV). Early work implicated a cytochrome a_1 as the terminal oxidase which was believed to be directly reduced by NO_2^- without the mediation of cytochrome c (Aleem 1968; Sewell and Aleem 1969). However, a cytochrome aa_3 with properties very similar to the mitochondrial enzyme was identified (Sewell et al. 1972) and purified (Sect. 3.4). More recent schemes therefore place cytochrome aa_3 at the site of oxygen reduction and implicate cytochrome a_1 as a component of the nitrite oxidase (Aleem 1977; Ferguson 1982a). This is supported by the solubilisation and purification of a NO_2^--cytochrome c oxidoreductase activity which contains both heme a_1 and a c-type cytochrome of M_r 29 K (Tanaka et al. 1983). The latter is of a similar size to cytochrome c_1 (Sect. 3.2).

A small basic cytochrome has been purified (Table 3.6) which is active both as electron donor to the cytochrome aa_3 and as electron acceptor from the nitrite-oxidising enzyme (Tanaka et al. 1983). Thus the oxidative electron transport system of *N. agilis* is a simple one involving only the three components represented in Fig. 3.10. Energy conservation in this oxidative arm is used to drive reversed electron transport to form NADH.

A thermodynamic peculiarity of the system deserves some comment. According to the midpoint potentials, cytochrome c-550 ($E_{m,7}$ 274 mV) should be poorly reduced by the more oxidising NO_2^-/NO_3^- couple ($E_{m,7}$ 430 mV). This puzzle was elegantly resolved by a scheme proposed by Cobley (1976a, b) involving a transmembrane electron transfer between nitrite oxidase, believed to be attached to the cytoplasmic surface of the membrane, and cytochrome c, attached to the periplasmic surface. Cobley suggested that this electron flow (moving as a hydride ion) would be enhanced by any membrane potential ($\Delta\psi$), positive outside. This scheme therefore explained the inhibitory effects on NO_2^- oxidation of agents (such as ADP or valinomycin and K^+) which act to collapse the membrane potential. Also the reducing potential of the electrons from NO_2^- should be increased by an amount equal to $\Delta\psi$. The effect is to partly overcome the thermodynamic imbalance between the cytochrome c and NO_2^- and neatly explains the observation of Ingledew and Chappell (1975) that in inverted vesicles energised by ATP hydrolysis, the apparent E_m of cytochrome c-550 trapped within the vesicles is raised from 274 to 360 mV if NO_2^- is used as electron donor.

Recent interest in the proton pumping ability of cytochrome aa_3 type oxidase (Sect. 3.4) led Ferguson (1982a) to replace the hydride transfer model of Cobley with a simple electron transfer across the membrane, accompanied by proton movement through the terminal oxidase (Fig. 3.10). However, initial experiments indicate that the purified oxidase cannot pump protons (Sone et al. 1983).

In addition to cytochrome c-550, *N. agilis* contains a cytochrome c-549, 554 which bears some resemblance to *Pseudomonas* cytochrome c peroxidase and is discussed in Section 3.4 and a cytochrome c-553 (M_r 11.5 K) which has an unusually positive midpoint potential of >450 mV (Chaudhry et al. 1981).

3.3.5 The Oxidation of Ferrous Iron

The oxidation of Fe^{2+} by O_2 catalysed by *Thiobacillus ferrooxidans* represents one of the narrowest thermodynamic spans used in biological energy conservation. The organism is strictly acidophilic, maintaining a cytoplasmic pH near neutrality in the presence of a bulk phase around pH 2 (Ingledew 1982). Under these conditions, the standard oxidative energy available is given by

$$E_{m,\ pH\,2,\ O/H_2O} - E_{h,\ pH\,2,\ Fe^{2+}/Fe^{3+}} = 1120 - 770 = 350\ \mathrm{mV}.$$

Components of the oxidative electron transport chain are represented in Fig. 3.10. A cytochrome c-552 ($E_{m,\ pH\,3.2}$ 620 mV) and a blue copper protein, colourfully named rusticyanin ($E_{m,\ pH\,3.2}$ 680 mV), can be released by EDTA washing of membranes (Cobley and Haddock 1975; Ingledew and Cobley 1980). By analogy with the plastocyanin-cytochrome c-553 pair in algae (Chap. 4.1.3), these may be functionally interchangeable. Both are reduced directly by Fe^{2+} (Cox and Boxer 1978) but cell-free extracts carry out Fe^{2+} oxidation at a small fraction of the rate of intact cells leading to a, so far unsuccessful, search for a Fe^{2+}-cytochrome c oxidoreductase. One suggestion is that superoxide is formed from the reaction of Fe^{2+} with oxygen mediated by a polynuclear phosvitin-like complex (Ingledew and Cobley 1980). Superoxide would then act as e^- donor to cytochrome c (Dugan and Lundgren 1965).

The cytochrome c oxidase is described as cytochrome a_1 with an α-peak at 597 nm and two potentiometric components ($E_{m,\ pH\,3.2}$ 720, 610 mV). Like the spectroscopically similar *Nitrosomonas* enzyme, it is probably a cytochrome aa_3 with a slightly shifted α-peak (Sect. 3.4). Experiments on azide inhibition suggest that the oxygen-binding site is accessible from the cytoplasmic side of the membrane (Ingledew et al. 1978). As a consequence of this topography of redox reactions (Fig. 3.10), protons taken up by the reduction of oxygen in the cytoplasm can be replaced by flow through the ATP synthase from the acidic bulk enviroment thus allowing energy conservation. In keeping with the other chemolithotrophs, the throughput of electrons in the oxidative arm has to be rapid to support the reversed electron transport required for NADH synthesis, and the concentrations of cytochrome c-550, rusticyanin and cytochrome oxidase are therefore very high (Tikhonova et al. 1967).

3.4 Oxygen as a Terminal Electron Acceptor

3.4.1 Introduction

Bacterial cytochrome oxidases that reduce molecular oxygen to water can be of the aa_3, a_1, *o* or d type. Cytochrome cd_1, which is also able to reduce oxygen is thought to function primarily as a nitrite reductase (Sect. 3.5). A major way in which heterotrophic aerobic bacteria differ from the mitochondrion is the complexity of their interaction with oxygen. Linear chains terminating in a single oxidase may be the exception rather than the rule and in an individual species, the quantitative contribution of two or three terminal oxidases may vary depending on the growth conditions. Thus oxygen limitation may induce cytochrome *o* (Sapshead and Wimpenny 1972; Froud and Anthony 1984): media in which CN^- production may occur may induce CN^-- insensitive cytochrome d (Jones 1977), although this is disputed (Akimenko and Trutko 1984). In most cases we do not know whether this adaptive branching at the end of the electron transport chains requires synthesis of different electron donor cytochromes or whether a single donor suffices. Since such donors are often c-type cytochromes this constitutes an important area of cytochrome c function which remains to be clarified.

The cytochrome c oxidases which are the subject of this section are transmembrane enzymes which oxidise cytochrome c in the periplasmic space (Wood 1983), a compartment topographically equivalent to the intermembrane space of the mitochondrion (Chap. 2). They reduce oxygen to water using protons taken up from the cytoplasmic side of the membrane. In the mitochondrion, this constitutes the third site of energy conservation and Jones (1977) has suggested that this site in bacterial respiration requires the presence of cytochrome c. We have seen that most chemolithotrophic respirations involve periplasmic oxidations that donate electrons at the level of cytochrome c (Fig. 3.2d) thus making them totally dependent on energy conservation at site III. Much current interest centres around whether the chemiosmotic properties of such systems derive from the topography of the redox reactions (Fig. 3.10), the proton-pumping activity of the cytochrome oxidase or a combination of the two.

Identification of a particular oxidase is most often on the basis of the visible absorption spectrum of reduced respiratory membranes and the effect of carbon monoxide on such spectra. The distinctive features of these spectroscopic analyses are shown in Fig. 3.11. It should be emphasised that CO will bind to and cause spectral changes in proteins that are not oxidases (Wood 1984). This is probably the case with the so-called cytochrome *o* of *Vitreoscilla* which is a soluble protein capable of only slowly reducing O_2 to H_2O_2 (Liu and Webster 1974; Webster 1975). In addition, several c-type cytochromes bind CO but show no oxidase activity (Wood 1984). Two more rigorous approaches to the identification of an oxidase are less frequently applied. One is to establish kinetic competence using rapid reaction techniques (Jones 1977). The

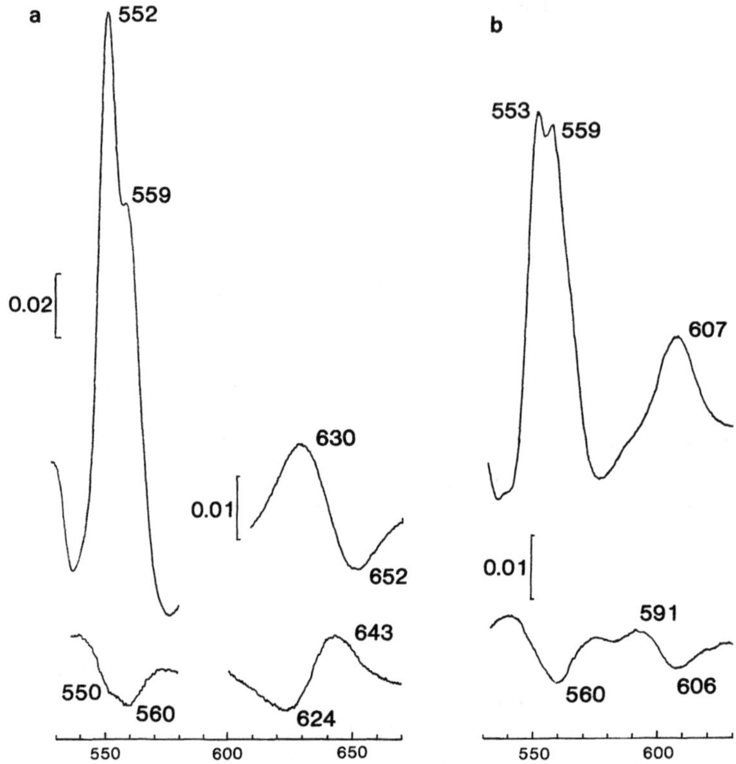

Fig. 3.11a, b. Spectroscopic features of cytochrome oxidases. **a** Difference spectroscopy using spheroplast membranes of *Azotobacter vinelandii. Upper spectra:* dithionite-reduced against air-oxidised. *Lower spectra:* dithionite-reduced+CO against dithionite-reduced. **b** Difference spectroscopy using French Press membranes of *Pa. denitrificans. Upper spectra:* dithionite-reduced against ferricyanide-oxidised. *Lower spectra:* dithionite reduced+CO against dithionite-reduced. In both organisms there is a large contribution of b-type cytochromes to the spectral region 550–560 nm, only a small proportion of which is cytochrome *o* (detected as a trough at 560 nm on the CO difference spectra). In the case of *A. vinelandii,* CO binding to c-type cytochrome is also evident (subsidiary trough at 550 nm). Cytochrome d of *A. vinelandii* shows a characteristic 630-nm peak and 652-nm trough in reduced minus oxidised spectra which is strongly perturbed by CO to give a peak at 643 nm and trough at 624 nm. Cytochrome aa_3 of *Pa. denitrificans* is identified by a peak at 607 nm in reduced minus oxidised spectra which is shifted to the blue on CO binding to yield a small peak at 591 nm and a trough at 606 nm in CO difference spectra

second is the reversal of CO inhibition of respiration using light of energy required to dissociate the CO complex (photochemical action spectra; Castor and Chance 1955). However, as Wood (1984) pointed out, light relief of CO inhibition of respiration would still be observed if the CO binding component were part of the electron transport system rather than the terminal oxidase.

3.4.2 Cytochrome aa_3

Spectra. Cytochromes aa_3 are characterised by α-bands at 600–607 nm (Fig. 3.11) and Soret bands at 440–445 nm in reduced minus oxidised difference spectra. CO difference spectra show maxima near 590 nm (Fig. 3.11) and 425 nm. Thus the α-band differs from that of cytochrome a_1 (585–595 nm) but the CO difference spectra of the two are similar and photochemical action spectra cannot be used to distinguish them. The a-type cytochrome of *Nitrosomonas europaea* was originally described as a_1 on the basis of its α-peak at 597 nm but its structure resembles that of the aa_3 group (Yamanaka et al. 1985). This may also be so for the "cytochrome a_1" of *Thiobacillus ferrooxidans* (Sect. 3.3.5).

Structure. The bacterial cytochromes aa_3 appear to be similar to the mitochondrial enzyme with respect to ligand-binding properties, the presence of potentiometrically and magnetically distinct heme and copper centres and in the rapid kinetics of the oxygen reaction (Poole 1983). In structure however they are simpler, often with two or at the most, three subunits. Several lines of evidence indicate that they contain relatives of subunits I and II of the mitochondrial enzyme. Thus all but the *Thiobacillus novellus* enzyme have a large subunit of molecular weight 45–57 K with atypical electrophoretic behaviour in SDS gels (Table 3.5). Amino acid compositions (Yamanaka and Fukumori 1981) and a limited amount of sequence information (Steffens et al. 1983) support homology with subunit I of mitochondrial cytochrome aa_3. The oxidases of *Rhodopseudomonas sphaeroides, Paracoccus denitrificans* and the thermophilic bacterium PS3 contain a second subunit of M_r 28–37 K which cross-reacts immunologically with yeast subunit II.

A distinctive feature of the oxidases from the two thermophilic bacteria is the presence of heme c associated with subunit II (Table 3.5). Some authors describe these oxidases as c_1 aa_3 because of the similarity in molecular weight of the subunit II and cytochrome c_1. An alternative explanation for the origin of the c-type cytochrome (Nicholls and Sone 1984), which we favour, is that subunit II of these oxidases derives from the genetic fusion of a small c-type cytochrome gene to an existing subunit II gene. Whatever the affinities of this cytochrome c subunit, it appears to be functionally competent in electron transfer (discussed below) and is therefore unlikely to be a preparative artefact.

Oxidase Activity. We have seen (Chap. 2) that mitochondrial cytochrome oxidase contains a high affinity site for cytochrome c which allows TMPD oxidation without dissociation of the cytochrome. In the oxidases of the thermophilic bacteria, a cytochrome c is resident on the molecule and purifies with the oxidase after solubilisation. The heme c can be oxidised at $-80\,°C$ after flash photolysis of the heme a_3/CO complex (Sone and Yanagita 1982) and TMPD oxidation does not require added cytochrome c (Yoshida et al. 1984; Nicholls and Sone 1984). Thus it seems probable that the resident heme c of

Table 3.5. Bacterial cytochromes aa_3

Organism	Subunits (K)	Redox centres[a]	H^+ Pumping	e^- Donor (Table 3.6)	References
Paracoccus denitrificans	45, 28[b]	2.0 a 2.3 Cu	0.6 H^+/e^-[c]	c-550[d]	Ludwig and Schatz (1980), Solioz et al. (1982), Davies et al. (1983)
Rhodopseudomonas sphaeroides	45, 37, 35[e]	1.6[f] a (Cu ND)	0	c_2[g]	Gennis et al. (1982)
Nitrobacter agilis	51, 31[h,k]	1.6 a 2.6 Cu	ND	c-550	Yamanaka et al. (1981 b), Chaudhry et al. (1980), Yamanaka et al. (1979b), Yamanaka et al. (1982)
Thiobacillus novellus	32, 23	1 a 1 Cu	ND	c-550	Yamanaka and Fujii (1980)
Nitrosomonas europaea	50, 33**	ND	ND	ND	Yamanaka et al. (1985)
Pseudomonas AMl	50, 32	1.2 a 1.4 Cu	ND	c_H (Table 3.3)	Yamanaka et al. (1985), Fukumori et al. (1985)
Thermophilic bacterium PS3[j]	56, 38*, 22[i,k]	1.8 a 2.1 Cu 1 c	0.6 – 1.4 H^+/e^-[m]	c-551[l]	Sone et al. (1979), Sone and Yanagita (1982), Sone and Hinkle (1982), Yanagita et al. (1983), Baines et al. (1984), Sone and Yanagita (1984)
Thermus thermophilus HB8	55, 33*[n]	1.8 a 1.9 Cu 1 c	1 H^+/e^-	c-552	Fee et al. (1980), Yoshida and Fee (1984), Honnami and Oshima (1980)
Bacillus subtilis W23	57, 37, 21	ND	ND	c-550[o]	Vrij et al. (1983)
Bacillus firmus	56, 40, 14*	2.1 a (Cu ND) 0.9 c	ND	ND	Kitada and Krulwich (1984)
Mitochondria	57, 26, 30, 17, 12, 11, 10, 9, 8, 6, 6, 6, 5	2 a 2 Cu	1 – 2 H^+/e^-	c	Chapter 2.2.1

* Subunit containing heme c; ND, not determined.

** See appendix note 14.

[a] Figures are calculated from nmol heme a or Cu per mg protein using the sum of the subunit molecular weights assuming single copies of each.

[b] The smaller subunit is cross-reactive with mitochondrial subunit II and shows sequence homology to it (Steffens et al. 1983). The larger subunit shows homology to mitochondrial subunit I. The native enzyme is a monomer (Ludwig et al. 1982).

[c] This figure is obtained by extrapolation to zero turnover (see text). Dicyclohexylcarbodiimide DCCD does not inhibit this proton movement (Puttner et al. 1983).

[d] Cytochrome c-550 is an effective donor to the solubilised enzyme. However, a tightly bound cytochrome c of unknown properties is rapidly oxidised in membrane preparations (Davies et al. 1983).

[e] 37 and 35-K subunits cross-react with antiserum against subunit II of *Para. denitrificans* cytochrome oxidase.

these oxidases is the functional equivalent of cytochrome c in the high affinity site of mitochondrial oxidase.

Indeed, tightly associated cytochrome c may be a common feature of bacterial oxidases. For example, that from the alkalophile, *Bacillus firmus* contains a cytochrome c subunit of M_r 14 K. Also, the washed membranes from both *Pa. denitrificans* and *Rps. sphaeroides* show high ascorbate-TMPD oxidase activity indicating retention of a tightly bound c-type cytochrome (Davies et al. 1983; Gennis et al. 1982). In the former case, this activity is not affected by a monoclonal antibody raised against the soluble cytochrome c-550 nor does added cytochrome c-550 enhance the NADH oxidase activity (Kuo et al. 1985). Solubilisation of the membrane with dodecylmaltoside releases a particle which exhibits quinol oxidase activity. This particle contains the bc_1 complex, cytochrome aa_3 and a previously uncharacterised c-type cytochrome (Berry and Trumpower 1985). Figure 3.2c, which shows a c-type cytochrome bound to the cytochrome oxidase may therefore represent a common situation among the bacterial cytochromes aa_3. As we shall see it also appears to be a common feature of the cytochromes *o*.

The functional significance of the bound c-type cytochrome is uncertain. It may allow direct reduction of the oxidase by cytochrome c_1 in complex III without the need for mediation by a soluble cytochrome c. If this is so, the role of the soluble cytochromes c, also found in these systems, may be solely to mediate electron flow from soluble cytochrome c reductases to the membrane and from the membrane to soluble terminal cytochrome c oxidising systems.

Cytochrome c oxidase activity can be measured by either the polarographic or spectrophotometric methods described in Chapter 2.2.3. In view of the complication of the intrinsic ascorbate-TMPD oxidase activity in certain bacterial cytochromes aa_3, the spectrophotometric assay should be preferred. Either yeast or horse cytochrome c have often been used as convenient substrates for the bacterial cytochromes aa_3 and information on the identification and reactivity of the natural substrates is sparse. Table 3.6 is a summary of the proposed

[f] Protein is only 80% pure and this probably represents 2 mol heme a/mol.

[g] Membranes containing no cytochrome c_2 exhibited high ascorbate activity TMPD oxidase-but Triton-solubilised enzyme had none. In the latter, activity was restored by added cytochrome c.

[h] Chaudhry et al. (1980) found a third subunit of M_r 13 K.

[i] Sone et al. (1979) originally found only one subunit but later work showed that the missing pair of subunits was susceptible to aggregation in SDS.

[j] Some spectroscopic contamination by cytochrome *o* was observed which was more severe in preparations from poorly aerated cells.

[k] Molecular weight derived from slopes of Ferguson plots.

[l] Cytochrome c-551 is the best donor of those tested and is membrane-attached. It is not included in Table 3.6 as its properties have not yet been published.

[m] High H^+/e^- figures are obtained at low turnover number (Sone and Yanagita 1984). Proton pumping is inhibited by DCCD.

[n] HonNami and Oshima (1980) found a third subunit of M_r 29 K.

[o] Cytochrome c-550 was isolated from unstirred cultures (Miki and Okunuki 1969b) but has not yet been shown to be present in aerobic cultures.

Table 3.6. Cytochromes c which donate to aa_3 type oxidases

Cytochrome	Size**	Sequence type	$E_{m,7}$	pI	Reactivity with			References
					P.a. cd_1	T.n. aa_3t	Bovine aa_3t	
Tuna cytochrome c	103	c_2	260	10.5	9	93	93	a, b
Pa. denitrificans c-550	129	c_2	256	4.5	3	3	4	b, c, d, e
Th. novellus c-550	13.3 K	ND	276	7.5	6	100	23	a, b, f, g
N. agilis c-550	109	c_2	271	8.5	7	ND	15	b, c, g, h
Rps. sphaeroides c_2*	124	c_2	352	5.5	1	ND	1	c, i, j, k, l
B. subtilis c-550*	12.5 K	ND	210	8.7	73	ND	"low"	m
The. thermophilus c-552	131	c_2	190	10.8	123	ND	4	n, o, p
Pseudomonas AMI	11 K	ND	294	8.8	ND	ND	ND	q, r
Ps. aeruginosa c-551	82	$c_7$tt	286	4.7	100	1	0	a, c, s

* These cytochromes have not been tested with their own purified oxidase. They are included in the Table because they are the only cytochromes in *B. subtilis* and *Rps. sphaeroides* with properties similar to the well-studied cytochromes that are known to donate to a cytochrome aa_3.
** The figure for size is the number of amino acids if the sequence is known or the estimated molecular weight if it is not.
t Bakers' yeast cytochrome c gave 100% activity tt see Table 1.6.
[a]Yamanaka and Fukumori 1977; [b]Yamanaka and Fukumori 1981; [c]Meyer and Kamen 1982; [d]Scholes et al. 1971; [e]Yamanaka et al. 1979b; [f]Yamanaka et al. 1971; [g]Yamanaka et al. 1982; [h]Chaudhry et al. 1981; [i]Pettigrew et al. 1975; [j]Bartsch 1978; [k]Yamanaka and Okunuki 1968; [l]Errede and Kamen 1978; [m]Miki and Okunuki 1969b; [n]Hon-nami and Oshima 1977; [o]Yoshida et al. 1984; [p]Titani et al. 1985; [q]O'Keefe and Anthony 1980a, b; [r]Fukumori et al. 1985; [s]Horio et al. 1960.
Ps. aeruginosa c-551 is included for comparison.

cytochrome c donors to the bacterial cytochromes aa_3 considered in this chapter and a few general comments can be made.

1. All are water-soluble proteins with a probable location in the periplasmic space (this has been directly demonstrated in some cases by spheroplast formation: Scholes and Smith 1968; Lorence et al. 1981).
2. They are members of the "large" class I cytochromes c with 100–130 residues, in contrast to the "small" class I cytochromes c exemplified by *Pseudomonas aeruginosa* c-551 (82 residues). Amino acid sequence information indicates homology to the cytochrome c_2 group (Vol. 2, Chap. 3).
3. Midpoint redox potentials fall within a narrow range and are comparable to those of the acceptor redox centres on the cytochrome oxidase.
4. The kinetics of the "natural" cytochrome c oxidase reactions (i.e. with substrate and oxidase from the same organism) have not been studied systematically (and in some cases have not been studied at all!). We can only

comment generally that in those instances studied, the "natural" reaction proceeds with relatively low K_m and high turnover. Although under saturating conditions mitochondrial cytochromes c from yeast or horse may appear more efficient substrates to bacterial cytochromes aa_3, the K_m for the natural substrate can be much lower (Hon-nami and Oshima 1980; Sone and Yanagita 1982; Yoshida and Fee 1984).

5. Studies of the cross-reactivity of a cytochrome c donor from one source and a terminal oxidase from another are frequently carried out but interpretation is fraught with difficulty. Because different pairings may have different pH and ionic strength optima and different K_m values, comparisons under fixed conditions and a single substrate concentration are virtually meaningless. Table 3.6 does show that *Thiobacillus novellus* c-550 and *Nitrobacter agilis* c-550 react moderately well with bovine cytochrome oxidase while mitochondrial cytochrome c reacts well with the *Thiobacillus* oxidase. The general rule holds that cytochromes reactive with the bovine enzyme are poorly reactive with cytochrome cd_1 (nitrite reductase) and vice versa (see also Sect. 3.5).

6. Since basic cytochromes c are relatively uncommon in bacteria, the predominance of such proteins in Table 3.6 suggests that the basic character may often be associated with donor activity to cytochrome aa_3. This is not a general rule because the cytochromes c of *Pa. denitrificans* and *Rps. sphaeroides* are acidic. Reactivity with mitochondrial cytochrome oxidase does seem to require basic character (although basic character itself does not ensure good reaction).

7. In many organisms there is more than one soluble c-type cytochrome and their relative contributions to respiration are poorly understood. The cytochromes c of *Thiobacillus* and *Nitrobacter* have been discussed in Section 3.3. In methanol-grown *Pa. denitrificans* there is a CO-binding cytochrome c (Van Versefeld and Stouthammer 1978) that has not been further characterised. By analogy with other methylotrophs (Sect. 3.3) we would expect it to mediate electron transfer between methanol dehydrogenase and cytochrome aa_3. Cytochrome c-550 of *B. subtilis* (Table 3.6) was isolated in very low yield from cells grown in unstirred peptone culture (Miki and Okunuki 1969a). Also present in soluble extracts was a cytochrome c-554 (pI 4.4, E_m −80 mV, α/β 1.3, M_r 14 K; Miki and Okunuki 1969b). Neither of these cytochromes has yet been reported in cells grown aerobically for the isolation of cytochrome aa_3 (Vrij et al. 1983).

8. Respiratory oxygen uptake in cyanobacteria also seems to be associated with cytochrome aa_3 (Peschek 1981) although no purification of the enzyme has yet been achieved. The cyanobacterial oxidase activities show interesting parallels to mitochondrial oxidase in their specificity. Oxygen uptake by membranes of *Anabaena variabilis* is stimulated by basic cytochromes such as horse cytochrome c and *Anabaena* cytochrome c-553 but not by acidic cytochromes such as *Euglena* c-552 (Lockau 1981). Since *Anabaena* cytochrome c-553 also reacts well with the photo-oxidised reaction centre, this cytochrome and the accompanying basic plastocyanin may perform a dual function in the photosyn-

thetic and respiratory chains (Davis et al. 1980; Lockau 1981; see also Sects. 3.2 and 3.6). In contrast, in *Anacystis nidulans*, the acidic cytochrome c-553 shows no reactivity with the membrane oxidase (Kienzl and Peschek 1982) which, like *Anabaena* membranes, prefers basic cytochromes. Early studies found a basic cytochrome c-552 in *Anacystis* (Holton and Myers 1967 a and b) and this should be tested as the physiological donor to the oxidase activity. In the light of these results, Davis et al. (1980) and Peschek and Schmetterer (1982) proposed that the extant, dual-purpose basic cytochrome of *Anabaena* represents a primitive link between respiratory and photosynthetic electron transport chains. They suggested that functional separation of the two chains at the level of cytochrome c occurred with the appearance of an acidic cytochrome c donor to the reaction centre. This leads us to speculate on the nature of the respiratory activity demonstrated in choroplasts (Bennoun 1982). Will a distinct cytochrome c be found in these eukaryotic organelles which donates electrons to a low level cytochrome oxidase?

Proton Translocation. Mitochondrial cytochrome oxidase is generally believed to couple electron transfer to oxygen with proton transfer from the matrix to the intermembrane space (Chap. 2.2.1). Analogous proton translocation in bacteria would be from the cytoplasm to the periplasmic space. The ability of a purified cytochrome c oxidase to translocate protons is tested by reconstitution in liposomes. Pulses of reduced cytochrome c (c_r) are used to initiate electron transfer and pH changes in the supporting medium are monitored (Fig. 3.12). Since only oxidase having a right-side-out orientation can interact

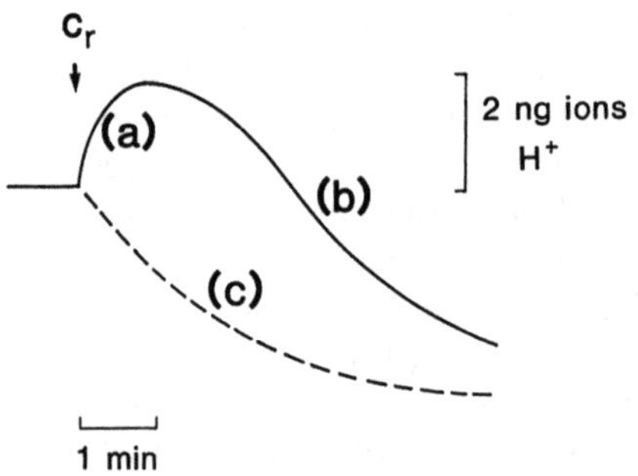

Fig. 3.12. Proton translocation by reconstituted vesicles of cytochrome c oxidase. Yeast ferrocytochrome c (c_r) (3.8 nmol) was added to phospholipid vesicles containing cytochrome c oxidase from the thermophilic bacterium PS3 in 0.1 mM MOPS pH 6.6 containing 25 mM K_2 SO_4, 2.5 mM $MgSO_4$ and 0.1 μg ml^{-1} valinomycin. pH changes were measured. An initial acidification phase (*a*) is due to proton translocation by the oxidase and this decays (*b*) due to leakage of protons across the membrane. Phase (*b*) results in a net alkalinisation of the medium due to the H^+ uptake involved in the oxygen reduction reaction within the vesicles. This alkalinisation is observed immediately (*c*) if the uncoupler 3,5-ditertbutyl-4-hydroxybenzylidine malononitrile is present. A H^+/e^- ratio is obtained by relating the initial acidification to the amount of cytochrome c oxidised. (After Sone and Hinkle 1982)

with c_r, the protons involved in water formation derive from the internal space of the liposomes and their uptake is not initially detected in the supporting medium. Thus, in the absence of scalar proton effects, acidification of the external medium is due to proton translocation. Production of the proton gradient is maximised by charge compensation using valinomycin and K^+ and a point is reached where the impermeability of the liposome membrane to protons begins to break down. As a result, H^+/e^- ratios decrease with increasing number of turnovers of the system and are often extrapolated to zero turnover (Table 3.5).

Cytochrome aa_3 from *Pa. denitrificans* was found to translocate 0.6 H^+/e^- at zero turnover, this being approximately half as effective as the mitochondrial enzyme under the same conditions (Solioz et al. 1982). Cytochrome aa_3 from the thermophilic bacterium PS3 was more effective, with H^+/e^- figures of greater than one at low turnover (Sone and Hinkle 1982; Sone and Yanagita 1984). The hydrophobic carbodiimide DCCD which is proposed to specifically block the proton channel associated with subunit III in the mitochondrial enzyme also inhibits proton translocation in the PS 3 oxidase (Sone and Hinkle 1982) but is relatively ineffective with the *Paracoccus* enzyme (Puttner et al. 1983).*

The cytochromes aa_3 of *Rps. sphaeroides* and *Nitrobacter agilis* could not translocate protons in reconstituted liposomes (Gennis et al. 1982; Sone et al. 1983). This is particularly surprising in the latter case where the oxidase may be the only mechanism for generation of a proton motive force in this organism. Further work is required to determine whether essential components have been lost during purification.

3.4.3 Cytochrome *o*

Spectra. Cytochrome *o* was originally defined on the basis of photochemical action spectra as a b-type cytochrome with CO-sensitive oxidase activity (Chance et al. 1953). In CO-difference spectra, peaks are observed near 416, 540 and 578 nm with troughs at 432 and 560 nm (Fig. 3.11). In reduced minus oxidised spectra, it contributes to the cytochrome b peak but its midpoint redox potential is raised by CO binding and can therefore be resolved by redox potentiometry.

Purified cytochromes *o* fall into two groups: those containing only heme b and those containing both heme b and heme c. In all, a heme b centre is believed to be the site of oxygen reduction and the additional redox centres probably act as electron buffers comparable to their counterparts in cytochrome aa_3. Froud and Anthony (1984) proposed that the name cytochrome *o* should be expanded according to the nature of these associated redox centres. Those oxidases containing c-type heme would be known as cytochrome *co* and those containing additional b-type heme as cytochrome *bo*. We will follow this recommendation.

---

* See appendix note 15

The purified cytochromes *co* have characteristic contributions of low-spin heme c and heme b to the α-peak. In all those studied (see next section), carbon monoxide binds to both heme types and this is reflected in carbon monoxide difference spectra which typically show a double trough near 550 nm and 560 nm (Fig. 3.11). Many authors ignore the clear contribution of c-type cytochrome to this spectrum and its significance is not understood.

Structure. Cytochrome *o* has been purified from *Escherichia coli* (Kita et al. 1984), *Azotobacter vinelandii* (Yang et al. 1979; Jurtshuk et al. 1981), *Pseudomonas aeruginosa* (Matsushita et al. 1982a; Yang 1982), *Methylophilus methylotrophus* (Carver and Jones 1983; Froud and Anthony 1984), *Rps. palustris* (King and Drews 1976) and *Rps. capsulata* (Hudig and Drews 1982).

The cytochrome *bo* of *E. coli* contains two heme b and two Cu in a proposed complex of 33-K and 55-K subunits (Kita et al. 1984). It is considered only briefly here because the respiratory chain does not involve cytochrome c. Cytochromes *co* purified from *Ps. aeruginosa*, *Mph. methylotrophus* and *Rps. palustris* contain subunits of 29 and 22 K and most preparations have additional bands of lower molecular weight. The subunit composition of *A. vinelandii* cytochrome *co* has not been reported. The oxidases contain heme c and heme b but the relative amounts vary widely from 1.2 c:1 b quoted for *Mph. methylotrophus* oxidase (Froud and Anthony 1984) to 10 c:1 b for *A. vinelandii* oxidase (Jurtshuk et al. 1981). This may be due to progressive loss of the heme b during purification. Peroxidase activity is associated with the two larger subunits after SDS gel electrophoresis. This would normally be taken to indicate the presence of covalently bound heme c but Froud and Anthony (1984) concluded that heme c is bound to the 22-K subunit and heme b to the 29-K subunit in the enzyme from *Mph. methylotrophus*.

The identity of the c-type cytochrome component is uncertain. The oxidase of *A. vinelandii* is frequently referred to as cytochrome c_4:o but although the respiratory membranes do contain cytochrome c_4, it has not been positively identified as part of the purified oxidase.* Similarly, membranes from *Ps. aeruginosa* contain cytochrome c_4 (Pettigrew, unpublished observation) but this has not been directly compared with the 22-K heme c band of the purified oxidase and other heme c bands of similar M_r are present in the membrane (Matsushita et al. 1982b). The cytochrome c in cytochrome *co* of *Mph. methylotrophus* was originally proposed to be a membrane counterpart of the soluble cytochrome c_L (Carver and Jones 1983) but this has been disputed by Froud and Anthony (1984).

A cytochrome oxidase from *Rps. capsulata* is a dimer of M_r 130 K containing one heme b (although this may again be a consequence of heme loss during purification; Hudig and Drews 1982). It does not bind CO and if it is to be described as a cytochrome *o*, then some relaxation of the original definition would be required. Although the purified enzyme does not contain heme c, a CO-binding cytochrome c of M_r 13 K persisted in early stages of the purification (Hudig and Drews 1983). This cytochrome c is immunologically

* See appendix note 16

distinct from the well-characterised soluble cytochrome c_2 and differs from it in isoelectric point (basic) and midpoint potential (234 mV).

Oxidase Activity. Unlike mitochondrial cytochrome aa_3, the cytochromes *co* can oxidise TMPD without the requirement for added soluble cytochrome c. Removal of the c-type component results in loss of activity. Cytochrome c-dependent dye oxidation finds a practical use in the "NADI" test for oxidase activity in bacteria which utilises a solution containing α-naphthol and dimethyl-p-phenylenediamine (Cooper 1970). The broad applicability of this test is probably a reflection of the widespread occurrence of the cytochromes *co* (Yang 1982).

An odd feature of these oxidases is that they lack ascorbate oxidase activity even though the c-type cytochrome is reducible with ascorbate (Yang et al. 1979; Matsushita et al. 1982a). This suggests that most of the c-type cytochrome molecules are not functionally associated with the cytochrome *o*. A further peculiarity is the low midpoint redox potential (−30 to +40 mV) for the heme b centres (Yang et al. 1979; Yang 1982) which is almost 300 mV more negative than that of the c-type cytochrome. However, the measurement of midpoint potentials for cytochrome oxidases under anaerobic conditions may give a misleading impression of their operating potentials. The preferential binding of oxygen to the ferrous heme will act to raise the potential by 60 mV for each factor of ten difference in the binding affinity. In these respects the cytochrome *o* from *Rps. capsulata* differs from the cytochromes *co*. The purified preparation retains ascorbate-TMPD oxidase activity in the absence of heme c and the midpoint redox potential of the heme b is 413 mV.

The physiological significance of the very high ascorbate-TMPD oxidase activity of *A. vinelandii* membranes is disputed. Using selective CN^- inhibition, Jones and Redfearn (1967) proposed that the cytochrome *co* branch took a small but significant portion of the total electron flux. However, mutants lacking cytochrome c and unable to oxidise ascorbate-TMPD grow normally using the respiratory branch terminating in cytochrome d (Hoffman et al. 1980). Only under severe oxygen limitation was growth poor suggesting that the high O_2 affinity of cytochrome *co* may be important at low oxygen tensions. This may also be true for *Mph. methylotrophus* in which methanol oxidation proceeds mainly via cytochrome aa_3 when oxygen is not limiting (Carver and Jones 1983).

The natural electron donors to the cytochromes *co* are poorly characterised. With the exception of that from *A. vinelandii*, all cytochromes *co* oxidise horse cytochrome c (membranes from *A. vinelandii* have this activity but the purified oxidase does not) (Yang et al. 1979). Both *A. vinelandii* and *Ps. aeruginosa* contain the set of cytochromes c-551, c_4 and c_5 which should be tested as donors to their respective cytochromes *co*. Cytochrome c_H (Table 3.3) is the most effective electron donor to cytochrome *co* of *Mph. methylotrophus* (Froud and Anthony 1984) and cytochrome c_2 is a likely candidate to donate to cytochrome *co* from *Rps. palustris*.

Table 3.7. Properties of two respiratory mutants of *Rps. capsulata*

Property	Mutant		Reference
	M6	M7	
(a) Sensitivity to antimycin A	High	Low	Zannoni et al. 1976
cyanide (I_{50})	2 μM	mM	La Monica and Marrs 1976
CO	No	Yes	Zannoni et al. 1976
(b) Presence of b_{413}*	Yes	No	Zannoni et al. 1974
(c) c_2 oxidase activity	Yes	No	Marrs and Gest 1973
(d) Anti-c_2 inhibition of respiration	Yes	No	Baccarini-Melandri et al. 1978
(e) Steady-state reduction of c_2	30%	85%	Zannoni et al. 1976

* a cytochrome b with E_m of 413 mV

The terminal oxidation pathways of *Rps. capsulata* have been extensively investigated using two key respiratory mutants (M6 and M7) isolated by Marrs and Gest (1973) (Table 3.7). The former contains a high potential cytochrome b (E_m 413 mV) which catalyses cytochrome c_2 oxidation in a cyanide-sensitive but CO-insensitive fashion. The latter contains an alternative oxidase which does not oxidise cytochrome c_2 and is CO-sensitive. In wild-type cells grown either photosynthetically (La Monica and Marrs 1976) or heterotrophically (Baccarini-Melandri et al. 1973) both oxidases are present and can be demonstrated by titration of respiration with cyanide (an advantage of studying the photosynthetic mode is that respiration is then gratuitous and is not susceptible to compensatory changes; La Monica and Marrs 1976). These studies are compatible with those of Zannoni et al. (1980) and Michels and Haddock (1980) who isolated a different type of mutant, lacking cytochrome c_2 and incapable of phototrophic growth. This mutant has a fully functional respiratory pathway with the characteristics of mutant M7.*

Thus present evidence favours a dual role for cytochrome c_2 as electron donor to both the high potential b-type oxidase and the photo-oxidised reaction centre. A complication of this scheme is the presence of the basic c-type cytochrome (described earlier) which is associated with the oxidase at early stages in the purification (Hudig and Drews 1983). One possibility, by analogy to cytochrome caa_3 of *Thermus thermophilus*, is that the membrane cytochrome c is tightly bound to the oxidase in situ and cytochrome c_2 is a diffusible donor.

H^+ Translocation and Energy Conservation. Little information is available on the role of cytochromes *co* in energy conservation and few vesicle reconstitution experiments have yet been performed with purified enzymes. The general suggestion of Jones (1977) that the third site of energy conservation is associated with the presence of cytochrome c is supported by studies on *Rps. capsulata* where the cytochrome c_2:b-type oxidase pathway constitutes such a site but the alternative cytochrome c-independent pathway does not (Baccarini-Melandri et al. 1973). Consistent with this is the proton transloca-

* See appendix note 17

tion by the former enzyme reconstituted in liposomes (Hudig and Drews 1984). Zannoni et al. (1976) proposed that the alternative low yield pathway may serve to oxidise cytoplasmic reducing equivalents even in the presence of the high energy charge maintained by the cyclic electron flow during photosynthesis.

In *Mph. methylotrophus* membranes, the proton concentration changes and charge transfer associated with ascorbate-TMPD oxidation can be accounted for by internal uptake of protons in the oxygen reaction without proton translocation (Dawson and Jones 1981). Selective inhibition of cytochrome *o* by low concentrations of CN^- had no effect on these measurements suggesting that both cytochrome aa_3 and *co* have similar contributions to energy conservation.

We have seen that cytochrome *co* of *A. vinelandii* plays little part in oxygen consumption under fully aerobic conditions. Perhaps the highly active cytochrome d pathway is required for respiratory protection of the oxygen-sensitive nitrogenase. Under O_2 limitation however the cytochrome *co* contribution may increase and allow use of the third site of energy conservation.

In summary, the cytochromes *o* are at an early stage of characterisation. By analogy with cytochrome aa_3, we might expect a functional unit capable of carrying four electrons on four separate redox centres. In the *E. coli* enzyme this seems to be accomplished by the presence of two hemes b and two Cu; in the cytochromes *co*, two heme b and two heme c may be present but quantitative analyses are not yet available. The redox centres of the cytochromes *co* are associated with the two larger subunits (approx. 29 K and 22 K). The small subunits may be genuine components, impurities or artefacts of preparation (for example of proteolytic origin). The relationship of the c-type cytochrome component with known, purified cytochromes c is uncertain but the electron donor to the cytochromes *co* is probably one of the small soluble class I cytochromes and the system may be described by Fig. 3.2c.

3.4.4 Cytochrome c Peroxidases

Hydrogen peroxide may be formed as a result of incomplete reduction of oxygen by flavoproteins or by terminal oxidases. It may be removed by catalase in a dismutation reaction or be reduced to water by a peroxidase. Some peroxidases (such as horse radish peroxidase) are very non-specific with respect to their electron donor; others obtain electrons from the respiratory chain via cytochrome c.

Such a process not only results in detoxification of the hydrogen peroxide but also may allow energy conservation by the electron transport chain. In the catalase-negative micro-aerobe *Campylobacter sputorum* subspecies mucosalis the flow of electrons through cytochrome c peroxidase is a major respiratory route (Fig. 3.13; Goodhew and Pettigrew, unpublished observations). Most aerobic bacteria are, however, catalase-positive and the presence of both catalase and cytochrome c peroxidase in some may reflect a requirement for a peroxidatic mechanism in both the cytoplasm and the periplasmic space.

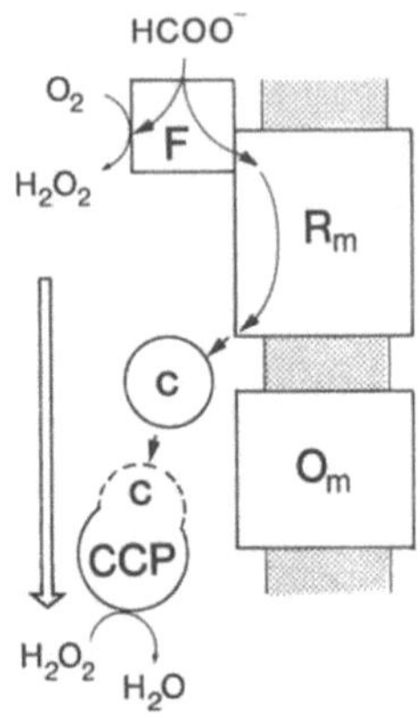

Fig. 3.13. Hydrogen peroxide production and removal in *Campylobacter mucosalis.* The diagram has been constructed on the basis of unpublished results of Goodhew and Pettigrew and according to the model of Fig. 3.2 although the topography of the reactions has not yet been established. It is proposed that formate oxidation may lead either directly to reduction of O_2 to H_2O_2 or to electron transport. The H_2O_2 produced by the formate oxidase reaction (*F*) is then further reduced to H_2O by a cytochrome c peroxidase (*CCP*) which accepts electrons from the electron transport system (R_m = membrane-bound cytochrome c reductase) via cytochrome c (*c*). The *CCP* is represented like that of *Ps. aeruginosa* (Fig. 3.14) but there is no evidence for this. The system results in the respiratory reduction of O_2 to H_2O via H_2O_2 without the involvement of a terminal oxidase (O_m). → e^- Flow; ⇒ diffusion of H_2O_2

With the exception of yeast cytochrome c peroxidase considered in Chapter 2, most work has been done on the enzyme from *Ps. aeruginosa.* It is markedly induced in culture of low oxygen tension (Lenhoff and Kaplan 1956). This suggests, either that more hydrogen peroxide is formed under these conditions, or perhaps more likely, that some cellular constituents such as heme d, which are susceptible to peroxide-induced damage, only appear at low oxygen tension (Sect. 3.5). The enzyme is purified from acetone powders of cells (Ellfolk and Soininen 1970) or from sonicated cells (Foote et al. 1983). Discrepancies in native M_r values (54 K, Ellfolk and Soininen 1971; 77 K, Singh and Wharton 1973) probably reflect a rapid monomer-dimer equilibrium since all workers agree on a polypeptide M_r of 40 K containing two hemes. The presence of covalently bound heme distinguishes the bacterial enzyme from that of yeast and is in agreement with the general proposal of Wood (1983) that periplasmic heme proteins contain heme c. The peptide chain is susceptible to endogenous protease action which yields a diheme N-terminal fragment of 29 K and a fragment of 11 K with a ragged N-terminus. An early acid precipitation stage in the purification appears to remove this protease (Soininen and Ellfolk 1975; Foote et al. 1983). We might note in passing that a 29-K heme peptide was identified in respiratory membranes of *Ps. aeruginosa* (Matsushita et al. 1982; Sect. 3.3).

For some years the *Pseudomonas* enzyme was regarded as very different in nature (and by implication, action; Ellfolk et al. 1973) from the well-studied high-spin, protoheme IX-containing yeast cytochrome c peroxidase (Chap. 2). In that enzyme, H_2O_2 binds to the peroxidatic heme and removes two reducing equivalents to leave a ferryl (Fe IV) oxene intermediate and a protein radical. Restoration of the resting ferric state requires two successive one electron transfers from cytochrome c. It now appears that the peroxidatic centre of *Pseudomonas* cytochrome c peroxidase is essentially similar; the difference is in the site of the second oxidising equivalent which derives from a second heme group in the bacterial enzyme and from the polypeptide chain in the yeast enzyme.

By analogy with cytochrome aa_3, the cytochrome c peroxidase of *Ps. aeruginosa* has an electron-accepting pole and a ligand-binding pole (Fig. 3.14). The electron-accepting pole is a high potential heme (Ellfolk et al. 1983) which can be reduced by azurin and cytochrome c-551 or non-physiologically by ascorbate (Ronnberg et al. 1981a). In the fully oxidised enzyme, the ligand-binding pole which is formed by a low potential heme does not bind hydrogen peroxide productively (Araiso et al. 1980; Ronnberg et al. 1981a) nor does it bind anionic ligands such as CN^- (Ellfolk et al. 1984). Reduction of the high potential heme has the effect of "opening" the low potential heme to ligand binding. Thus, like cytochrome oxidase (Palmer et al. 1976), the peroxidase contains a structural mechanism for preventing binding of substrate unless the enzyme has the reducing power to form water.

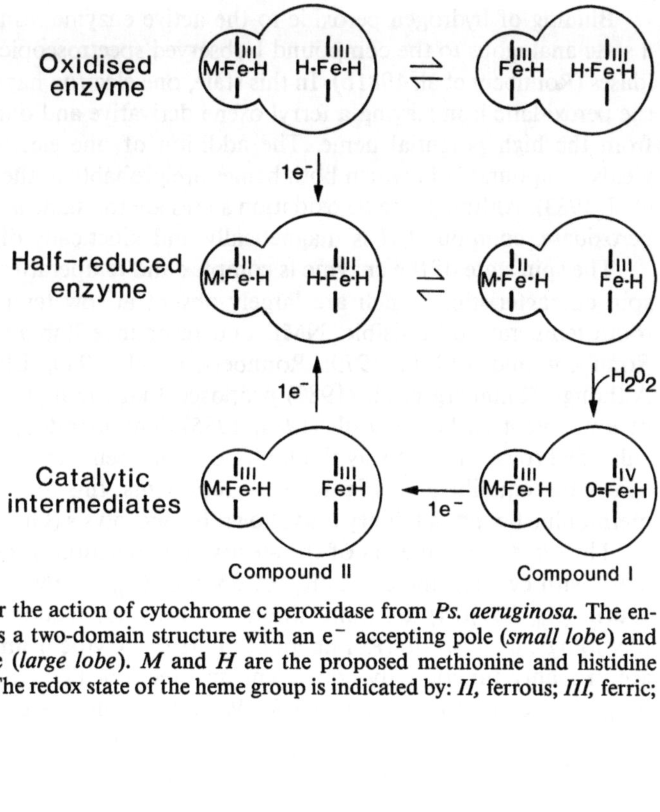

Fig. 3.14. A model for the action of cytochrome c peroxidase from *Ps. aeruginosa.* The enzyme is represented as a two-domain structure with an e^- accepting pole (*small lobe*) and a ligand-binding pole (*large lobe*). *M* and *H* are the proposed methionine and histidine ligands of the heme. The redox state of the heme group is indicated by: *II,* ferrous; *III,* ferric; *IV,* ferryl

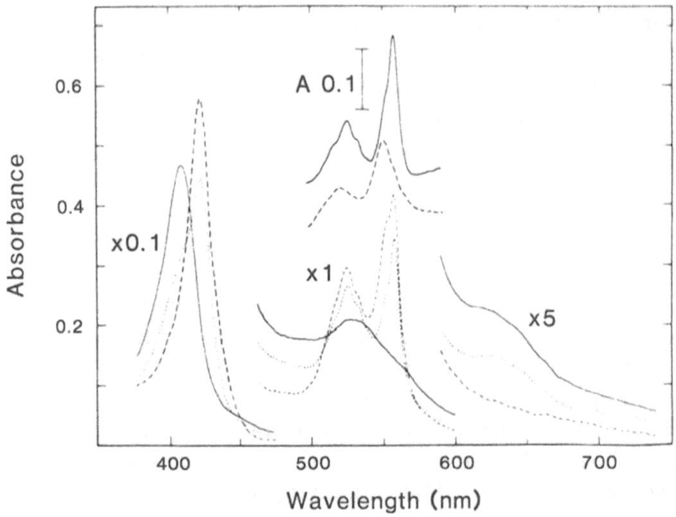

Fig. 3.15. The spectra of *Ps. aeruginosa* cytochrome c peroxidase. *Solid line:* ferricyanide oxidised; *dotted line:* ascorbate reduced; *dashed line* dithionite reduced. The *insert* shows the difference spectra (500–600 nm) of ascorbate reduced minus oxidised (*solid line*) and dithionite reduced minus ascorbate reduced (*dashed line*) (Ellfolk et al. 1983)

Binding of hydrogen peroxide to the active enzyme results in formation of a state analogous to the compound I observed spectroscopically in other peroxidases (Ronnberg et al. 1981b). In this state, one electron has been removed from the peroxidatic iron leaving a ferryl oxene derivative and one has been removed from the high potential heme. The addition of one electron to compound I yields compound II in which both hemes are probably in the ferric state (Ellfolk et al. 1983). Although the Fe oxidation states are the same as those in the resting peroxidase, compound II is magnetically and kinetically distinct.

The spin state of the enzyme is complex and temperature-dependent. High-spin characteristics, which are largely absent at low temperatures appear at room temperature in visible, NMR and resonance Raman spectra (Fig. 3.15; Ronnberg and Ellfolk 1979; Ronnberg et al. 1980; Ellfolk et al. 1984). Although Ronnberg et al. (1980) proposed that the high-spin heme was that of low potential, Foote et al. (1984, 1985) demonstrated, using magnetic circular dichroism, that it was the high potential heme that existed in a thermal spin-state equilibrium. Figure 3.14 represents this by the lability of the methionine ligand, a lability also shown by photolysis (Greenwood et al. 1984).

The kinetic parameters of the steady-state reaction of cytochrome c peroxidase with cytochrome c-551 (K_m 69 μM), c_5 (K_m 90 μM), c_4 (K_m 69 μM) and azurin (K_m 120 μM) are quite similar and do not readily justify a choice of physiological substrate (Ronnberg et al. 1975). Cytochrome c-551 and azurin are probably functionally equivalent and reciprocally synthesised depending on growth conditions (Chap. 4.1.3). Whatever donor is physiologically domi-

Table 3.8. Properties of *Ps. aeruginosa* cytochrome c peroxidase and related cytochromes

Source	Cytochrome	Growth conditions	M_r (native) (kD)	M_r (peptide) (kD)	Hemes per peptide	Spectra ox	Spectra red	E_m (mV)	Reference
Pseudomonas[a] *aeruginosa*	c-557 (552)	5% $O_2+NO_3^-$	54	41	2	407 524 (640) (705)	420 524 557 (552)	−330 +320	Ellfolk and Soininen 1970, 1971; Singh and Wharton 1973
Pseudomonas stutzeri	[b]	Anaerobic	ND	37	2	[b]		[c]	Villalain et al. 1984[d]
Alcaligenes faecalis	c-557 (551)	Aerobic	65	ND	[e]	408 530 620	420 (440) 525 557 (551)	[f]	Iwasaki and Matsubara 1971
Nitrobacter agilis	c-554 (549)	Aerobic	90	46	2	402	419 (430) 523 549 554	[f]	Chaudhry et al. 1981

Figures in parentheses are shoulders on spectral peaks.
[a] The *Ps. fluorescens* of Lenhoff and Kaplan (1956), used by Ellfolk and co-workers, has been shown to be *Ps. aeruginosa*. An enzyme possibly similar to the *Ps. aeruginosa* enzyme was isolated from *Ps. denitrificans* (Coulson and Oliver 1979).
[b] Spectral maxima were not quoted. A 620-nm band was present in oxidised and half-reduced forms and a 705-nm band was present in the oxidised enzyme. The α-peak of the half-reduced form was approx. 555 nm with a shoulder on the short wavelength side. The fully reduced α-peak was approx. 551 nm with a shoulder on the long wavelength side.
[c] One heme was reducible with ascorbate.
[d] Kodama and Mori (1969) purified a cytochrome c-558 (552) from a strain of *Ps. stutzeri* which was spectrally similar to the *Ps. aeruginosa* enzyme but was not reducible with ascorbate and appeared to have only 2 hemes per 74 kD.
[e] Two hemes per 65 kD.
[f] Not reducible with ascorbate.

nant, the cytochrome c peroxidase is a periplasmic enzyme (Goodhew, Hunter and Pettigrew, unpublished observations) acting according to Fig. 3.2d. Like the reactions of mitochondrial cytochrome c, activity is inhibited by increasing salt concentration consistent with an electrostatic component in complex stabilisation (Soininen and Ellfolk 1972). In the bacterial systems, both redox partners are acidic but it is noticeable that a ring of lysine residues around the "front face" of the molecule is a feature of both basic mitochondrial cytochrome c and acidic *Ps. aeruginosa* cytochrome c-551 (Vol. 2, Chap. 4).

Probable relatives of the *Ps. aeruginosa* enzyme on the basis of size and spectra are shown in Table 3.8 but only the protein from *Ps. stutzeri* has been tested for peroxidase activity (Villalain et al. 1984). The similarities appear too great to be fortuitous although differences do exist in the lower heme contents and the apparent absence of a high potential heme in the proteins purified by Kodama and Mori (1969) and Iwasaki and Matsubara (1971). These features should be re-examined in the light of the present model of the *Ps. aeruginosa* enzyme.

Cytochrome c peroxidases which contain heme c have been isolated from *Thiobacillus novellus* (Yamanaka 1972) and *Nitrosomonas europaea* (Anderson et al. 1968). Both enzymes can oxidise horse cytochrome c and the former can oxidise the basic cytochrome c-550 (Table 3.6) from the same organism. In the enzyme from *Thiobacillus novellus*, high-spin character was apparent in the short wavelength Soret maximum at 398 nm and a small 630 nm peak but, unlike the proteins of Table 3.8, the α-band is at 550 nm and is not characteristically split.

3.5 Nitrate Respiration and Denitrification

3.5.1 The Biology of Denitrification

Denitrification is defined as the respiratory (dissimilatory) reduction of nitrite to gaseous products (N_2, NO and N_2O) which are lost to the atmosphere (Delwiche and Bryan 1976). The process is of particular ecological importance because some of the products may contribute to the destruction of the ozone layer and it may be economically important, given the high cost of nitrate fertiliser.

Most denitrifiers are non-fermentative facultative gram-negative chemoheterotrophic bacteria which use NO_3^- as a less preferred respiratory substitute for oxygen (Payne 1981). Distribution of the ability to denitrify is erratic without any obvious evolutionary pattern but it is not clear whether this is as a consequence of sporadic loss of a commonly held metabolic facility or of lateral gene transfer.

Usually nitrate is the initial respiratory acceptor for denitrification but intermediates in the reduction pathway can often be used and in some cases (e.g.

NO_2^- in *Alcaligenes*) are required. Most denitrifiers can also carry out assimilation of nitrate nitrogen and therefore contain two enzyme systems which operate under different controls. Assimilatory nitrate reductases are repressed by NH_3 but are unaffected by oxygen while the respiratory enzymes show the converse pattern of regulation. In some organisms (e.g. *Achromobacter fischeri*) complete reduction of nitrite to NH_3 can take place but without the NH_3 repression which is symptomatic of the assimilatory process.

C-type cytochromes are found in large amounts in denitrifying bacteria. Their role is generally poorly defined but several probably act as redox mediators between the electron transport system and the terminal enzymes of denitrification. In addition, c-type cytochromes are involved as components of some of these terminal enzymes, a role corresponding to that of Fig. 3.2e.

3.5.2 The Reduction of Nitrate to Nitrite

The respiratory nitrate reductases are membrane-bound Mo, Fe-S proteins (Thauer et al. 1977; Haddock and Jones 1977; Stouthamer et al. 1980). In the best studied examples of *E. coli* and *Pa. denitrificans*, the enzymes accept electrons from the periplasmic side of the membrane via a specific b-type cytochrome and reduce nitrate at the cytoplasmic surface (reviewed in Stouthamer et al. 1980; Ferguson 1982b). Thus c-type cytochromes are not usually involved either in the terminal enzyme or the electron transport chain and we will not consider this process further. However, *Thiobacillus denitrificans* is an exception in that nitrate respiration is blocked by antimycin A and a c-type cytochrome acts as the donor to nitrate reductase.

3.5.3 The Reduction of Nitrite

In contrast to the nitrate reductases, those enzymes reducing nitrite are not generally membrane-bound. A variety of types have been isolated, the best characterised of which is cytochrome cd_1.

3.5.3.1 Cytochrome cd_1

Purification. Cytochrome cd_1 is a homodimer of M_r 120 K containing one heme c and one heme d_1 per monomer (Newton 1969; Kuronen et al. 1975; Henry and Bessieres 1984). It has been purified from *Ps. aeruginosa* (Horio et al. 1961b; Kuronen and Ellfolk 1972; Gudat et al. 1973; Parr et al. 1976; Silvestrini et al. 1983), *Pa. denitrificans* (Newton 1969; Lam and Nicholas 1969; Timkovich et al. 1982), *Alcaligenes faecalis* (Matsubara and Iwasaki 1972), *Thiobacillus denitrificans* (Sawhney and Nicholas 1978; Le Gall et al. 1979) and *Ps. perfectomarinus* (Liu et al. 1983). Losses of protein and heme

d_1 can occur on ion exchange columns, a problem especially pronounced for the *Pa. denitrificans* enzyme for which a hydrophobic Sepharose method was developed (Timkovich et al. 1982). Immuno-affinity chromatography, although at first sight an attractive alternative to ion exchange, suffers from the problem of heme d_1 loss in the 6 M urea required to break the antibody: antigen interaction (Silvestrini et al. 1983).

Enzyme Activity. The *Pseudomonas* enzyme was initially described as a cytochrome oxidase (Horio et al. 1961b) but it was shown that the enzyme could also act as a nitrite reductase (Yamanaka et al. 1961). The spectroscopic absence of cytochrome cd_1 in mutants, their inability to grow anaerobically on nitrite and the failure of NO_2^- to oxidise their cytochrome system is convincing evidence for the physiological importance of the cytochrome cd_1 as a nitrite reductase (Hartingsveldt and Stouthamer 1973). However, there is disagreement as to the product of the reaction. A long-standing assumption, but based on indirect evidence, has been that NO is the product of nitrite reduction. The central finding of Yamanaka and Okunuki (1963b) extended by Shimada and Orii (1975, 1978) and Orii and Shimada (1978) was that the spectrum of ferri-cd_1 + NO is identical to that obtained after reaction of ferro-cd_1 with NO_2^-. However, a number of considerations cast doubt on the relevance of these NO-binding studies. The ability to grow on NO is rare among denitrifiers and NO is not generally released during respiration (Payne 1981). In an elegant experiment, St. John and Hollocher (1977) trapped $^{15}N_2O$ from $^{15}NO_2^-$ in a pool of added N_2O but could not do so with NO. N_2O thus showed the properties of a free obligatory intermediate but NO did not. The initial observation that a spectroscopically detectable NO-binding protein which acted to sequester NO was present in extracts of *Ps. aeruginosa* (Rowe et al. 1977) was later shown to be due to oxidation of cytochrome cd_1 by NO (Zumft et al. 1979). That such an oxidation can occur with the release of N_2O has been demonstrated directly with purified cytochrome cd_1 and gas chromatography (Wharton and Weintraub 1980; Sawhney and Nicholas 1978; Matsubara and Iwasaki 1972). Indeed, in the case of the reduction of nitrite by the *Ps. aeruginosa* enzyme, NO and N_2O have a precursor: product relationship with the former being maintained at a low steady-state level while the latter accumulates (Wharton and Weintraub 1980). Thus the balance of evidence favours the reduction of nitrite to nitrous oxide catalysed by cytochrome cd_1.

The physiological significance of the oxygen reaction is uncertain. In *Pa. denitrificans*, cell membranes contain on *o*-type oxidase, with K_m two orders of magnitude lower than that of cytochrome cd_1, which could account for 90% of the total oxidase activity (Lam and Nicholas 1969). Furthermore the cytochrome cd_1 only appears in the presence of nitrate or nitrite and under lowered oxygen tension (Kodama 1970; Sapshead and Wimpenny 1972; Parr et al. 1976; Liu et al. 1983). However, since under these conditions the cytochrome oxidase activity of a wild-type culture of *Ps. aeruginosa* was twice that of a mutant lacking nitrite reductase activity (Hartingsveldt and Stouthamer 1973) it appears that cytochrome cd_1 can contribute significantly to the overall oxygen-reducing ability of normal cells. The reaction with oxygen shows a stoichiometry of cytochrome oxidation of 4 mol/mol O_2 consistent with water as the reduced product. Neither catalase nor superoxide dismutase affect the rate of reaction (Timkovich and Robinson 1979).

Spectroscopy and the Spin State of the Hemes. Although different preparations of cytochrome cd_1 have spectra similar to that of Fig. 1.11c the exact pattern and intensity between 600 and 700 nm is quite variable. Parr et al. (1974) have proposed that oxy-sulfur anions formed in dithionite solutions can bind to the protein and cause a shift in the ferro-heme d_1 spectrum from 655 to 635 nm, an interpretation contested by Shimada and Orii (1976) who suggested acidification by the dithionite as the cause of the spectral change due to the pH dependence of the ferro-heme d_1 spectrum seen by Gudat et al. (1973). The spectrum of the ferricytochrome cd_1 of *Pa. denitrificans* contains an additional band at 700 nm (Newton 1969).

Removal of heme d_1 by treatment with acid-acetone results in loss of spectral bands at 460 nm and 600–700 nm (Horio et al. 1961b; Newton 1969) and loss of enzymic activity. Reconstitution with heme d_1 (Yamanaka and Okunuki 1963a) gave full recovery of activity (Hill and Wharton 1978) and only heme a was a partially effective substitute (50% recovery).

On the basis of EPR spectroscopy (Gudat et al. 1973; Muhoberac and Wharton 1983), magnetic circular dichroism at low temperature (Walsh et al. 1979; Foote et al. 1984), NMR spectroscopy (Timkovich and Cork 1983) and Mossbauer spectroscopy (Huynh et al. 1982), heme c is hexacoordinate (Met-His) and low-spin in both redox states while heme d_1 changes from a hexacoordinate, low-spin, bis-His ferric state to a pentacoordinate, high-spin ferrous state. This "opening" of the heme d_1 on reduction is reminiscent of both cytochrome P450 and cytochrome aa_3 and may have important functional implications (see Fig. 3.16).

Ligand Binding and Heme:Heme Interactions. Cytochrome cd_1 was originally seen as a soluble, easily purified and potentially simpler terminal oxidase than cytochrome aa_3, the study of which might aid the understanding of that complex mitochondrial enzyme. This proved to be mistaken on two counts: firstly, the enzyme is now accepted as a nitrite rather than an oxygen reductase and secondly, its kinetics and ligand-binding properties are very complex. The effects of ligand binding potentially illuminate two important areas of molecular function – the interaction between the heme groups and the site of nitrite reduction – but in neither of these areas has a completely clear picture yet emerged. The site of nitrite binding and reduction is probably heme d_1. Thus NO_2^- reacts preferentially with ferro-heme d_1 to produce a visible spectrum resembling that obtained by adding NO to the ascorbate-reduced enzyme (Silvestrini et al. 1979; Muhoberac and Wharton 1983). Addition of NO to the

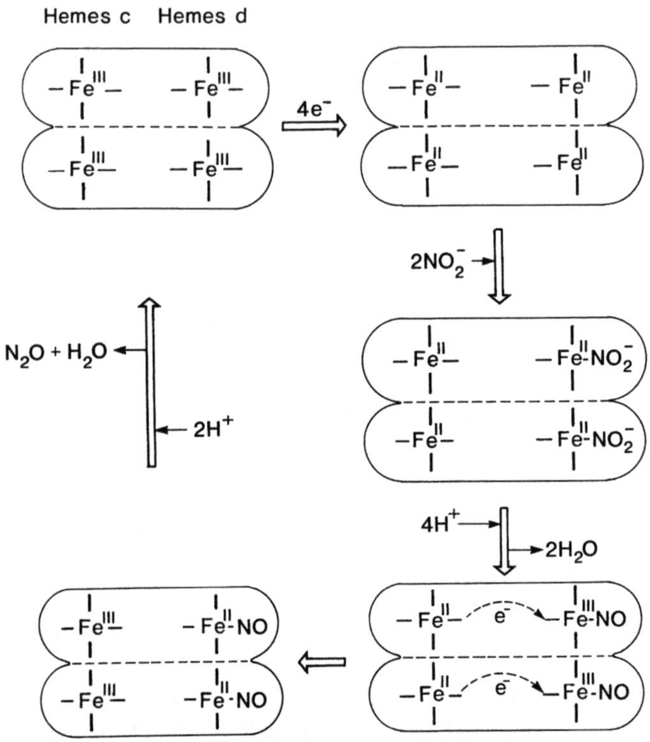

Fig. 3.16. A structural model for the nitrite reductase activity of cytochrome cd_1. The molecule is represented as a dimer with a subunit interface (*broken lines*). The two hemes c which constitute the electron-accepting pole are placed to the *left* of each molecule while the ligand binding pole (*hemes d*) are on the *right*. The oxidised molecule (*III*, ferric hemes) accepts $4e^-$ and then binds a NO_2^- at each ferro heme d (*II*, ferrohemes). Each NO_2^- is reduced to NO in one e^- steps and then two more e^- from the hemes c (*broken lines*; intramolecular electron transfer) are used to reduce the two bound NO to a single N_2O. Heme c remains hexacoordinate throughout but the reduction of heme d results in displacement of the sixth ligand and a high-spin state (Huynh et al. 1982) thus allowing ligand access. The net equation for the reaction is

$$2NO_2^- + 6H^+ + 4e^- \rightarrow N_2O + 3H_2O \ .$$

The model satisfactorily accounts for the data implicating NO as an intermediate and the spin-state switch of heme d_1. Although both hemes c and d_1 can bind NO under physiological conditions, heme d_1 will bind NO preferentially. The dimeric state is an essential feature of the model

reaction mixture shows the effects of product inhibition (Dhesi and Timkovich 1984). However, two lines of evidence complicate this interpretation. Firstly, heme c can also bind NO at pH values below 7 (Shimada and Orii 1975, 1978; Silvestrini et al. 1979; Johnson et al. 1980) and the pH of the periplasmic space is likely to be slightly acidic. Secondly, CO inhibits the cytochrome oxidase but not the nitrite reductase reaction (Yamanaka et al. 1961) and NO_2^- can oxidise the ferro-cd_1 : CO complex faster than the rate of dissociation of CO (Parr et al. 1974). Since it is accepted that CO binds only to ferro- heme d_1 (Parr et al. 1975), this implies that O_2 binds at heme d_1 and NO_2^- binds elsewhere in the molecule. However, other explanations are possible. Thus Barber et al. (1978) proposed that O_2 (K_A 10^4 M^{-1}) is simply a much poorer ligand than NO_2^- (K_A 1.2×10^6 M^{-1}) or CO (K_A $0.05-5 \times 10^6$ M^{-1}). In contrast, CN^-, which inhibits both enzymic reactions, binds strongly (K_A 5×10^{11} M^{-1}). A further possibility is that the dimer can bind only one O_2 and reduce it in a singe $4e^-$ transfer process while NO_2^- may bind at each heme d_1 to allow two single electron reductions to NO followed by reductive ligation of the two NO to form N_2O. In such a model (Fig. 3.16), the binding of one CO may block the reaction with O_2 but allow the binding of a single NO_2 and partial reduction to NO. Consistent with this scheme is the monophasic binding of O_2 (Greenwood et al. 1978) and the kinetic model for CO binding of Parr et al. (1975) which incorporates a species in which a single bound CO prohibits the binding of a second.

A variety of studies suggest that cooperative interactions exist between heme groups in cytochrome cd_1, a phenomenon which may be of importance in understanding how the molecule can carry out its reductive function. A positive cooperativity between hemes d_1 is found on binding CO and CN^- to the ferrocytochrome (Parr et al. 1975; Barber et al. 1978) and is also seen in reductive titrations (Blatt and Pecht 1979). Reaction with oxygen is however monophasic (Greenwood et al. 1978) and may reflect binding of a single molecule per molecule of cytochrome cd_1 so that a synchronised $4e^-$ reduction may take place.

Silvestrini et al. (1982) found that CO ligation to heme d_1 lowers the redox potential of heme c by at least 80 mV but such a strong interaction between the two heme types was not found in the analysis of reductive titrations (Blatt and Pecht 1979). No perturbation of the magnetic properties of heme c occurred on binding CO to heme d_1 (Vickery et al. 1978) and the conversion of high-spin to low-spin heme d_1 on binding cyanide did not affect the Raman spectra of heme c (Ching et al. 1982).

Blatt and Pecht (1979) proposed a negative cooperativity of 26 mV between hemes c, and the biphasic reduction of heme c by chromous ions (Barber et al. 1977) or by azurin (Parr et al. 1977) is consistent with heme c-heme c interaction although other explanations are possible.

Electron Transfer Activity and Redox Potentials. Heme c can be identified as the site of entry of electrons by using a fast reductant such as reduced azurin (Wharton et al. 1973; Parr et al. 1977), reduced cytochrome c-551 (Silvestrini et al. 1982) or chromous ions (Barber et al. 1977) under anaerobic conditions. In such experiments the kinetics of heme c reduction involve fast and slow phases and the latter is independent of reductant and matches the monophasic kinetics seen for wavelengths at which changes in heme d_1 only are observed. These results are consistent with a model of fast reduction of heme c followed

(a) Kinetic model

$$A_r + c_o d_{1,o} \underset{k_2}{\overset{k_1}{\rightleftharpoons}} A_r{:}c_o d_{1,o} \underset{k_4}{\overset{k_3}{\rightleftharpoons}} A_o{:}c_r d_{1,o} \underset{k_6}{\overset{k_5}{\rightleftharpoons}} A_o + c_r d_{1,o}$$

$$A_r \underset{k_8}{\overset{k_7}{\rightleftharpoons}} A_r^*$$

$$c_r d_{1,o} \underset{k_{10}}{\overset{k_9}{\rightleftharpoons}} c_o d_{1,r}$$

$$(cd_1)_r + O_2 \xrightarrow{k_{11}} (cd_1)_o + 2H_2O$$

(b) Thermodynamic and kinetic parameters

	k_3 (s^{-1})	k_4 (s^{-1})	$K_{3/4}$	ΔE_m (mV) (c minus A)
Wharton et al., 1973	29	120	0.24	−36
Parr et al., 1977	120	200	0.6	−13

	k_9 (s^{-1})	k_{10} (s^{-1})	$K_{9/10}$	ΔE_m (mV) (d_1 minus c)
Wharton et al., 1973	0.2−2	10	0.02−0.2	−40 to −100
Parr et al., 1977	0.2	0.35	0.6	−13

Fig. 3.17 a, b. A kinetic model for the azurin: oxygen oxidoreductase activity of cytochrome cd_1. A_r and A_o denote the reduced and oxidised states of azurin. A^* is a conformational isomer. The two hemes of cytochrome cd_1 can exist in either oxidised (*o*) or reduced (*r*) states. In both investigations k_3 was taken as the limiting rate constant at high [A] for the fast phase of azurin oxidation by cytochrome cd_1. k_4 was taken as the limiting rate constant for the fast phase of azurin reduction by the reduced cytochrome cd_1 : CO complex. The variation in the determination of k_9 by Wharton et al. (1973) arises from the very small amplitude of the spectroscopic changes observed. The conditions were 0.1 M phosphate pH 6.6 for Wharton et al. (1973) and 0.1 M phosphate pH 7 for Parr et al. (1977)

by slow intramolecular electron transfer, a view confirmed by studies on the reverse reaction. The model is presented in Fig. 3.17 for azurin as reductant and gives the conflicting rate constants obtained by Wharton et al. (1973) and Parr et al. (1977). The reason for the differences in results is not known. The kinetic measurements allow calculation of the difference in the midpoint redox potential between the two heme types and indicate that heme c is more oxidising than heme d_1 under these conditions. This is in agreement with most of the potentiometric results (Horio et al. 1961a; Blatt and Pecht 1979). CO binding to heme d_1 lowers the midpoint potential of heme c by 80 mV (Silvestrini et al. 1982). This thermodynamic pattern suggests a device reminiscent of that proposed for cytochromes aa_3 by Palmer et al. (1976) whereby the ligand-binding centre only becomes reduced and accepts the ligand when the enzyme contains at least two electrons.

The oxygen reactions of cytochrome cd_1 have been studied by two research groups who differ in their findings. Wharton and Gibson (1976) found a fast O_2-dependent reaction at 460 nm with $k = 5.7 \times 10^4\ M^{-1}\ s^{-1}$ and a slower reaction at 549 nm with a limiting $k = 8\ s^{-1}$ consistent with rapid reaction of oxygen at heme d_1 followed by slower intramolecular electron transfer. On the other hand, Greenwood et al. (1978) found that oxidation of both heme c and heme d_1 occurred at the same fast rate implying a dramatic increase in the rate of intramolecular electron transfer over that found under anaerobic conditions. This resembles the effect of oxygen on electron transfer between hemes a and a_3. Spectra of intermediates in the reaction of oxygen with reduced cytochrome cd_1 are neither coincident nor isobestic with the oxidised or reduced species and have been taken as an indication of the presence of an oxygenated heme d_1 (Shimada and Orii 1976; Greenwood et al. 1978).

Structure. Although spectroscopically similar to heme d_1 of *Aerobacter aerogenes* (thought to be an iron chlorin with a hydroxyethyl side chain, Barrett 1956), Newton (1969) and Lemberg and Barrett (1973) emphasised that heme d from cytochrome cd_1 is chemically distinct having a much greater solubility in water and a pyridine ferrohemochrome α-peak, 5 nm to the red. On the basis of NMR spectroscopy, Timkovich et al. (1984) proposed a chlorin structure with the ten substituents (four methyls, two hydroxymethyls, two formates, a propionate and an acrylate) shown in Fig. 1.10. However, Chang (1985) has offered an alternative interpretation (Fig. 1.10) and suggested a dioxo-isobacteriochlorin with four methyls, two acetyls, two oxo groups, a propionate and an acrylate.

The highly stable protein dimer is formed by the interaction of two elongated monomers to form a cross-structure. Both electron microscopy (Saraste et al. 1977) and small angle X-ray scattering (Berger and Wharton 1980) suggest cylindrical monomers of 40×100 Å but low resolution X-ray crystallography indicates a more compact structure (Akey et al. 1980; Takano et al. 1979). The absence of fluorescent quenching of certain internal tryptophans and added fluorescent probes by the heme groups indicates a distance of > 80 Å between the hemes and some points in the molecule (Mitra and Bersohn 1980). This distance implies that the two heme groups of the monomer may be located at one end of the cylindrical structure. Furthermore, the ability of the enzyme to carry out coordinated 4 e^- reduction of an oxygen or two NO_2^- (Fig. 3.16) implies that the four hemes of the dimer may be clustered at one end of the molecule. On the basis of polarised single crystal spectroscopy, the heme c and d_1 appear to be perpendicularly orientated (Makinen et al. 1983) in contrast to the coplanar orientations proposed for cytochrome c electron transfer complexes (Chap. 2.5). This feature may explain the relatively slow intramolecular electron transfer rate (Vol. 2, Chap. 8).

Proteolysis of cytochrome cd_1 (subunit M_r 62 K) yields a quite stable 58-K product with increased enzyme activity (Horowitz et al. 1982a). In the presence of CN^- this can be further cleaved to give 48-K, 28-K, and 11-K products with the largest peptide retaining heme d_1 and the smallest, heme c (Horowitz et al. 1982). Thus heme c may be bound to a separate domain of the monomer which is made susceptible to proteolysis by CN^- binding. Further investigation of this heme c domain and its relationship to the class I cytochromes will be of considerable interest.

3.5.3.2 Other Nitrite Reductases

In *Achromobacter cycloclastes* (Iwasaki and Matsubara 1972), Alcaligenes sp. (Miyata and Mori 1969; Suzuki and Iwasaki 1962), *Rps. sphaeroides* (Sawada et al. 1978) and possibly in *Corynebacteria nephridii* (Renner and Becker 1970) the dissimilatory nitrite reductase is a non-heme, Cu-containing enzyme which can catalyse the reaction:

$$NH_2OH + HNO_2 \rightarrow N_2O + 2H_2O,$$

in contrast to cytochrome cd_1, which is strongly inhibited by hydroxylamine. NO_2^- can also be reduced to NO using an appropriate ferrocytochrome c as a donor (Suzuki and Iwasaki 1962; Iwasaki and Matsubara 1972).

A number of heme c-containing nitrite reductases are known which release NH_3 as a final product. These should probably be considered dissimilatory enzymes because they are repressed by oxygen but not by NH_3. The properties of these enzymes from *Achromobacter fischerii, Desulfovibrio desulfuricans* and *Escherichia coli* are summarised in Table 3.9.

There is controversy as to the physiological role of the *E. coli* cytochrome, which has been variously described as a cofactor in formate hydrogen lyase, a detoxifying nitrite reductase, a respiratory nitrite reductase or an electron transport component involved in nitrite respiration. The first possibility is thought to be unlikely because low levels of nitrite strongly inhibit formate hydrogen lyase (Fujita and Sato 1967) and cytochrome c-552 can be absent when this activity is high (Douglas et al. 1974). A large amount of evidence is consistent with a role in nitrite reduction. The cytochrome is induced by nitrite (Fujita and Sato 1966) in parallel with nitrite reductase (Cole et al. 1974). Spheroplasts have lost both cytochrome c-552 and their ability to couple CO_2 production with NO_2^- reduction (Fujita and Sato 1967; Cole et al. 1974) and purified cytochrome c-552 can restore this activity. Mutants lacking nitrite reductase also lack cytochrome c-552 (Cole and Ward 1973). The cytochrome is rapidly oxidised by NO_2^- in both whole cells (Fujita and Sato 1967) and in

Table 3.9. Heme containing nitrite reductases other than cytochrome cd_1

Source	Extract	Mol. wt.	Subunits	Hemes	E_m	Spectra	e^- Donor	Reaction	Reference
Achromobacter fischerii	Water-soluble	70–95 K	2×39 K[a]	2	ND	α 551 α/β 1.6 γ_r/α 5.1 γ_r/γ_o 1.3	Fp reductase[d], not horse cyt.c or *A. fischerii* c-551	$NO_2^- \rightarrow NH_3$ $NH_2OH \rightarrow NH_3$	Prakash and Sadana 1972, Husain and Sadana 1974
Desulfovibrio desulfuricans	Cholate	66 K[b]	–	6	[c]	α 553 α/β 1.5 γ_r/α 5.2 γ_r/γ_o 1.3	Viologens or hydrogenase + FAD, not cyt. c_3 or Fd or flavodoxin	$NO_2^- \rightarrow NH_3$ $NH_2OH \rightarrow NH_3$ S oxyions $\rightarrow S^{2-}$ in dithionite solution	Liu et al. 1980, Liu and Peck 1981
Escherichia coli	Water-soluble	136 K	–	10	~200	α 552 α/β 1.8 γ_r/α 4.76 γ_r/γ_o 1.4	Fp reductase	$NO_2^- \rightarrow NH_3$	Fujita 1966a, b
		70 K		6					Liu et al. 1981b

[a] The subunits are linked by S-S bridges.
[b] Molecular weight is from SDS electrophoresis; the native protein is highly aggregated.
[c] The cytochrome was reduced to a small extent by ascorbate.
[d] Fp = flavoprotein, FD = ferredoxin

the purified state, suggesting that it may be the nitrite reductase rather than act as donor to it.

Coleman et al. (1978) found that the NADH-nitrite oxidoreductase of *E. coli* could be separated from cytochrome c-552 on Sephadex and no cross-stimulation of activity was observed. However, a large loss of activity was found, consistent with removal of a cofactor. Liu et al. (1981b) showed that a NADH-nitrite oxidoreductase could be reconstituted using NADH oxidase+FAD+cytochrome c-552 with activity almost as high as that observed with cytochrome c-552 and reduced methyl viologen as donor. Thus FAD may be the missing cofactor in the studies of Coleman et al., and the balance of evidence very much favours a role for cytochrome c-552 as the terminal enzyme of nitrite respiration.

3.5.4 The Reduction of Nitric Oxide and Nitrous Oxide

There has been considerable debate as to whether NO is an intermediate in denitrification. We have concluded in Section 3.5.3.1 that N_2O rather than NO is probably the product of cytochrome cd_1 under physiological conditions. However, separate NO reductases do seem to be present in some denitrifiers and nitric oxide does stimulate proton translocation in oxidant pulse experiments (Shapleigh and Payne 1985). This disagreement is not yet resolved. Study of the terminal N_2O reductase has been hindered by its lability and assay difficulties and only the enzyme from *Ps. perfectomarinus* is well-characterised. The protein is a dimer of 60-K subunits each containing four Cu (Zumft and Matsubara 1982). Reduction with dithionite converts the pink form into a blue form which retains paramagnetic character. This remarkable result is to be contrasted with the reduction of azurin (absent in this organism) from a blue paramagnetic copper II state to a colourless diamagnetic copper I state. Although Snyder and Hollocher (1984) found that the chromatographic behaviour of the Cu protein and N_2O reductase in cell extracts were not congruent, Zumft et al. (1985) showed that preparation under anaerobic conditions solved this problem and yielded a more active enzyme.

A similar enzyme is probably present in *Rps. capsulata* (McEwan et al. 1985) but the enzyme from *Pa. denitrificans* differs in lacking Cu and accepting electrons from the electron transport chain via an antimycin-sensitive pathway (see discussion on energy conservation).

3.5.5 Electron Transport and the Role of Cytochrome c

A scheme for the interaction of the terminal denitrifying enzymes with the electron transport system of *Pa. denitrificans* is shown in Fig. 3.18. This scheme contains features of importance to energy conservation which will be discussed in the following section. In other denitrifying organisms variations

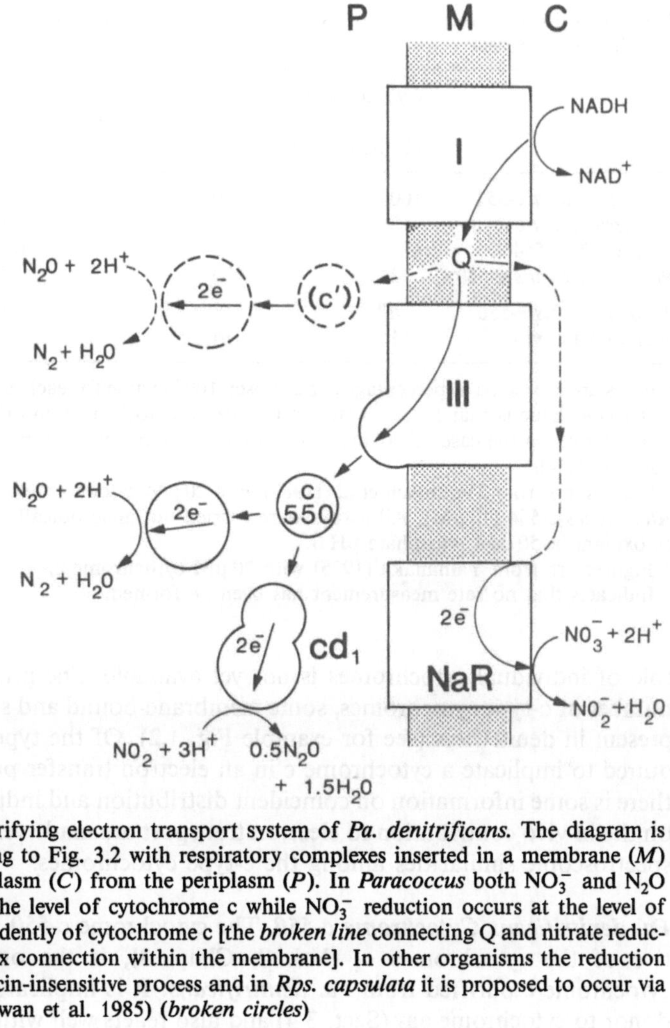

Fig. 3.18. The denitrifying electron transport system of *Pa. denitrificans.* The diagram is constructed according to Fig. 3.2 with respiratory complexes inserted in a membrane (*M*) separating the cytoplasm (*C*) from the periplasm (*P*). In *Paracoccus* both NO_3^- and N_2O reduction occur at the level of cytochrome c while NO_3^- reduction occurs at the level of coenzyme Q independently of cytochrome c [the *broken line* connecting Q and nitrate reductase (*NaR*) is a redox connection within the membrane]. In other organisms the reduction of N_2O is an antimycin-insensitive process and in *Rps. capsulata* it is proposed to occur via cytochrome c′ (McEwan et al. 1985) (*broken circles*)

in this pattern of electron flow may exist although most are less well-characterised. In general, however, nitrate reductase receives electrons at the level of reduced quinone in an antimycin-insensitive process (an exception is *Thiobacillus denitrificans*); the cytochrome cd_1 type of nitrite reductase receives electrons at the level of cytochrome c and N_2O reductase may receive electrons either at the level of cytochrome c or at the level of reduced quinone depending on the organism.

The role of c-type cytochromes in this scheme is as soluble redox mediators linking the electron transport system with the terminal enzymes of NO_2^- and probably N_2O reduction. However, in most cases, detailed knowledge of the

Table 3.10. Relative reactivities of cytochromes c with cytochrome cd_1 and mitochondrial cytochrome c oxidase

	Cytochrome cd_1 [a]			aa_3 [b] (Bovine)
	Ps. aeruginosa	*Pa. denitrificans*	*A. faecalis*	
Ps. aeruginosa c-551	100	0	0	0
Ps. aeruginosa azurin	3	0	0	[c]
A. faecalis c-554	420	3	100	[c]
Pa. denitrificans c-550	8	100	2	4
Thio. novellus c-550	6[b]	[c]	[c]	23
Mammalian cyt.c	1	14	1	73

Figures are expressed as percentages of a chosen 100% value for each enzyme. In most cases this 100% value is that for the reaction between a cytochrome and the enzyme from the same source. In the case of bovine cytochrome aa_3, yeast cytochrome c is the most active donor (100%).

[a] Figures are from Timkovich et al. (1982) with 20 μM cytochrome c or azurin except for *Alcaligenes* c-554 (10 μM). All assays were carried out anaerobically with 0.5 mM nitrite as oxidant in 50 mM phosphate pH 6.

[b] Figures are from Yamanaka (1975) with 20 μM cytochrome c.

[c] Indicates that no rate measurement has been performed.

role of individual cytochromes is not yet available. The problem is the large number of c-type cytochromes, some membrane-bound and some free, that are present in denitrifiers (see for example Fig. 1.2). Of the types of evidence required to implicate a cytochrome c in an electron transfer process (Sect. 3.1), there is some information on coincident distribution and induction but studies on the in vivo contribution to electron transport are made very difficult by the spectroscopic similarities among the c-type cytochromes.

***Pa. denitrificans* Cytochrome c-550.** The cytochrome c-550, a structural relative of the cytochromes c_2 (Vol. 2, Chap. 3), is the only small soluble cytochrome c purified from *Pa. denitrificans.* It is implicated as the electron donor to cytochrome aa_3 (Sect. 3.4) and also reacts well with cytochrome cd_1, the nitrite reductase (Table 3.10). There is little sequence similarity between cytochrome c-550 and the cytochromes c-551 which are the electron donors to cytochrome cd_1 of the Pseudomonads and this is reflected in the poor cross-reactivity of Table 3.10.

Cytochrome c-551. The cytochromes c-551 are monoheme, monomeric cytochromes of small size (8–9 K), positive E_m (200–265 mV) and ferroheme spectra with $\alpha/\beta > 1.7$ and α(551–553 nm) (Chap. 1, Table 1.5). Of the species discussed in this chapter they are found during anaerobic growth in most actively denitrifying Pseudomonads (Ambler 1977a), in *Alcaligenes faecalis* (Timkovich et al. 1982) and possibly in *Thiobacillus denitrificans* (Sawhney and Nicholas 1978) and halophilic *Paracoccus* (Hori 1961). However, *Ps.*

fluorescens biotype B, D and F, *Alcaligenes* sp. (formerly called *Ps. denitrificans*) and *Pa. denitrificans* denitrify but do *not* contain the cytochrome, while it *is* found in the strict aerobe *Azotobacter vinelandii*. Although cytochrome c-551 is markedly induced by anaerobic conditions in *Ps. aeruginosa* (Matsushita et al. 1982b) it is present in similar amounts in aerobic and anaerobic *Ps. perfectomarinus* (Liu et al. 1983) and *Ps. stutzeri* (Fig. 1.2).

In spite of these ambiguities in its distribution and induction, cytochrome c-551 has long been assumed to be the physiological donor to *Ps. aeruginosa* cytochrome cd_1 (Yamanaka et al. 1961). It is proposed that it performs this role interchangeably with the copper protein, azurin, and indeed that the presence of both azurin and cytochrome c-551 genes may allow an adaptive response to changes in copper and iron in the environment by analogy with the established reciprocal relationship between algal cytochrome c-553 and plastocyanin (Wood 1978b; Bohner et al. 1980; see Chap. 4.1.3). We should note that, in spite of similar maximum turnover rates ($100-200\ min^{-1}$), the apparent K_m for cytochrome c-551 ($2-19\ \mu M$) is less than that for azurin ($15-49\ \mu M$ with different authors) (Yamanaka et al. 1961; Gudat et al. 1973; Barber et al. 1976; Tordi et al. 1985) and at substrate concentrations close to K_m for c-551, azurin appears to be a very poor alternative (Timkovich et al. 1982; Table 3.10). However, the electron exchange between azurin and c-551 is very fast (Antonini et al. 1970) and may result in effective equilibration of a mixed electron donor pool before transfer to cytochrome cd_1 (Meyer and Kamen 1982). It is interesting that the redox potentials of *Ps. aeruginosa* azurin and cytochrome c-551 show a matching pH dependence (Silvestrini et al. 1981; Pettigrew et al. 1983; Vol. 2, Chap. 7). The former authors have suggested that this may constitute a control over the rate of electron transfer at the bioenergetic membrane where proton concentration will be variable.

The steady-state kinetics of cytochrome cd_1 with ferrocytochrome c-551 are poorly characterised. The progress curves are usually complex, particularly with nitrite as acceptor (Robinson et al. 1979). Like mitochondrial cytochrome oxidase their appearance can be explained by product inhibition (Barber et al. 1976; Robinson et al. 1979) but unlike the mitochondrial enzyme, simple first-order curves are seldom obtained and this may be due to unequal binding of the ferro- and ferricytochrome c (Robinson et al. 1979). Tordi et al. (1985) suggested that only one site of the dimeric cytochrome cd_1 is active at low substrate concentration giving rise to the K_m values quoted above. They proposed that azurin and cytochrome c-551 compete for binding to this first site and interfere with binding to the second identical site.

Cytochrome c_4 and c_5. The trio of cytochrome c-551, c_4 and c_5 appears together in the denitrifying Pseudomonads and also in the strict aerobe *Azotobacter vinelandii* from which c_4 and c_5 were first purified (Swank and Burris 1969). The cytochromes c_4 are diheme monomeric cytochromes of M_r approx. 20 K, positive E_m ($200-300$ mV) and ferroheme spectra with α/β = $1.2-1.3$ and a usually symmetrical α-peak at $550-552$ nm (Chap. 1, Table 1.5).

All but a small fraction of cytochrome c_4 is tightly bound to the cell membrane (Fig. 1.2) but is released in a monomeric, water-soluble form after butanol treatment. This unusual behaviour may reflect the two-domain structure of the molecule (Vol. 2, Chap. 4). Each domain may have a hydrophobic surface region which interacts with a membrane protein in situ and with each other in the water-soluble state. The function of cytochrome c_4 is not known. It is only one of a number of membrane c-type cytochromes in the Pseudomonads and it is present in both aerobic and denitrifying cells. Although it has been suggested that it is a part of the cytochrome *co* oxidase of *Azotobacter vinelandii* (Sect. 3.4) and may be so in the Pseudomonads also, this has not been directly demonstrated.

One intriguing possibility regarding the mode of action of cytochrome c_4 is that it may interact at the same time with a reductase and oxidase and conduct electrons across itself. Another possibility is that its diheme nature might allow a synchronous and perhaps cooperative delivery of a pair of electrons to a terminal oxidase. These require investigation conducted in the light of the structural (Vol. 2, Chap. 4) and potentiometric (Vol. 2, Chap. 7) information that is available.

It is probable that relatives of cytochrome c_4 are present in other organisms. The *Alcaligenes* sp. originally described as a "*Ps. denitrificans*" contains a membrane-bound cytochrome c with the spectra, heme content and size of cytochrome c_4 (Shidara 1980). This cytochrome stimulated release of N_2 from NO_2^- in cell extracts (Iwasaki 1960) and was found to rise in cells grown on N_2O, while nitrite reductase levels fell. In *Pa. denitrificans*, a cytochrome c of M_r 20 K is associated with a membrane particle with quinol oxidase activity (Berry and Trumpower 1985; Sects. 3.4.2 and 3.2.5).

The cytochrome c-552 of *Ps. perfectomarinus* was found only under anaerobic conditions (or at least was only soluble under these conditions; Liu et al. 1983). In its heme content and size (1.7 mol heme c per 28 K) and its amino acid composition (Liu et al. 1981) it resembles cytochrome c_4. However, the α/β-peak ratio of 2 and the small shoulder on the short wavelength side of the reduced Soret band (Liu et al. 1981a) are more reminiscent of a cytochrome c_3 (Chap. 1, Fig. 1.6). In addition, there is a remarkable separation of 350 mV in the midpoint potentials of the two heme groups and the possibility of a c_3-like hemochrome ($E_m = -180$ mV) and a high potential cytochrome c hemochrome ($E_m = +174$ mV) present on the same structure must be considered. With the loss of energy involved in electron transfer across such a soluble diheme system having this extreme separation of redox potentials, it is difficult to conceive of a role for such a system in a phosphorylating electron transport process. Perhaps enzymic activity (such as peroxidase) should be examined. The cytochrome c (551 nm) ($\alpha/\beta = 1.26$) found in aerobic cells (Liu et al. 1983) may be the genuine cytochrome c_4.

Cytochrome c_5 usually occurs as a dimer of a monoheme cytochrome of M_r approx. 10 K (although that from *Ps. mendocina* is monomeric; Ambler and Taylor 1973) and the α-peak of the ferrocytochrome is shifted to the red near 555 nm with a decreased extinction (α/β = 1.4). These properties appear to be characteristic of the class but have been reported in detail only for the *Azotobacter* cytochrome (Swank and Burris 1969; Campbell et al. 1973). The N-terminal raggedness observed in amino acid sequence work (Ambler and Taylor 1973) has given rise to a persistent worry that cytochrome c_5 is the fragment of a larger protein derived by endogenous protease action. Both cytochrome cd_1 (Horowitz et al. 1982) and cytochrome c peroxidase (Soininen and Ellfolk 1975) appear to be susceptible to limited proteolysis although the low cysteine content of the former (Nagata et al. 1970) appears to rule out a precursor relationship with cytochrome c_5 which is unusual in containing four cysteines. The sequence of a heme fragment of cytochrome cd_1 confirms this conclusion (Kalkinnen and Ellfolk 1978).

The function of cytochrome c_5 is not known. A cytochrome c-554 with the spectral characteristics and size of cytochrome c_5 was found in both aerobic and anaerobic cells of *Ps. perfectomarinus* (Liu et al. 1983) and *Ps. stutzeri* (Fig. 1.2). In halophilic *Paracoccus* a cytochrome c-554 fell to a low level during anaerobic growth (Hori 1961) but paradoxically reacted very well with *Ps. aeruginosa* cytochrome cd_1. On the other hand, the cytochrome c-554 of *Thiobacillus denitrificans* is unique in its association with nitrate reductase and does not react with cytochrome cd_1 (Amminuddin and Nicholas 1973; Sawhney and Nicholas 1977, 1978).

Cytochromes c′. The cytochromes c′ form a homogeneous class (IIa, Chap. 1, Table 1.4) with respect to spectra (Fig. 1.6), midpoint redox potential (−10 to +100 mV), M_r (14−16 K monomers) and amino acid composition (Meyer and Kamen 1982). Amino acid sequences (Vol. 2, Chap. 3) show homology between the cytochromes c and certain spectroscopically distinct cytochromes c′ (low spin, class IIb) exemplified by cytochrome c-556 of *Rps. palustris*. This structural knowledge is not matched by an understanding of function. We consider them in this section because some evidence suggests a role in denitrification.

Cytochrome c′ is found in the photosynthetic bacteria – the Chromatiaceae and the Rhodospirillaceae – the aerobe, *Azotobacter vinelandii* and the facultative anaerobes, *Alcaligenes* sp. and halotolerant *Paracoccus* (Meyer and Kamen 1982). Although widespread in the Rhodospirillaceae it is absent in *Rhodomicrobium vannielii* (Morita and Conti 1963), *Rps. viridis* (Olson and Nadler 1965) and the one strain of *Rps. palustris* that contains the low-spin cytochrome c-556 (DeKlerk and Kamen 1966). In *Rhodospirillum rubrum* and *Rps. capsulata* it is present in both photosynthetic and respiring cells (Taniguchi and Kamen 1965; Prince et al. 1975), while in *Rps. sphaeroides* it is present only in the former (Kikuchi et al. 1965).

This erratic distribution makes discussion of function difficult. Further difficulty arises from the probability that cytochrome c′ is altered upon solubilisation from the membrane. Thus in *Chromatium*, although purified cytochrome c′ binds CO, intact cells bind much less CO than expected from their c′ content (Cusanovich et al. 1968). Taniguchi and Kamen (1965) could not detect cytochrome c′ by immunological methods in membrane particles of *Rsp. rubrum*, yet Kakuno et al. (1971) were able to solubilise cytochrome c′ from such membranes using detergent. Finally, the distinctive EPR signals of purified cytochrome c′ are not observed in intact cells or membrane fractions (Dutton and Leigh 1973; Prince et al. 1974a; Corker and Sharpe 1975). Kakuno et al. (1971) have suggested that cytochrome c′ in situ may have a b-type optical spectrum and in this context it is noteworthy that the low-spin sequence homologues of cytochrome c′ have red-shifted absorption bands. It is possible that this represents the native chromophoric structure of the cytochrome c′ group and an X-ray structure for cytochrome c-556 is needed.

Although a role for cytochrome c′ in photosynthetic electron transport has been proposed on the basis of light-induced absorption changes (Olson and Chance 1960; Morita 1968; Cusanovich et al. 1968), the problems in identifying the cytochrome in situ cast doubt on these interpretations. A different possibility is that the cytochrome is involved as part of the denitrification process. Thus cytochrome c′ was isolated from *Alcaligenes* sp. (Suzuki and Iwasaki 1962) and Shidara (1980) purified a cytochrome c-556 from the same organism. For the latter, the red shift of all spectral peaks, the low $Soret_{red}/Soret_{ox}$, the broad δ-band and the midpoint redox potential of 291 mV suggest that this monoheme cytochrome might be a member of the low-spin class II cytochromes. This cytochrome c-556 stimulated the release of N_2 from NO_2^- using dimethylphenylenediamine as reductant.*

Denitrification as a complete sequence of reactions is rare amongst the Rhodospirillaceae but the ability to reduce nitrous oxide may be more common (McEwan et al. 1985). In *Rps. capsulata* and *Rps. sphaeroides*, N_2O reduction is insensitive to antimycin A and the purified reductase can oxidise cytochrome c′. McEwan et al. (1985) suggested that cytochrome c′ may mediate the transfer of electrons to N_2O reductase at the level of reduced quinone (Fig. 3.19). The occurrence of redox proteins associated with denitrification is summarised in Table 3.11.

3.5.6 Energy Conservation

Growth of denitrifying organisms is possible on NO_3^-, NO_2^- or N_2O indicating that all the stages of denitrification allow energy conservation by oxidative phosphorylation. Table 3.12 indicates that considerable oxidative energy is available at each stage of denitrification and indeed that the process of nitrite reduction to nitrogen is thermodynamically more favourable than oxygen reduction. This however is not reflected in the efficiency of oxidative phosphorylation or the growth yield.

In *Ps. denitrificans*, Koike and Hattori (1975) found similar phosphorylation yields for each stage which were 40% lower than for oxygen. The origin of this lower efficiency appears to be the topographical design of the denitrifying electron transport system. In *Pa. denitrificans* nitrate is reduced at the cyto-

* See appendix note 18

Table 3.11. Distribution of electron transport proteins in denitrifying and related bacteria

Organism	Mode of growth	c-551	c_4	c_5	c′	Azurin	cd_1	Reference
Ps. aeruginosa	Denitrifying[a]	+	+	+	−	+	+	g, h
Ps. fluorescens (Biotypes ABDEFG)[b]		−	ND	+	−	+	ND	g
Ps. fluorescens (Biotype C)[b]	Denitrifying	+	ND	ND	−	+	ND	g
Ps. stutzeri	Denitrifying	+	+	+	−	−[c]	+	g, i
Ps. mendocina	Denitrifying	+	+	+	−	ND	ND	g
Ps. denitrificans (NCIB 9496)	Denitrifying	+	ND	+	−	+	ND	g
Ps. perfectomarinus	Denitrifying	+	+	+	−	ND	+	j
Ps. putida	Aerobic	−	ND	ND	−	+	ND	g
Az. vinelandii	Aerobic	+	+	+	−	−	ND	g, k
Alc. faecalis	Denitrifying	ND	ND	ND	ND	+	+	l, m
Alc. species[d]	Denitrifying	ND	(+)	ND	+	+	ND	n, o
"Halotolerant *Paracoccus*"	Denitrifying	(+)	(+)[e]	(+)	+	ND	+	p, q
Thio. denitrificans	Denitrifying	(+)	ND	(+)	ND	ND	+	r
Pa. denitrificans	Denitrifying	[f]	ND	ND	−	ND	+	s

+ Indicates that a redox component has been shown to be present; (+) indicates that a redox component is present which is not structurally characterised but which has some of the properties characteristic of the group; − indicates that the component was specifically looked for but not found; ND indicates that a component was not reported but was not proved to be absent. In fact, proof that a protein is absent, rather than present in small amounts or under particular growth conditions, is difficult to obtain.

[a] "Denitrifying" indicates a potential ability to grow anaerobically in the presence of nitrate but not all studies were carried out under denitrifying conditions. All denitrifiers can grow aerobically and prefer to do so.

[b] Biotypes B, D and F can denitrify; A, E and G cannot (Stanier et al. 1966). Ambler (1977) emphasised that biotype C is the only "very actively denitrifying group".

[c] Azurin is reported to be absent in *Ps. stutzeri* but in view of its probable reciprocal relationship with cytochrome c-551 its presence should be investigated under Fe-limitation and Cu-supplementation.

[d] This is the species studied by Iwasaki and co-workers and originally described as a "*Ps. denitrificans*". The amino acid sequence of the azurin differs from that of a known culture of *Alcaligenes denitrificans* and thus the species assignment remains in doubt.

[e] A cytochrome c-554 (548) resembles cytochrome c_4 in amino acid sequence but is only half the size.

[f] *Para. denitrificans* c-550 is a "large" class I cytochrome c with much greater similarity to the cytochrome c_2/mitochondrial cytochrome c group than to the *Pseudomonas* cytochromes c-551.

References:
[g]Ambler (1977); [h]Yamanaka and Okunuki (1963b); [i]Kodama (1970); [j]Liu et al. (1983); [k]Swank and Burris (1969); [l]Iwasaki and Matsubara (1971); [m]Timkovich et al. (1982); [n]Suzuki and Iwasaki (1962); [o]Shidara (1980); [p]Hori (1961); [q]Meyer and Kamen (1982); [r]Sawhney and Nicholas (1978); [s]Newton (1969).

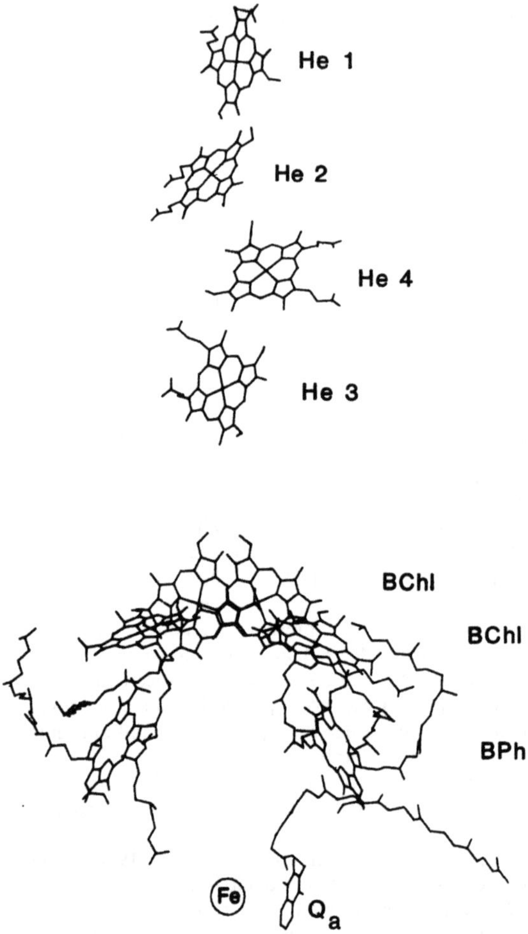

Fig. 3.19. The arrangement of the redox centres in the reaction centre of *Rps. viridis.* The four heme groups (*He 1-He 4*), the bacteriochlorophylls (*BChl*), the bacteriophaeophytin (*BPh*), the quinone (Q_a) and the iron (*Fe*) are shown (Deisenhofer et al. 1985)

plasmic side of the membrane (Sect. 3.5.2) but Alefounder and Ferguson (1980), in agreement with earlier studies on *Ps. aeruginosa* by Wood (1978a), found that cytochrome cd_1, the nitrite reductase, was released from spheroplasts. This is consistent with the location of its donor, cytochrome c-551, and the general concept of periplasmic c-type cytochromes (Wood 1983; Sect. 3.1). Quantitative analysis of H^+ translocation during electron transport confirms the periplasmic location of nitrite reduction and shows that this is also the site for the reduction of nitrous oxide (Meijer et al. 1979; Boogerd et al. 1981) and nitric oxide (Shapleigh and Payne 1985). As a consequence of this location, an electrogenic effect opposite to that required for energy conservation is generated by the oxidation of cytoplasmic substrates and the reduc-

Table 3.12. Energetics of N-oxide reduction

	$E_{m,7}$
$NO_3^- + 2e^- + 2H^+ \rightarrow NO_2^- + H_2O$	+ 430 mV
$NO_2^- + 2e^- + 3H^+ \rightarrow 0.5N_2O + 1.5H_2O$	+ 762 mV
$N_2O + 2e^- + 2H^+ \rightarrow N_2 + H_2O$	+ 1355 mV

Overall $2NO_3^- + 10e^- + 12H^+ \rightarrow N_2 + 6H_2O$.

tion of periplasmic acceptors (Fig. 3.18). The diminution of the H^+ gradient can quantitatively explain the lower phosphorylation yields (Boogerd et al. 1981).

In other organisms, the phosphorylation yields may vary. For example, nitrous oxide reduction may be an antimycin-insensitive process (e.g. in *Rps. capsulata*, McEwan et al. 1985) and therefore allows poorer energy conservation. A strain of *Desulfovibrio desulfuricans* can reduce nitrate to ammonia during anaerobic growth on lactate (Liu and Peck 1981) and under these conditions a hexaheme cytochrome c with nitrite reductase activity is induced (Table 3.9). The enzyme is transmembrane, accepting cytoplasmic electrons and reducing periplasmic nitrite (Steenkamp and Peck 1981). Again the cytoplasmic proton release and periplasmic proton uptake associated with the terminal reactions work against the establishment of the chemiosmotic proton gradient but nevertheless acidification transients were observed with whole cells given nitrite pulses (Steenkamp and Peck 1981).

In the case of *E. coli* cytochrome c-552, the potential span between cytoplasmic NADH and the periplasmic cytochrome (E_m −200 mV; Table 3.9) seems very short to support oxidative phosphorylation. It may be that *E. coli* cytochrome c-552 is a detoxifying enzyme for nitrite or that it simply acts to allow reoxidation without energy conservation of the NADH produced in fermentative growth.

3.6 The Oxidation of Cytochrome c by the Photosynthetic Reaction Centre

3.6.1 Photosynthetic Reaction Centres

Events Following Light Absorption. The central event in all photosynthetic systems is the ejection of an electron from a special pair of chlorophyll molecules (P) after excitation to the singlet state by energy transmitted from light-harvesting pigments. The electron comes to reside on a primary acceptor (A) via fast intermediate electron transfers resulting in a charge separation ($P^+ A^-$) across the bioenergetic membrane. This charge separation gives rise to a membrane potential, positive on the side opposite the cytoplasm (in bacteria) or the stroma (in the chloroplast) (Blankenship and Parson 1978; Crofts and Wraight 1983).

In the cyclic photosynthetic mode, the electron on A^- returns to P^+ via the bc_1/bf complex with associated energy conservation (Sect. 3.2). In the non-cyclic mode, the electron may be used in the production of NAD(P)H for biosynthesis. As noted in Section 3.3.3, photosynthetic systems differ in the reducing power of A^-. In the green sulfur bacteria (Chlorobiaceae) and photosystem I of the chloroplast, the primary acceptor can reduce NAD(P) via ferredoxin as a redox mediator. In this non-cyclic mode, electrons must be supplied from an exogenous donor which is water in the case of oxygenic photosynthesis (via photosystem II) and sulfur compounds in the Chlorobiaceae. In the purple bacteria A^-, which is believed to be an iron-quinone centre, is not of low enough redox potential to achieve reduction of NAD(P) and energy-dependent reversed electron transport from organic substrates or reduced sulfur compounds is required.

These different photosynthetic systems are represented in Fig. 3.5 and compared to the mitochondrial respiratory system. This diagram emphasises the role of c-type cytochromes as the immediate electron donors to the photo-oxidised chlorophyll P^+. Thus they play a crucial role in cyclic photosynthetic electron transport, mediating electron transport from the bc_1/bf complex and also in the oxidation of sulfur compounds in the photosynthetic bacteria (Sect. 3.3.3).

Structure of the Reaction Centre. The best studied reaction centres are those of the purple bacteria. They are of two types, the distribution of which has no obvious phylogenetic significance. In one, exemplified by *Rhodopseudomonas sphaeroides*, there are three polypeptide chains (H 34 K, M 32 K, L 28 K; Okamura et al. 1974) and four bacteriochlorophylls, two bacteriophaeophytins, two ubiquinones and one Fe atom. The H and M subunits are accessible at the cytoplasmic surface to appropriate antisera conjugated with ferritin while all three subunits are accessible at the periplasmic face of the membrane (Valkirs and Feher 1982). Analysis of the distribution of hydrophobic residues in the sequence of the M subunit suggests a pattern of several transmembrane α-helices (Williams et al. 1983). Some similarity was detected with a chloroplast polypeptide of similar size implicated in quinone binding. Study of orientated multi-layers of the reaction centre by neutron and X-ray diffraction confirms its transmembrane nature and shows an asymmetric mass distribution with 2/3 at the cytoplasmic side (Pachence et al. 1981, 1983).

The second type of reaction centre exemplified by *Rps. viridis* and *Chromatium vinosum* also has H, L and M subunits and a similar array of prosthetic centres but, in addition, contains a hydrophobic c-type cytochrome of $M_r \sim 40$ K containing four heme groups (Thornber et al. 1980). The functional relationships of these prosthetic groups, which had been established by laser flash kinetic analysis, has been elegantly confirmed by X-ray crystallography of the reaction centre from *Rps. viridis* (Deisenhofer et al. 1984; Fig. 3.19). A plausible electron transfer route leads from the bacteriochlorophyll dimer through closely associated bacteriochlorophyll and bacteriophaeophytin molecules to a quinone close to an iron. A second symmetrically related route has no quinone at its terminus perhaps through loss of one of the two quinones on crystallisation. The membrane orientation of this domain cannot yet be deduced from the crystallography but much independent evidence supports a location for P^+ on the side opposite the cytoplasm and a process of quinol reduction and proton uptake at the cytoplasmic side (Dutton and Prince 1978; Crofts and Wraight 1983).

One of the c-type heme groups lies close to the bacteriochlorophyll dimer and is likely to be the electron donor to P^+. Thus the two types of reaction centres differ in the nature of the immediate donor to P^+, in one case, a mobile, soluble cytochrome c and in the other, a fixed, membrane-associated cytochrome c. This distinction is further examined in the following sections.

3.6.2 Cytochromes c_2 in the Rhodospirillaceae

The cytochromes c_2 are a well-studied group of low-spin monoheme cytochromes with spectra (α-peak symmetrical, 549–552 nm; α/β 1.6), size (97–124 amino acids) and amino acid sequence (Ambler et al. 1979) closely resembling the mitochondrial cytochromes c (Vol. 2, Chap. 3). At present, cytochromes from the Rhodospirillaceae dominate the group but sequence homologues are also found in *Paracoccus denitrificans* (Timkovich et al. 1976; Ambler et al. 1981), *Agrobacterium tumefaciens* (van Beeumen et al. 1980), *Thermus thermophilus* (Titani et al. 1985) and *Nitrobacter agilis* (Tanaka et al. 1982) (Vol. 2, Chap. 3). They may be a feature of organisms synthesising a cytochrome aa_3 type oxidase (Table 3.6). The cytochromes c_2 from non-photosynthetic sources may be sufficiently distinctive (for example in redox potential) that a subdivision of the group will be justified in the future. This section deals only with cytochromes c_2 from the Rhodospirillaceae. *

These cytochromes are distinguished from their mitochondrial relatives in showing poor reactivity with mitochondrial cytochrome oxidase and their more positive midpoint potentials ($E_{m,\,pH\,7}$ 290–390 mV; Table 3.13). In fact the constraint on redox potential for a cytochrome c donor to P^+ may be narrower than this since most of the cytochrome c_2 midpoint potentials measured at pH 5 fall within the range 350–395 mV) (Pettigrew et al. 1978). Because of the known proton-translocating properties of the membrane (Hochman et al. 1975) and the periplasmic location of cytochrome c_2 (Prince et al. 1975), the operating environment of c_2 may be acidic.

The exceptions which fall outside this range are interesting. The iso-II-cytochromes c_2 of *Rhodospirillum molischianum* and *R. fulvum* are accompanied by higher potential homologues (Pettigrew et al. 1978), while the cytochrome c_2 of *Rhodopseudomonas viridis* has been shown to donate, not

* See appendix note 19

Table 3.13. Properties of the cytochromes c_2

Source	$E_{m,7}$ (mV)	$E_{m,5}$ (mV)	pI	Net charge[a]	Rate with cyt.aa_3[b]	Rate with NADH cyt.c reductase[c]
R. photometricum	341	348		+5	0.2	14
R. rubrum	323	357	6.2	$+1\frac{1}{2}$	0.9	48
Rps. sphaeroides	352	368	5.5	+1	1	102
Rps. capsulata	368	368	7.1	+3	0.6	129
Rps. palustris 37	366	383	9.7	$+5\frac{1}{2}$	0.6	55
Rps. acidophila	370	370		+3		
Rps. viridis	296	309		$+3\frac{1}{2}$	0.4	6
Rm. vanniellii	355	392	7.9	+3	0.3	56
R. molischianum I	388	396	9.8	$+5\frac{1}{2}$	0	13
R. molischianum II	305	305	9.4	$+4\frac{1}{2}$	0	25
R. fulvum I	369	382		$+5\frac{1}{2}$	0	9
R. fulvum II	291	291		$+4\frac{1}{2}$		

The organisms are grouped according to those having better than 48% sequence identity.
[a] Net charge was calculated assuming Lys = Arg = 1+; Asp = Glu = 1−; His = 1/2+ and ignoring the charges on the heme. Thus this should be regarded as a measure of relative rather than absolute charge at a given pH close to neutrality.
[b] Rate with cytochrome oxidase from bovine heart mitochondria. The figures are relative to the rate with mitochondrial cytochrome c at 5 μM cytochrome c, 0.1 M MES buffer pH 6.0 (Errede and Kamen 1978).
[c] Rate with NADH cytochrome c reductase from bovine heart. The figures are relative to the rate with mitochondrial cytochrome c at 5 μM cytochrome, 50 mM HEPES pH 7.5 (Errede and Kamen 1978). The reaction follows a zero order time course followed by a first-order portion of variable magnitude. The figures here are for the first-order rate.

directly to P^+, but via a cytochrome c bound to the reaction centre in this organism (Shill and Wood 1984).

Kinetics of Electron Transfer: (1) in Chromatophores. Several of the Rhodospirillaceae contain extensive cytoplasmic vesicular membranes which may be continuous with the plasma membrane.

In many forms of cell breakage these vesicles are nipped off to form chromatophores which contain trapped components, including cytochrome c_2, which were originally periplasmic in the intact cell. Because of this tendency to retain periplasmic proteins, chromatophores are fully active in cyclic electron transport driven by light absorption and carry out coupled proton uptake (Hochman et al. 1975). Anti-c_2-antiserum is only effective in blocking photooxidation of cytochrome c_2 if the chromatophore membrane is first solubilised with detergent (Prince et al. 1975).

In contrast, treatment of cells with lysozyme-EDTA results in formation of spheroplasts which lose most of their c-type cytochrome and their ability to catalyse cyclic electron transport (Prince et al. 1975). Membrane vesicles of the same membrane orientation as spheroplasts could be prepared by mild cell breakage (Hochman et al. 1975). They readily lost both photophosphorylation and electron transport activities but these functions could be reconstituted by added cytochrome c_2 (Hochman and Carmeli 1977).

Most work has been performed on the chromatophores of *Rps. sphaeroides* and *Rps. capsulata* and the cyclic electron transport system has been extensively characterised (Dutton and Prince 1978). The initial charge separation across the membrane after light absorption is very fast giving rise within 1 ns to a reducing centre (E_m −20 mV) on the cytoplasmic side, believed to be an association of iron and quinone, and an oxidising centre on the periplasmic side in the form of the radical cation of the bacteriochlorophyll dimer (E_m 450 mV). This charge separation will decay within 0.1 s and to prevent this, a rapid reduction of the oxidising centre is required. Cytochrome c_2 is the immediate e^- donor to the photo-oxidised bacteriochlorophyll molecule but the kinetics are quite complex and require further explanation.

Until recently, the photo-oxidation of c-type cytochrome in *Rps. sphaeroides* chromatophores was considered solely to reflect the role of cytochrome c_2 as the immediate donor to the activated reaction centre. In dark equilibrium titrations 30% of the c-type cytochrome behaves as a species with E_m 340 mV (close to the value for the purified protein) and the rest with E_m 295 mV (Dutton et al. 1975). It was proposed that the major part of the cytochrome c_2 of the chromatophore is bound to the reaction centre resulting in a lowering of E_m and that it is the rapid oxidation of this functionally competent fraction that is observed after flash illumination of chromatophores poised at chosen ambient redox potentials. A Nernst plot with E_m approx. 40 mV more negative is obtained for the oxidation produced by a second flash of light and this was interpreted according to the statistical model of Case and Parson (1971) whereby a system oxidising a pair of identical redox centres with average reduction level x obeys a Nernst plot defined by $2x-x^2$ for the first flash and x^2 for the second (Dutton et al. 1975; Prince and Dutton 1977b) (see also Sect. 3.6.3). The two theoretical curves differ by 44 mV and the true E_m is equidistant between them. Consistent with this model Dutton et al. (1975) found that although methods for making chromatophores varied in the proportions of total cytochrome c_2 to reaction centre, the amounts of cytochrome c oxidation that occurred after the first, and after combined subsequent flashes were approximately equal. This indicated that all reaction centres associated with one cytochrome could oxidise a second. The time course of oxidation after a single flash was biphasic with half-times of 4 μs and 300 μs (Dutton et al. 1975; Bowyer et al. 1979). A model was proposed in which a two-state equilibrium exists for cytochrome c_2 bound to the reaction centre, one state requiring a re-orientation before electron transfer could take place and giving rise to the slow phase of oxidation (Dutton et al. 1975; Dutton and Prince 1978).

Two discoveries necessitated a re-evaluation of these experiments on the thermodynamics and kinetics of cytochrome c_2 oxidation in *Rps. sphaeroides.*

The first was the demonstration that antimycin A did not completely block the re-reduction of cytochrome c_2 during flash-induced photo-oxidation (Bowyer et al. 1979). These authors found that if the quinone analogue UHDBT was added as well as antimycin A almost all the cytochrome c was oxidised after the first flash and matched the amount of P^+ that was reduced. In contrast to Dutton et al. (1975) they further found that chromatophores contained only 1 mol cytochrome c_2 per mole photo-oxidisable reaction centre. The second discovery was that of Wood (1980a, b) who found a membrane cytochrome with E_m 295 mV and α-peak at 552 nm. This cytochrome resembled mitochondrial cytochrome c_1 in molecular weight (30 K) and was present in approximately equal amounts with cytochrome c_2 in chromatophores. Catalytic cytochrome c_2 can mediate its photo-oxidation in washed membranes and since 80% of the total cytochrome c in chromatophores was photo-oxidisable, this membrane cytochrome c must be contributing, at least in part, to the oxidation kinetics. This is borne out by the finding of Bowyer et al. (1981) that the difference spectrum of the α-peak obtained after photo-oxidation was shifted 1.5 nm to the red compared to that of cytochrome c_2 and probably contains contributions from both c-type cytochromes. Similarly, the midpoint potential of 305 mV obtained by Bowyer et al. (1979) for the oxidation after a single flash in the presence of UHDBT is probably an average of the midpoint potentials of the cytochrome c_1 (295 mV) and the cytochrome c_2 (340 mV).

A possible model to incorporate these findings (Fig. 3.20) involves a serial arrangement of cytochrome c_1 and c_2 in association with a pair of reaction centres. After a saturating light flash the cytochrome c_2 will go oxidised in the fast phase and then mediate electron transfer from the cytochrome c_1 to a second P^+ in the slow phase. This model requires that the difference spectrum after the fast phase be that of cytochrome c_2 while that after the slow phase be cytochrome c_1. Unpublished results of Bowyer et al. quoted in Bowyer et al. (1981) support this requirement. The model is also consistent with the aver-

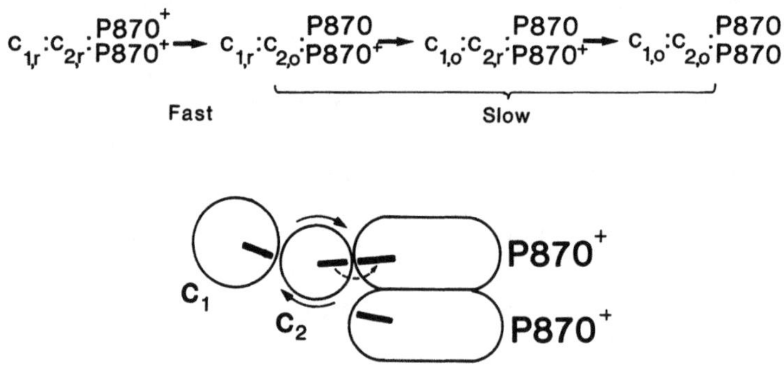

Fig. 3.20. A model for the photo-oxidation of cytochrome c_2 in chromatophores from *Rps. sphaeroides*. $c_{1,r}$ and $c_{2,r}$ are the reduced states of cytochromes c_1 and c_2. $c_{1,o}$ and $c_{2,o}$ are the oxidised states. P 870 is the photooxidisable reaction centre bacteriochlorophyll. (+) denotes its photooxidised state

aged value of E_m obtained after the first flash and the stoichiometry of cytochromes and reaction centres in the chromatophores.

Whether this is also true in *Rps. capsulata* remains uncertain. Cyclic electron transport in this organism is not sensitive to antimycin A (Evans and Crofts 1974b), there is no difference between the midpoint potential of c-type cytochrome in chromatophores and purified cytochrome c_2 (Evans and Crofts 1974a; Prince and Dutton 1977b) and no membrane cytochrome with the properties of cytochrome c_1 has yet been detected (but see Sect. 3.2). However, in view of the close similarity between the cytochrome c_2 sequences of *Rps. capsulata* and *Paracoccus denitrificans* (Vol. 2, Chap. 3) and the fact that *Rps. capsulata* cytochrome c_2 is the most reactive of its class with mitochondrial complex III (Table 3.13), a thorough search for a cytochrome c_1 would seem advisable. Bowyer et al. (1981) found the same red shift in photo-oxidised difference spectra of *Rps. capsulata* and *sphaeroides*.

A genetically-engineered c_2^- genotype in a wild-type *Rps. capsulata* background was constructed by Daldal et al. (1986). Unlike the mutants discussed in Section 3.4.3 which lack all c-type cytochromes and could not grow photosynthetically, the c_2^- construct grew at a similar rate to the wild-type in saturating light intensity. Only under reduced light was there poor growth. Thus inefficient cyclic electron transport may be possible in the absence of cytochrome c_2. But under natural conditions of variable, and often low light intensity, there would be strong selection against such a mutation. Similar mutations in *Rps. sphaeroides* are unable to grow photosynthetically under any conditions (T. Donohue and S. Kaplan, personal communication).

Kinetics of Electron Transfer: (2) with Detergent-Solubilised Reaction Centres. Reaction centres solubilised and purified from *Rps. sphaeroides* using detergents have a relatively simple molecular composition and lack bound c-type cytochrome (Clayton and Wang 1971). Added reduced cytochrome c can be photo-oxidised by these preparations and the time course of the reaction is dependent on the ionic strength. Thus Ke et al. (1970) at 10 mM Tris Cl found a fast, zero-order oxidation of cytochrome c_2 ($t_{1/2} = 12\ \mu s$) while at 10 mM phosphate, 60 mM NaCl the reaction is much slower and first-order in cytochrome c_2 (Rickle and Cusanovich 1979). Overfield et al. (1979) showed that biphasic kinetics are obtained at intermediate ionic strengths incorporating both the zero-order fast phase and the first-order slow phase and under certain conditions the time course closely resembles that seen in chromatophores or whole cells ($t_{1/2}$ 4 µs and 300 µs).

The fast phase is believed to represent oxidation of cytochrome c_2 bound prior to flash activation (K_D 10^{-5} M at low ionic strength) while the slow phase has a rate limit at high cytochrome c concentration which is interpreted as due to a re-orientation step prior to electron transfer (Rickle and Cusanovich 1979; Overfield et al. 1979). Similar kinetic models are proposed by both groups and can be written:

$$c_{2,r} + P^+ \underset{k_2}{\overset{k_1}{\rightleftharpoons}} c_{2,r}..P^+ \xrightarrow{k_3} c_{2,r}.P^+ \xrightarrow{k_4} c_{2,o}.P\ ,$$

where $c_{2,r}$ is reduced cytochrome c_2 and $c_{2,o}$ is oxidised. P^+ is the photo-oxidised reaction centre and (..) and (.) indicate distal and proximal orientations respectively. The conversion of proximal to distal is the proposed re-orientation step required before electron transfer can occur. k_3 is the limiting rate constant at high [cyt.c] and has a value of $3-7\times10^3\ s^{-1}$ (Overfield et al. 1979; Rickle and Cusanovich 1979). k_1 is the second-order rate constant for complex formation and has a value of $8\times10^8\ M^{-1}\ s^{-1}$ while the electron transfer rate constant k_4 is taken as that for the fast phase ($10^6\ s^{-1}$; Overfield et al. 1979).

The fall in rate of the slow phase with rising ionic strength is consistent with electrostatic interactions stabilising the complex although these must be local because the cytochrome c_2 and the reaction centre of *Rps. sphaeroides* have similar acidic isoelectric points (Prince et al. 1974b).

Kinetics of Electron Transfer: (3) with Reaction Centres in Phospholipid Vesicles. As with the soluble system, photo-oxidation of *Rps. sphaeroides* cytochrome c_2 by reaction centres incorporated within phospholipid vesicles involves a fast first-order phase and a slow second-order phase. In neutral vesicles the contribution of the fast phase ($t_{1/2} = 2\ \mu s$) is very dependent on ionic strength and disappears at 0.1 M KCl (Overfield and Wraight 1980a). The fast phase was again proposed to be due to a prior association of reduced cytochrome with reaction centres ($K_D\ 5\times10^{-6}$ M) and the slow phase to collisional interaction (which, in the case of the neutral vesicles does not appear to be enhanced by the possibility of two-dimensional diffusion). Important differences were found when negatively charged vesicles (containing phosphatidyl serine) were used (Overfield and Wraight 1980b; Riegler et al. 1984). The contribution of the fast phase was independent of ionic strength and the rate of the slow phase *increased* with ionic strength to an optimum at 0.1 M KCl. Furthermore, although the slow phase was a second-order collisional process, a rate limit at a small excess of RC over cytochrome was reached. These authors proposed that a cytochrome c molecule is retained on a single negatively charged vesicle with $K_D\ 10^{-7}$ M and that its lateral diffusion to encounter the small number of reaction centres per vesicle is restricted by tight electrostatic interactions with the membrane which are loosened by higher ionic strength. The calculated diffusion constant at low ionic strength resembled that of integral membrane proteins and they suggest that cytochrome c_2 may move with a cluster of lipids associated with it.

However, the situation is complicated by the fact that when the dissociation rate of reduced cytochrome from unactivated reaction centres is considered [as measured by the partial saturation flash technique of van Grondelle et al. (1976), see below], low ionic strengths favour mobility of cytochrome between reaction centres as long as the temperature is high enough to allow rapid diffusion across the lipid. Overfield and Wraight suggested that low ionic strength may stabilise a reaction centre state which permits rapid dissociation of cytochrome and pointed out that the natural ionic composition of chromatophores is similar to that of freshwater, in contrast to the high ionic strengths typically used for chromatophore preparation.

Structural Studies: the Interaction of Cytochrome c_2 with the Reaction Centre. The complex formation predicted from the kinetic studies using purified reaction centres has been confirmed by direct binding experiments. However, there are important points of disagreement regarding the number of binding sites involved and the relative affinities of reduced and oxidised cytochrome c_2 (Table 3.14).

The larger number of binding sites obtained by Pachence et al. (1983) cannot be explained by non-specific binding to phosphatidyl choline since no binding to vesicles in the absence of reaction centres was observed. However, since the concentration of reaction centres was measured by a single turnover photo-oxidation experiment, an overestimate of binding sites would be obtained if damaged reaction centres which could bind, but not oxidise, cytochrome c were present.

Purified reaction centres scarcely discriminate between the two redox states of either horse cytochrome c or *Rps. sphaeroides* cytochrome c_2 and any effect on redox potential would be predicted to be small (Rosen et al. 1980). We have seen that the original conclusion that the midpoint potential of cytochrome c_2 within the chromatophore is 50 mV lower than the free

Table 3.14. Binding of cytochrome c_2 and horse cytochrome c to reaction centres (RC)

State of RC	Method	Cyt.	Sites/RC	ΔE_m [a]	$K_{D,ox}$	$K_{D,red}$	Reference
In detergent solution	Equil. dialysis, 10 mM Tris pH 8, 0.1% Triton	c_2	1	+10[b]	1.4 µM	0.87 µM	Rosen et al. 1980
In detergent solution	Equil. dialysis, 10 mM Tris pH 8, 0.1% Triton	HHC	1		0.41 µM	0.37 µM	Rosen et al. 1980
In phospholipid vesicles	Centrifugation, 10 mM Tris pH 8[d]	HHC	3	−40[c]	1 µM		Pachence et al. 1983; Moser et al. 1982

[a] ΔE_m is the difference between the midpoint potential of the cytochrome free and bound to reaction centres. A negative value indicates a fall in potential on binding.

[b] This figure was not measured directly but was calculated using the equation $E_{m,f} - E_{m,b} = RT/nF \ln K_{D,red}/K_{D,ox}$.

[c] This was measured directly. A similar effect was predicted from the difference between the apparent E_m in situ and that measured in solution for *Rps. sphaeroides* cytochrome c_2. However, this is now in doubt as discussed in the text.

[d] The $K_{D,ox}$ at 10 mM Tris was obtained according to Fig. 1 of Pachence et al. (1983).

cytochrome must be re-examined with the discovery of cytochrome c_1. Nevertheless, direct measurements with horse cytochrome c and phosphatidyl choline vesicles containing reaction centres do demonstrate a lowered E_m for this cytochrome to a degree consistent with a five times stronger binding of the oxidised state (Moser et al. 1982) although no studies have been performed with cytochrome c_2.

There is agreement that binding is strongly dependent on ionic strength and Pachence et al. (1983) found 23 times weaker binding in 0.1 M NaCl. However, unlike the case of mitochondria (Chap. 2.8) the ionic strength of the periplasmic space (and therefore the chromatophores) of these freshwater organisms is probably low and thus the binding constants obtained at lower ionic strengths may be physiologically relevant. Since the cytochrome c concentration within chromatophores may be higher than mM, binding sites should be saturated (Rosen et al. 1980). The existence of a phase in the photo-oxidation kinetics of whole cells as fast as that seen with isolated reaction centres at low ionic strength indicates binding of c_2 prior to the light flash. The slower phase may also be due to bound cytochrome c_2 but in an orientation distal to the reaction centre active site (Dutton and Prince 1978).

The kinetic order of the two phases for cytochrome c_2 in situ cannot be readily assessed in the usual way, but the method of partial saturation flash analysis (van Grondele et al. 1976) can be used to determine whether a cytochrome c_2 molecule is confined to a single reaction centre. At a flash intensity that activates only 50% of the reaction centres and a c_2:reaction centre ratio of 0.5, all the c_2 will be oxidised only if able to diffuse off the unactivated centres. No such mobility was observed indicating a long-term association of a cytochrome c_2 with a reaction centre. Since the reduction of cytochrome c_2 was shown to be a second-order collisional process (Prince et al. 1978) cytochrome c_2 associated with the reaction centre may have rotational mobility to allow alternating interactions with its reductase and its oxidase (Dutton and Prince 1978).

Cytochrome c_2 can be covalently cross-linked to subunit L of reaction centres using the bifunctional cross-linking reagent dithiobissuccinimidyl propionate. A photo-oxidation rate similar to that for the unmodified complex ($t_{1/2}$ 0.8 μs) was found for low ionic strength and this rate persisted to higher ionic strengths while the half-time for the unmodified complex extended to 4 ms (Rosen et al. 1983). When horse cytochrome c was used, binding occurred to either subunits M or L and reaction rates were considerably slower in the cross-linked complex. This presumably reflects a lower specificity of horse cytochrome c for the binding site leading to greater mobility within the complex and suboptimal cross-linking.

Protection experiments of the type described in Chapter 2 have surprisingly indicated that cytochrome c_2 does not bind to the reaction centre at the front face but rather in the region of the C-terminal helix (Rieder et al. 1985). This represents a major upset to the general theory of electron transfer via the heme edge of cytochrome c and requires further investigation.

The interaction of mitochondrial cytochrome c with reaction centres has been studied by X-ray diffraction of orientated phospholipid multi-layers (Pachence et al. 1983) which allows a description of the changes in density across the multi-layer on addition of cytochrome c. Cytochrome c is accommodated in the multi-layer with little change in periodicity even though the intermembrane gap is smaller than the diameter of the protein. This implies some penetration of the cytochrome into the bilayer and consistent with this is a change in the lipid distribution due to an enforced "flipping" of phospholipids to the opposite side of the membrane.

3.6.3 Reaction Centre Cytochrome c

Properties and Distribution. In some photosynthetic bacteria, c-type cytochrome remains firmly associated with the reaction centre after detergent solubilisation of the membrane. This type of reaction centre is erratically distributed and is found in the purple sulfur bacteria *Chromatium vinosum* and *Thiocapsa pfennigii*, the Rhodopseudomonads *viridis* and *gelatinosa* and perhaps also in the green sulfur bacteria (Table 3.15). The c-type cytochromes are characterised by $M_r \sim 40$ K, the presence of high and low potential heme pairs, rapid photo-oxidation, which may persist to very low temperatures, and hydrophobic nature. The latter property has made purification difficult.

The cytochrome was isolated from *Chromatium* chromatophores after acetone powdering and extraction with 2% cholate (Kennel and Kamen 1971). It retained the spectroscopic and potentiometric features of the photo-oxidisable cytochrome c-553, 558 and contained four to five hemes per polypeptide. Thus both heme pairs appear to reside on a single chain. Attempts to extract the corresponding protein from membranes of *T. pfennigii* (Meyer et al. 1973) and *Rps. viridis* (Knaff and Kraichoke 1983) yielded spectroscopically altered products.

Kinetics of Photo-Oxidation (Chromatium). Steady-state illumination of *Chromatium vinosum* chromatophores resulted in the rapid photo-oxidation of a cytochrome c-555 at positively poised potentials and a c-553 at more reducing potentials (Cusanovich et al. 1968). The extent of photo-oxidation observed at different ambient potentials allowed the calculation of an E_m of 319 mV for the former and 10 mV for the latter and the kinetics of cytochrome oxidation matched those of P^+ reduction (Parson 1969). Subsequent work using short saturating single turnover flashes found that when cytochrome c-555 was fully reduced, the first flash oxidised all the P but only half the cytochrome, while the second flash oxidised the remaining cytochrome (Parson 1969; Case et al. 1970; Seibert and DeVault 1970; Parson and Case 1970; Dutton 1971).

When cytochrome c-555 was partially reduced, cytochrome oxidation in response to the first and second flashes at different ambient potentials could

Table 3.15. Properties of membrane-bound cytochromes implicated in photosynthetic electron transport

Source		$E_{m,7}$ (mV)	Hemes/ RC	Electron acceptor	$E_{m,7}$ of acceptor	Rate of photooxidation		Electron donor	Subunits in RC	References
						298 Kelvin	80 Kelvin			
Chromatium vinosum	555	319	2	$P883^+$	490	2.3 μs	None	Cyclic (c-551)[b]	45 K (heme) 30 K, 27 K, 22 K, 12 K	Cusanovich et al. 1968; Thornber 1969; Case and Parson 1971; Parson 1969; Dutton 1971; van Grondelle et al. 1977; Kennel and Kamen 1971; Halsey and Byers 1975
	553	10	2–7[a]			1.1 μs	2.5 ms	non-cyclic		
*Rps. viridis**	558	330	2	$P960^+$	500	~2 μs	ND	c_2	38 K (heme) 35 K, 28 K 24 K	Case et al. 1970; Thornber and Olson 1971; Cogdell and Crofts 1972; Thornber 1969; Trosper et al. 1977; Thornber et al. 1980; Pucheu et al. 1976; Prince et al. 1976; Shill and Wood 1984
	553	−12	2 or 3			300 ns	ND			
Thiocapsa pfennigii	555	340	2		490	<100 μs	ND			Prince 1978; Olson et al. 1969
	550	0	2							
Rps. gelatinosa	553[c]	128		$P600^+$	400	0.7 μs	4.6 μs			Dutton 1971; Kihara and Chance 1969
	548[c]	280					60 μs			
Chlorobium limicola	553	165	2	$P840^+$	250	Biphasic, 5 and 50 μs				Prince and Olson 1976; but see also Knaff et al. 1973
Chlorops. ethylica[d]	553			$P840^+$		90 μs	None			Swarthoff et al. 1981; see also Fowler et al. 1971

[a] Rapid flash experiments suggest two hemes while chemical determination suggests seven.
[b] The c-555 component is generally believed to accept cyclic electrons but its e^- donor (c-551) can be reduced by thiosulfate.
[c] These wavelengths were measured at 80 Kelvin.
[d] *Chloropseudomonas ethylica* has been shown to be a consortium of *Prosthecochloris aestuarii* and *Desulfuromonas acetoxidans* (Pfennig and Biebl 1976).
* See appendix note 20.

be fitted to a statistical model involving two identical hemes, either or both of which may be reduced prior to the flash (Case and Parson 1971; see also Sect. 3.6.2). It was proposed therefore that two identical c-555 hemes are associated with a single reaction centre. Similar experiments demonstrated that two c-553 hemes must be associated with the same reaction centre.

A peculiarity of this system is that only oxidation of c-553 occurs at potentials at which both types of heme are reduced. This cannot be fully explained by the different half-times for oxidation of the two types of heme (Table 3.15). Dutton and Prince (1978) proposed that the high potential c-555 can take up a proximal or distal orientation with respect to P and that reduction of c-553 favours the latter. Oxidation of the c-555 heme does not persist to very low temperature unlike that of the c-553 (Table 3.15) and these authors further suggested that low temperatures also favour the distal orientation. These studies must now be re-evaluated in the light of the X-ray crystallographic work.*

Electron Donor to Reaction Centre Cytochrome c (Chromatium and Rps. viridis). *Chromatium* cytochrome c-555 has long been proposed to act in a cyclic photophosphorylation system (Dutton and Prince 1978) but its re-reduction in chromatophores is too slow to support such a cellular role. Using whole cells of *Chromatium*, van Grondelle et al. (1977) showed that photo-oxidised cytochrome c-555 can be re-reduced by a c-551 with a half-time of 300 μs. They found that although c-551 is present in less than 1 mol/mol reaction centre it is mobile enough to visit at least two c-555 hemes on different reaction centres.

The re-reduction of cytochrome c-551 is biphasic, the faster phase being blocked by o-phenanthroline, a classic inhibitor of cyclic photophosphorylation, while the slower phase was enhanced by thiosulfate indicating an entry of non-cyclic electrons. van Grondelle et al. (1977) emphasised the similarity in the diffusible nature and oxidation kinetics of this component to the cytochrome c_2 of *Rps. rubrum* and *Rps. sphaeroides* in spite of its action at one step removed from the photo-oxidised P^+.

Using alanine uptake as a measure of membrane potential generated by cyclic photophosphorylation in *Chromatium* spheroplasts, Knaff et al. (1980) found that only a cytochrome c-551 among the known *Chromatium* cytochromes could restore HOQNO-sensitive uptake in washed spheroplasts. This cytochrome was isolated by Gray et al. (1983) in low yield. It has a midpoint potential of 240 mV in agreement with that estimated in whole cells by van Grondelle et al. (1977) and M_r 15 K.

Thus a cytochrome c-551 of unknown affinities to other cytochromes appears to act as the soluble periplasmic diffusible donor to the cytochrome c-555 bound to the *Chromatium* reaction centre. Its apparent ability to accept both cyclic and non-cyclic electrons calls into question the long held assumption that the low potential cytochrome c-553 attached to the reaction centre is active in non-cyclic electron transport (Hind and Olson 1968).

As with *Chromatium* the re-reduction of reaction centre cytochrome c is very slow in membrane preparations of *Rps. viridis*. Addition of cytochrome

---

* See appendix note 20

c_2 greatly increases the rate of re-reduction of the c-557 component of the reaction centre and the cytochrome c_2 is itself re-reduced in an antimycin-sensitive fashion (Shill and Wood 1984).

The reductase system which must link these c-type cytochromes and the primary acceptor if cyclic photophosphorylation is to occur, is poorly understood. The presence of b-type cytochromes and a membrane-bound c-type cytochrome have recently been demonstrated in *Chromatium* (Knaff and Buchanan 1975; Knaff et al. 1979) and *Rps. viridis* (Wynn et al. 1985). In *Chromatium*, cytochrome b reduction is inhibited by quinone analogues believed to act at the level of the primary or secondary acceptors, but is enhanced by antimycin A (Bowyer and Crofts 1980). These results are consistent with the presence of a "b Q cycle" (Fig. 3.5) and a membrane complex related to mitochondrial complex III (see Sect. 3.2) but further information is required.

3.6.4 The Chlorobiaceae

The photosynthetic apparatus of the green sulfur bacteria is less well characterised than that of the purple sulfur bacteria, partly because of the difficulty in observing light-induced absorbance changes in the presence of the very high concentrations of *Chlorobium* chlorophyll. This light-harvesting chlorophyll can be removed without the use of detergent to yield a reaction centre complex analogous to the chromatophore of the purple bacteria (Fowler et al. 1971; Prince and Olson 1976). In *Chlorobium limicola* Prince and Olson (1976) found that a pair of identical cytochromes c-553 (E_m 165 mV) reduce the photo-oxidised P^+ (E_m 250 mV) and Fowler et al. (1971) obtained a similar result for the consortium "*Chloropseudomonas ethylica*" which contains the green sulfur bacterium *Prosthecochloris aestuarii* (Pfennig and Biebl 1976). Swarthoff et al. (1981) purified a reaction centre complex from the latter and found that it had suffered damage both at the electron acceptor side of the reaction (only 50% of reaction centres could form P^+) and at the electron donor side (only about 35% of P^+ formed could be rapidly reduced by cytochrome c-553).

The midpoint potential of the cytochrome c-553 donor bears the same thermodynamic relationship to that of P in this organism as seen in the analogous system of the purple bacteria except that both have a negative displacement of 200 mV. No evidence of a low potential donor analogous to cytochrome c-553 in *Chromatium* was found and no cytochrome photo-oxidation was observed at 80 Kelvin.

There is controversy as to the relationship between this cytochrome and the well-characterised soluble cytochromes c-555 which have been isolated from both *Chlorobium limicola* (f. thiosulfatophilum) (Gibson 1961; Meyer et al. 1968; Yamanaka and Okunuki 1968; Steinmetz and Fischer 1981) and *Prosthecochloris aestuarii* (Olson and Shaw 1969; Shioi et al. 1972). The cytochromes c-555 have a characteristically red-shifted and asymmetric α-peak, a low α/β-peak ratio and high γ-red/α-peak ratio. In these respects, in size and, to a small extent, in amino acid sequence (Vol. 2, Chap. 3) they resemble the algal cytochromes c-553 (see below). However, they are distinguished by having much lower midpoint potentials (103 mV for *Pr. estuarii* (Shioi et al. 1972) and 140 mV for *Chl. limicola* (Gibson 1961). The light-induced absorbance changes observed in whole cells were proposed to be due to this cytochrome (Olson and Sybesma 1963; Sybesma 1967; Shioi et al. 1976). However, several lines of evidence suggest that this may not be so.

Most authors find that the spectrum of photo-oxidation in both species is that of a cytochrome c-(551-553) rather than a c-555 (Prince and Olson 1976; Knaff et al. 1973; Swarthoff et al. 1981; Fowler et al. 1971). Also the midpoint potentials measured for the photo-oxidisable cytochrome of both species are higher than those of the soluble cytochromes. Finally, the persistence of the photo-oxidisable cytochrome c-553 in the reaction centre preparation of Swarthoff et al. (1981) and the solubilisation with detergent of a distinct cytochrome c-553 from *Chl. limicola* (Bartsch 1978) suggests that the photo-oxidisable cytochrome may be an integral membrane protein analogous to the cytochromes associated with the reaction centre in the purple bacteria.

However, the interpretation of these lines of evidence themselves is not free of ambiguity. Exact peak maxima for photo-oxidation can be perturbed by the presence of other components (for example carotenoids). We have found (Leitch, Moore and Pettigrew, unpublished results) that the midpoint potential of the *Chl. limicola* cytochrome c-555 is pH-dependent and varies from 180 mV at pH 6 to 140 mV at pH 8. This is accompanied by a change in the appearance of the α-peak. In studies of intact systems it is difficult to assess the true pH at the bioenergetic membrane. Also the biphasic kinetics and differential thermodynamic behaviour on the first and second flash is reminiscent of photo-oxidation in *Rps. sphaeroides* where the original model of a pair of identical cytochromes c_2 associated with the reaction centre has been replaced by a series model involving both cytochrome c_2 and cytochrome c_1 (see Sect. 3.6.2).

Thus the relative roles of the membrane-bound cytochrome c-553 and the soluble cytochrome c-555 require further investigation. A possible role for the latter should be re-examined in the light of the identification of a small soluble cytochrome as reductant for the reaction centre cytochrome c in *Chromatium* and *Rps. viridis*. In view of the findings of Kusai and Yamanaka (1973) that cytochrome c-555 is reduced by sulfide via flavocytochrome c-553 (Sect. 3.3.3) and also stimulates the reduction of cytochrome c-551 by thiosulfate-cytochrome c reductase in *Chl. limicola*, a role for cytochrome c-555 in the non-cyclic supply of electrons to the photosystem seems likely in this organism. However, *Pr. estuarii* does not contain flavocytochrome c-553 and does not utilise thiosulfate.

3.6.5 The Algal Cytochromes c and Plastocyanin

The mistaken equation of cytochrome f of higher plants and cytochrome c-553 of algae and cyanobacteria led to a confusion which was resolved by Wood (1977) who demonstrated that the latter group of organisms contained both a soluble cytochrome c-553 and a membrane-bound cytochrome clearly related to plant cytochrome f in properties.

The algal cytochromes c-553 (used as a generic name, not as an exact indicator of α-peak) are members of the class I c-type cytochromes (Chap. 1, Table 1.5) having low-spin absorption spectra with a red-shifted and usually asymmetric α-peak (552–555 nm), high γ-red/α-peak ratio (usually 6–7.2) and positive midpoint potential (340–390 mV) (Yakushiji 1971). These cytochromes are "small" as defined in Chapter 1, having 83–89 residues, and their amino acid sequences bear no close resemblance to any other group (Vol. 2, Chap. 3). The cytochromes c-553 are a sequence class which clearly span the eukaryotic/prokaryotic division. They have been isolated from the eukaryotic algae, dinoflagellates and diatoms (Perini et al. 1964; Mitsui and Tsushima 1968; Yakushiji 1971; Laycock and Craigie 1971; Lach et al. 1973; Mehard et al. 1975; Shimazaki et al. 1978; Bohme et al. 1980a; Sugimura et al. 1981; Sugimura et al. 1984; Hochman et al. 1985) and from all groups of the cyanobacteria (Susor and Krogman 1966; Biggins 1967; Holton and Myers 1967a,b; Aitken 1976; Yamanaka et al. 1978; Stewart and Bendall 1980).

Plastocyanin is a small soluble blue copper protein isolated from chloroplasts of higher plants and some algae. It was found that cytochrome c-553 and plastocyanin were functionally interchangeable as donors to P^+ in membrane particles (see below) and that they were naturally interchangeable depending on growth conditions in several algae and cyanobacteria (Chap. 4.1.3). Because of this functional equivalence, studies on the mode of action of plastocyanin are relevant to understanding cytochrome c function and are included in the following discussion.

Kinetics of Electron Transfer: (1) in Algae and Chloroplasts. The work of Duysens (1955) established that cytochrome f of plants and cytochrome f + c-553 of algae operated between the two photosystems, and the same general functional position was demonstrated for plastocyanin (Malkin and Bearden 1973; Visser et al. 1974). However, the spectroscopic changes which can be followed after flash photo-oxidation of P have been the subject of widely different interpretations. It is now clear that this was a consequence of several experimental problems. Firstly, in algae, both cytochrome f and cytochrome c-553 contribute to absorbance changes in the cytochrome c region. Secondly, plastocyanin cannot be directly observed due to its broad spectrum and its contribution must be inferred by indirect means. Thirdly, the complexity of the intact photosynthetic apparatus leads to overlap of the spectroscopic contributions of different species. Previous claims that cytochrome f and plastocyanin act in parallel rather than in series as donors to P700 (e.g. Haehnel 1977;

Bouges-Bocquet 1977) were in contradiction to the results of in vitro reconstitution studies (see below) and have now been retracted (Haehnel et al. 1980). The reduction of P700 in spinach chloroplasts, eukaryotic algae and cyanobacteria occurs in two phases with $t_{1/2}$ 20 μs and 200 μs. In spinach chloroplasts the fast phase is assigned to plastocyanin on the basis of its sensitivity to cyanide (which removes Cu from plastocyanin). This half-time is unaffected by changes in the concentration of plastocyanin (perturbed by osmotic shrinkage or swelling) or changes in the ionic strength, and is therefore considered to reflect a prior long-lived complex with P700. In organisms in which plastocyanin is replaced by cytochrome c-553 the fast phase of P700 reduction corresponds to the kinetics of oxidation of this cytochrome.

In contrast the kinetic characteristics of the second phase are consistent with a second-order reaction and the oxidation of a pool of plastocyanin. The half-time is approximately equal to that observed for cytochrome f oxidation, which occurs after a lag of ~50 μs, consistent with the presence of an intervening electron carrier (Bouges-Bocquet 1977).

Thus plastocyanin is the primary donor to P700. A portion may be associated with the reaction centre and oxidised very rapidly while the remainder exhibits the properties of a mobile pool which transfers electrons from cytochrome f. In algae and cyanobacteria the plastocyanin may be replaced by cytochrome c-553 and the model then closely resembles that proposed for the Rhodospirillaceae and involving cytochrome c_2 (Fig. 3.5; Fig. 3.20).

It has been a point of emphasis in this chapter that the sidedness of reactions in the electron transport chain are an important consideration in chemiosmotic models of oxidative phosphorylation. It is accepted that photoreactions drive electrons from inside to outside the thylakoid membrane (Witt 1979). This requires an internal location for P700 and, by implication, for its primary donor (Fig. 3.5). The following investigations on osmotically shocked chloroplasts which have lost their outer envelope and their stromal contents, yet retain their thylakoid structure, tend to support this conclusion for plastocyanin and cytochrome c-553.

Thus antisera against plastocyanin did not agglutinate spinach chloroplasts nor block electron transport (Hauska et al. 1971; Haehnel et al. 1981) and parallel experiments found the same for *Euglena* cytochrome c-552 (Wildner and Hauska 1974b). As with mitochondria, sonication leads to inversion of the membrane and results in loss of plastocyanin (Hauska et al. 1971) or cytochrome c-552 (Wildner and Hauska 1974). Haehnel et al. (1981) isolated inverted and right-side-out vesicles and showed that only the former allowed good access to P by added plastocyanin. The rate of electron transport in chloroplasts using ascorbate/diaminodurol as e^- donor to plastocyanin and methyl viologen as acceptor from photosystem 1 is dependent on the osmotic strength of the supporting medium (Lockau 1979). The rate enhancement by hyperosmolarity can also be achieved by permeant ions and implies that plastocyanin is located behind an osmotic barrier and its interaction with P is influenced by ionic conditions.

Contradictory results were obtained by Schmidt et al. (1975) who found that anti-plastocyanin antibodies did inhibit electron transport in chloroplasts and Smith et al. (1977) who found that although plastocyanin was less accessible to hydrophilic labelling reagents in intact chloroplasts than in sonicated preparations, the degree of protection was not sufficient to justify location behind a lipid barrier. The latter authors concluded that plastocyanin may reside in a hydrophilic cleft on the outer surface. These anomalies have not yet been explained but may be a consequence of the extreme susceptibility of the thylakoids to membrane disruption. The present evidence favours an inner location.

Kinetics of Electron Transfer: (2) in Membrane Fragments. The reaction between plastocyanin or cytochrome c-553 and P^+ has been extensively studied in vitro using thylakoid fragments dispersed by detergent or sonication, and partially resolved or highly purified photosystem I particles. The pattern of reactivity with a donor from different sources appears to vary with the degree of resolution of the particle used and the activity decreases with purification to reach zero in the reaction centre preparation of Shiozawa et al. (1974).

With thylakoid fragments a specificity of reaction appropriate to the membrane of origin is observed. Using particles dispersed with digitonin, Wood and Bendall (1975) showed that spinach plastocyanin was by far the best donor to spinach P700 ($k = 8 \times 10^7\ M^{-1}\ s^{-1}$ at $I = 0.1$ M, pH 7) of those tested (including algal cytochrome c-553, cytochrome f and horse cytochrome c). Similarly, Tsuji and Fujita (1972) found that *Porphyra* cytochrome c-553 was only 20% as good as spinach plastocyanin in the reduction of spinach P^+ while *Anabaena* plastocyanin was unreactive. Conversely, *Anabaena* plastocyanin and cytochrome c-553 were equally effective as donors to *Anabaena* P^+ but spinach plastocyanin (2%) and *Porphyra* cytochrome c-553 (0.5%) were poorly reactive.

In contrast, the oxidation of basic donors such as *Anabaena* plastocyanin and cytochrome c-553 by highly resolved photosystem I particles from spinach chloroplasts was faster than that of acidic electron donors which included spinach plastocyanin itself (Davis et al. 1980). There was a strong correlation between K_m and isoelectric point and in the case of acidic donors the K_m was lowered by added Mg^{2+} while for basic donors it was raised. These authors proposed that the eukaryotic photosystem I may have a donor specificity that resides on an easily removable subunit that is lost in highly purified particles. They suggested that this component may be subunit III in the preparation of Bengis and Nelson (1977).

One possibility is that in the absence of subunit III, Mg^{2+} ions allow interaction of an acidic binding site on the donor molecule and a similarly charged site on the reaction centre by a bridging or charge-shielding process. A positively charged donor would not require Mg^{2+} and indeed could be inhibited by it. An alternative is that Mg^{2+} causes a conformational change in the reaction centre which facilitates electron transfer (Gross 1979).

Poplar plastocyanin does contain a ring of acidic residues close to the copper (Colman et al. 1978) and the acidic cytochromes c-553 may share this surface feature. Indeed, Wood (1978b) suggested that sequence similarity may exist between the two proteins and noted a negatively charged region of seven residues with conserved features. This region (residues 57–63) forms part of the ring of negative charge in the plastocyanin structure and it is interesting that of the nine negative residues that form this ring in spinach plastocyanin only one persists in *Anabaena* plastocyanin. Davis et al. (1980) proposed that plastocyanin (and cytochrome c-553) evolved from a basic to an acidic protein and that this evolution was accompanied by the appearance of a subunit on the reaction centre which facilitated the interaction of two negatively charged species.

3.7 Sulfate Respiration and the Production of Sulfide

3.7.1 The Biology of the Sulfate-Reducing Bacteria

The final mode of bacterial respiration to consider involves the reduction of sulfate. This possibly represents the most divergent respiratory mode in bacteria and there are few features in common with other types of respiration. Sulfate reduction is thus placed alone in Figs. 3.1 and 3.2. Two of the most striking differences from other respirations are the negative redox potentials involved and the presence of cytochrome c_3, a low potential cytochrome with no structural affinity to the other bacterial cytochrome c classes. Bacteria able to reduce sulfate to sulfide were originally grouped in two genera, *Desulfovibrio* and *Desulfotomaculum*. The former contained c-type cytochromes and the pigment desulfoviridin later shown to be associated with sulfite reductase. The latter lacked these redox components and was spore-forming. Recently identified genera include *Desulfobacter, Desulfosarcina, Desulfonema, Desulfobulbus, Desulfococcus* and *Thermodesulfobacter* but little is yet known of their respiration and redox proteins (Odom and Peck 1984).

The *Desulfovibrio* are ecologically versatile. In the absence of sulfate they can grow by fermentation on lactate, pyruvate or ethanol in syntrophic association with a methanogen to utilise the hydrogen released. In this growth mode the *Desulfovibrio* resemble the *Clostridia* in their utilisation of acetyl phosphate for substrate level phosphorylation (the phosphoroclastic reaction) and they have been classified as primitive on this basis. Lactate and pyruvate can also be used as respiratory substrates in the presence of sulfate, probably with the release of hydrogen as an intermediate reaction (see below). The *Desulfotomaculum* can also oxidise lactate with sulfate but here the process is not energy conserving and sulfate is probably acting as an accessory oxidant, like nitrate in *Clostridia*. Finally, *Desulfovibrio*, but not *Desulfotomaculum*, can grow chemolithotrophically in the presence of hydrogen and sulfate. Under natural conditions the hydrogen may be supplied by syntrophic association with a fermentative bacterium.

The *Desulfuromonas* cannot reduce sulfate but obligately require sulfur as respiratory electron acceptor using acetate as an oxidisable substrate. *Desulfuromonas acetoxidans* can live in syntrophic association with green sulfur bacteria which oxidise sulfide to sulfur in non-cyclic photosynthetic electron transfer (Sect. 3.3.3). Some of the *Desulfovibrio* can also oxidise acetate and reduce sulfur.

3.7.2 Electron Transport and the Roles of Cytochromes c

The Pathway of Sulfate Respiration. The pathway of sulfate reduction in *Desulfovibrio* is shown in Table 3.16. Adenosine phosphosulfate (APS) and SO_3^{2-} are free intermediates but there is controversy as to whether sulfite is reduced directly to sulfide ($6e^-$) or whether $3\times2e^-$ steps are involved requiring three separate enzymes. Sulfite reductase contains siroheme and is identified as desulfoviridin or desulforubidin depending on visible spectra. This enzyme will produce either sulfide or trithionate depending on the exact conditions. Thiosulfate reductase activity can be demonstrated but neither thiosulfate nor trithionate can be exchanged with external pools (Chambers and

Table 3.16. The pathway of sulfate reduction in *Desulfovibrio*

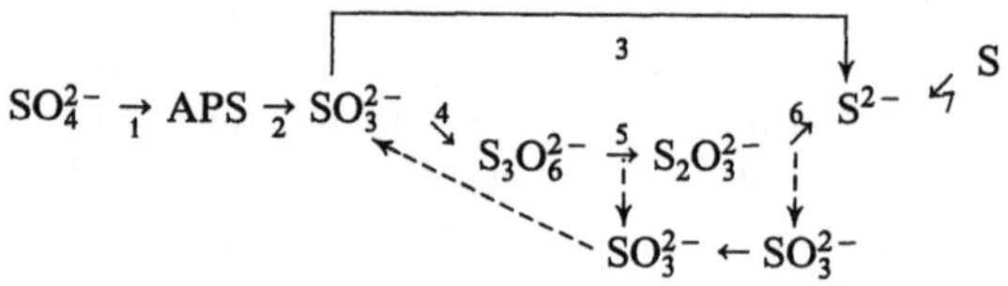

	$E_{m,7}$ [d]
1. $SO_4^{2-}+ATP\rightarrow APS+PP_i$	
2. $APS+2e^-\rightarrow SO_3^{2-}+AMP$	− 60[a]
3. $SO_3^{2-}+6e^-+6H^+\rightarrow S^{2-}+3H_2O$ [b]	−116
4. $3SO_3^{2-}+2e^-+6H^+\rightarrow S_3O_6^{2-}+3H_2O$	−173 [b]
5. $S_3O_6^{2-}+2e^-\rightarrow S_2O_3^{2-}+SO_3^{2-}$	+225 [b]
6. $S_2O_3^{2-}+2e^-\rightarrow S^{2-}+SO_3^{2-}$	−402 [b]
7. $S+2e^-\rightarrow S^{2-}$ [c]	−270

[a] This figure should be compared with the reaction $SO_4^{2-}+2e^-+2H^+\rightarrow SO_3^{2-}+H_2O$ which has an $E_{m,7}$ value of −516 mV (Thauer and Badziong 1980).
[b] There is controversy as to whether SO_3^{2-} is reduced directly to S^{2-} (Eq. 3) or by $3\times2e^-$ transfer steps (Eqs. 4−6).
[c] Some of the sulfate-reducing bacteria can reduce sulfur to sulfide (Biebl and Pfennig 1977) while in *Desulfuromonas acetoxidans* this is an obligate respiration.
[d] Values of $E_{m,7}$ are from Thauer et al. (1977).

Trudinger 1975). Thus there may be a branched production of sulfide under appropriate growth conditions (Thauer and Badziong 1980).

These features of the reduction of sulfate are shown in Table 3.16. The recycling sulfite pool indicated by the broken lines has been demonstrated using ^{35}S-thiosulfate (Findlay and Akagi 1970).

Electron Transport. There has been considerable confusion over the electron transport processes of sulfate respiration. This confusion arose from the following. Firstly, like the denitrifiers, there may be a multiplicity of terminal enzymes and donor reactions. Secondly, most assignments of the physiological role of a redox protein have been on the basis of in vitro stimulation of a reaction but such stimulation is not necessarily due to direct or physiological participation in a process. For example, cytochrome c_3 is an efficient oxygen scavenger which may stimulate oxygen-sensitive processes non-specifically (Le Gall and Postgate 1973). Thirdly, the compartmentation of redox processes between the periplasm and the cytoplasm has only recently been realised. This makes many proposed schemes redundant because components (such as cytochrome c_3 and ferredoxin), which react in vitro, are situated on opposite sides of the cell membrane.

In the hydrogen cycling scheme of Odom and Peck (1981 a), which we favour, the membrane separates a cytoplasmic hydrogen evolution system from a periplasmic, hydrogen-utilising system (Fig. 3.21). In the former, the oxida-

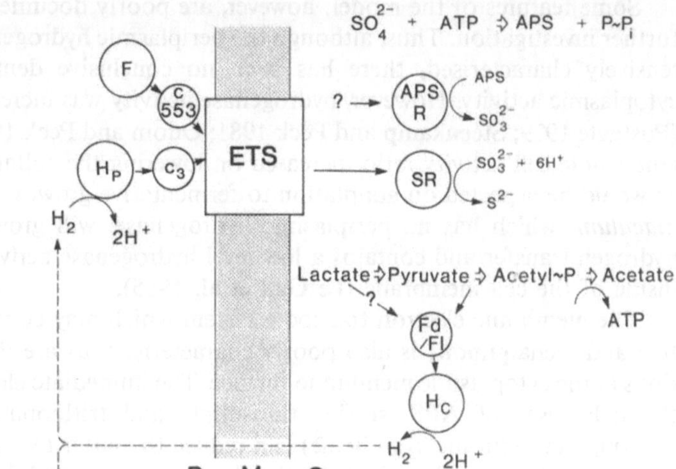

Fig. 3.21. Electron transport and energy conservation in *Desulfovibrio*. → Electron transfer; ⇒ metabolic conversion; - - -→ diffusion; ? the acceptor of lactate dehydrogenase and the donor to APS reductase are not known. *F* Formate hydrogenase; *Hp, Hc* periplasmic and cytoplasmic hydrogenase; *ETS* electron transport system; *APS-R* APS reductase; *SR* sulfite reductase; *Fd* ferredoxin; *Fl* flavodoxin. The diagram is constructed on the basis of the hydrogen-cycling scheme of Odom and Peck (1981 a) and the chemiosmotic model of Wood (1978 c)

tion of pyruvate and lactate is coupled to hydrogen production via ferredoxin (or flavodoxin) and the hydrogen can diffuse across the cell membrane. In the latter the hydrogen is oxidised via a periplasmic hydrogenase and cytochrome c_3 (4 heme). Electrons pass across the membrane and finally reduce cytoplasmic sulfite. The scheme elegantly explains the versatile life style of *Desulfovibrio*. The organism carries out a kind of internal hydrogen transfer on the same principle as that seen between a fermentative bacterium and a hydrogen-respiring bacterium. But, in the absence of sulfate as electron acceptor, the organism can revert to a purely fermentative mode while, if hydrogen and sulfate are supplied, it can operate in a purely respiratory mode.

Several arguments and experimental observations favour this scheme. Thus hydrogenase, formate dehydrogenase and cytochrome c_3 (4 heme) are released from spheroplasts while APS reductase, ferredoxin and sulfite reductase are cytoplasmic (Bell et al. 1974; van der Westen et al. 1978; Badziong and Thauer 1980; Odom and Peck 1981b). Lactate could not be oxidised by sulfate in spheroplasts but reconstitution was achieved with hydrogenase and cytochrome c_3 (4 heme) (Odom and Peck 1981a). Early stages of growth were associated with a burst of hydrogen production (Tsuji and Yagi 1980) and the subsequent disappearance of the hydrogen was accompanied by sulfide production. Finally, cytochrome c_3 (4 heme) is specifically reduced by hydrogenase (Yagi et al. 1968; Yagi 1970; Yagi and Maruyama 1971; Bell et al. 1978).

Some features of the model, however, are poorly documented and require further investigation. Thus, although the periplasmic hydrogenase has been extensively characterised, there has been no conclusive demonstration of a cytoplasmic activity. However, hydrogenase activity was increased on cell lysis (Postgate 1979; Steenkamp and Peck 1981; Odom and Peck 1981a) and the latent/whole cell activity ratio increased on lowering the sulfate concentration, as would be expected on adaptation to fermentative growth. Also, *Desulfotomaculum*, which has no periplasmic hydrogenase, will grow by interspecies hydrogen transfer and contains a low level hydrogenase activity bound to the inside of the cell membrane (Le Gall et al. 1965).

The membrane electron transport system which may contain cytochromes b, c and menaquinone is also poorly characterised, as are the terminal reactions at the cytoplasmic membrane surface. The immediate electron donors for the reductases of APS, sulfite, thiosulfate and trithionate are unknown. Although cytochrome c_3 (4 heme) can reduce the latter two enzymes in vitro, this cannot be a physiological reaction because of the periplasmic location of the cytochrome. A role for ferredoxin is a possibility, however, because that from *D. gigas* can exist as a trimer (FdI) with E_m −440 mV, which is active in H_2 production from pyruvate, and also as a tetramer (FdII) with E_m −130 mV, which stimulates the oxidation of hydrogen by sulfite (Bruschi et al. 1976; Moura et al. 1977; Huynh et al. 1980). Thus ferredoxin may play a dual role determined by its state of aggregation. Such a dual role is also accessible to flavodoxin which can undergo two single electron reductions with E_m values of −440 mV and −150 mV (Dubourdieu et al. 1975). A value for E_m around −150 mV would be appropriate for the $6e^-$ reduction of sulfite to sulfide (E_m −116 mV) but inappropriate for three $2e^-$ reductions (E_m −173, +225, −402 mV) (Table 3.16). In the second step a large amount of electrical potential energy would be unavailable for chemiosmotic energy conservation while the third step would be highly unfavourable.

Obligate sulfur reduction is a characteristic of *Desulfuromonas* which contains a triheme relative of cytochrome c_3 [c_3 (3 heme)], but the ability to grow on sulfur is also found in *D. gigas* and *D. desulfuricans* (Norway). Both cytochrome c_3 (4 heme) and cytochrome c_3 (3 heme) may be able to reduce sulfur directly and indeed the sulfur reductase activities of the different cytochromes c_3 correlate well with the ability of the organisms to grow on sulfur (Fauque et al. 1979a).

Properties of the Cytochromes c. The class of low potential multi-heme cytochromes c (class III, Chap. 1, Table 1.4) is divided into three groups, all of which are associated with the sulfidogenic bacteria. The *Desulfovibrio* all contain a tetraheme c_3 (M_r 14 K) and most contain an octaheme c_3 (M_r 26 K) (Table 3.17). A triheme c_3 (M_r 9 K) is found in *Desulfuromonas acetoxidans.*

Table 3.17. Distribution of electron transport proteins in *Desulfovibrio* and related bacteria

	Respiration with		c_3 (3 heme)	c_3 (4 heme)	c_3 (8 heme)	c-553	Reference
	SO_{2-4}	S					
Desulfovibrio							
gigas	+	+	−	+	+	−	a, g
vulgaris (H)	+	−	−	+	+	+	a, g
vulgaris (M)	+	ND	−	+	−	+	a, b
desulfuricans (EA)	+	ND	−	+	+	+	a, c
desulfuricans (N)	+	+	−	+	+	+	a, c, g
salexigens	+	ND	−	+	ND	ND	a
africanus	+	ND	−	+	ND	ND	a
Desulfotomaculum acetoxidans	+	ND	−	−	−	−	d
Desulfuromonas acetoxidans	−	+	+	−	−	−	e, f

Biebl and Pfennig (1977) tested for the ability to grow on sulfur. Those bacteria studied fell into two groups, the first producing ten times more sulfide than the second. In the Table, members of the first group are indicated by + and those of the second by − for respiration with S.

References:

[a]LeGall and Forget (1978); [b]Yagi (1979); [c]Guerlesquin et al. (1982); [d]Campbell and Postgate (1965); [e]Pfennig and Biebl (1976); [f]Probst et al. (1977); [g]Biebl and Pfennig (1977).

Cytochrome c-553 is a class I c-type (Chap. 1, Table 1.5) unrelated to the c_3 group in properties and structure.

Cytochrome c_3 (4 Heme). This cytochrome is uniformly present in the *Desulfovibrio* and is also found in *Thermodesulfobacter commune* (Hatchikian et al. 1983). It is characterised by size and tetraheme nature (Meyer et al. 1971; Yagi and Maruyama 1971). The optical absorption spectrum is also distinctive with a high γ_{red}/γ_{ox} ratio (1.5), a blue-shifted δ-peak (350 nm) and a shoulder on the short wavelength side of the γ-peak (Meyer and Kamen 1982; Chap. 1, Table 1.6). Although these properties define the group, the members show considerable variation in amino acid sequence (only 30–45% pairwise identity; Vol. 2, Chap. 3), isoelectric point (5.2–10.6) and antigenic cross-reactivity (Drucker and Campbell 1969; Singleton et al. 1982).

Cytochrome c_3 (4 heme) accepts electrons from the periplasmic hydrogenase and probably donates electrons to the membrane electron transport system. As noted above, cytochrome c_3 (4 heme) may also act as a sulfur reductase. A molecular description of the function of cytochrome c_3 (4 heme) requires consideration of the kinetic and thermodynamic relationships of the four heme groups. This has proved to be a difficult task and progress is discussed briefly here (Fig. 3.22) and more fully in the chapters on redox potential and electron transfer in Volume 2. Although the analysis is complicated by the need to distinguish macroscopic and microscopic potentials (Vol. 2, Chap. 7) and by the heme-heme interactions that are probably present, it is clear that the four hemes can adopt distinct midpoint potentials. The broadest range is seen for *D. desulfuricans* (Norway) with $E_{m,7.6}$ of −165, −305, −365 and −400 mV (Bianco and Haladjian 1981). NMR analysis of solutions at different degrees of reduction demonstrates that there are both positive and negative heme-heme interactions of up to 60 mV, that at least some of the hemes have pH-dependent properties and that the intramolecular

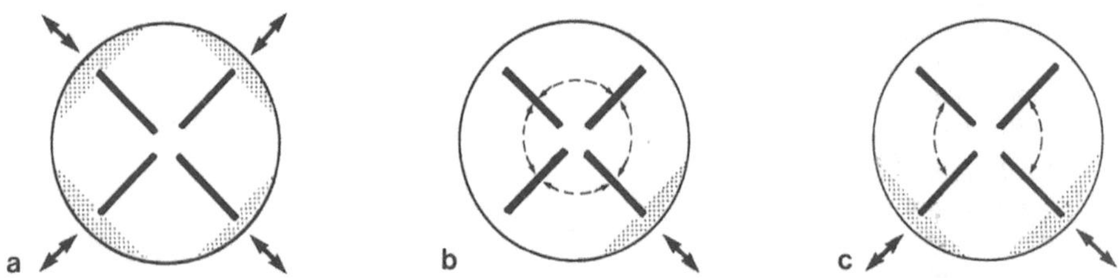

Fig. 3.22 a–c. Possible models for electron transfer in cytochrome c_3 (4 heme). In each model the central *solid bars* represent the 4 heme groups and the *shaded arcs* represent binding regions for a redox partner incorporating a site of electron transfer. - - -→ Intramolecular electron transfer; ↔ intermolecular electron transfer. In model **a** there are four sites of intermolecular electron transfer and no requirement for intramolecular equilibration. In **b** a single intermolecular electron transfer site allows electron entry and exit for four hemes which equilibrate internally. In **c** two surface sites allow electron transfer for two pairs of hemes

electron transfer rate is moderately fast, at least between some of the heme groups (Moura et al. 1982; Santos et al. 1984). All of these points may have physiological significance. Thus the cytochrome may allow cooperative delivery or uptake of more than one electron and may be responsive to pH changes across the bioenergetic membrane. However, these are speculative comments and there is no clear physiological rationale for a requirement for a multi-heme cytochrome in this particular electron transport system. Further comment must await investigation of the nature of the interaction of cytochrome c_3 with its redox partners (Yagi 1984).

Cytochrome c_3 (8 Heme). Cytochrome c_3 (8 heme) may be as generally distributed as its smaller relative (Table 3.17) but has received less attention, perhaps due to its membrane association (Bruschi et al. 1969; Hatchikian et al. 1972). Its octaheme nature, size and spectra (Bruschi et al. 1969; Guerlesquin et al. 1982) suggest a simple dimeric relationship with cytochrome c_3 and indeed the apocytochrome was 13 K on SDS gel electrophoresis (Guerlesquin et al. 1982). However, it is distinct from c_3 (4 heme) in amino acid composition and EPR spectra (Le Gall et al. 1971; Le Gall and Forget 1978).

The role of cytochrome c_3 (8 heme) is uncertain. The *D. gigas* cytochrome can stimulate thiosulfate reduction in a crude system (Hatchikian et al. 1972) and will efficiently mediate electron transfer between purified hydrogenase and thiosulfate reductase (Bruschi et al. 1977), whereas the c_3 (4 heme) is almost inactive. However, although the location of cytochrome c_3 (8 heme) is not known, it would be expected to be periplasmic or periplasmic-facing on the basis of Wood's general hypothesis (Wood 1983) and the reaction with thiosulfate reductase may not be physiological.

Cytochrome c_3 (3 Heme). Cytochrome c_3 (3 heme) was originally named c_7 but its obvious structural similarity to the cytochromes c_3 (4 heme) (Ambler 1971) makes this an unjustifiable distinction. It can be isolated either from the free-living *Desulfuromonas acetoxidans* (Probst et al. 1977) or from the *Chloropseudomonas ethylica* consortium (Olson and Shaw 1969; Meyer et al. 1971; Shioi et al. 1972). The cytochrome contains only three hemes (Meyer et al. 1971) and the heme content and size suggest that it is related to cytochrome c_3 (4 heme) by loss of about 30% of the molecule containing one heme group. Two of the heme groups have a midpoint potential of -177 mV and the third of -102 mV (Fiechtner and Kassner 1979).

c_3-Like Cytochromes. *Rps. sphaeroides* and *palustris* and *Anacystis nidulans* have cytochromes c which resemble c_3 (4 heme) in some properties (Meyer et al. 1971; Holton and Myers 1967a,b). Like c_3 (4 heme) they have negative E_m values (-150 to -254 mV) and a high γ_{red}/γ_{ox} ratio (1.5–1.6). However, they contain only 2 mol heme in proteins of M_r 20–24 K.

Cytochrome c-553. Cytochrome c-553 is present in *D. vulgaris* (Hildenborough and Miyazaki) and *D. desulfuricans* (Norway) (Table 3.17). The cytochrome is monoheme with M_r 9 K and its amino acid sequence has the features of a class I c-type cytochrome (Nakano et al. 1983; Bruschi and Le Gall 1972). The presence of a 695-nm band in the oxidised spectrum is indicative of a His-Fe-Met chromophore and this is confirmed by NMR spectroscopy (Vol. 2, Chap. 2). The midpoint potentials of the cytochrome c-553 from *D. vulgaris* (Hildenborough) and *D. desulfuricans* (Norway) are between 0 and 100 mV (Bertrand et al. 1982; Fauque et al. 1979b) and are consistent with their partial reducibility with ascorbate. However, the c-553 from the Miyazaki strain of *D. vulgaris* was found to have a midpoint potential of −260 mV (Yagi 1979). In view of the fact that it was also ascorbate reducible this requires re-examination.

The cytochrome c-553 from *D. vulgaris* (Miyazaki) was reactive with formate dehydrogenase but was not reduced by hydrogenase (Yagi 1979), while the cytochrome c_3 (4 heme) had the converse pattern of activity. This is consistent with a periplasmic role in the oxidation of formate (Fig. 3.21). However, c-553 is not as widespread as formate dehydrogenase and in *D. gigas*, the cytochrome c_3 (13 K) can act as acceptor (Riederer-Henderson and Peck 1970).

3.7.3 Energy Conservation

As noted above, the oxidation of lactate and pyruvate with release of hydrogen resembles the Clostridial life-style and ATP is generated by substrate-level phosphorylation from acetyl phosphate. Sulfate respiration, whether with lactate or hydrogen as electron donor, is complicated by the requirement for ATP in the sulfate-priming process. Since the product of this is AMP, the subsequent energy conservation must yield more than 2 ATP for growth to be possible on hydrogen. Growth yield calculations suggest 1 ATP/($SO_4^{2-} \rightarrow S^{2-}$) and 3 ATP/[$S_2O_3^{2-} \rightarrow S^{2-}$ (via SO_3^{2-})] with hydrogen as a substrate (Badziong and Thauer 1978). Thus the reduction of SO_3^{2-} to S^{2-} appears to yield 3 ATP, two of which are used in the sulfate-priming reaction.

We have noted the uncertainty in the pathway of sulfite reduction. Energy conservation may therefore follow one of two patterns. If sulfite is reduced directly to sulfide, an average E_m value of −116 mV applies (Table 3.16) and a ΔE_m between hydrogen and sulfite of 304 mV is obtained. This may permit of one ATP synthesis per electron pair by analogy to similar energy spans in the mitochondrion. On the other hand, if sulfite is reduced via a three-step process, not all of the three steps may be involved in energy conservation. Indeed, thiosulfate reduction by hydrogen can release very little if any free energy (Table 3.16), while trithionate reduction has a ΔE_m of 645 mV. Thus energy conservation may be concentrated in the $2e^-$ transfer to trithionate (2 ATP) and the $2e^-$ transfer to sulfite (1 ATP).

The oxidation of hydrogen by sulfite is a transmembrane process [vectorial electron transfer in the terminology of Kroger (1980)] which gives rise to a proton gradient simply as a result of the location of the terminal reactions. Such a gradient was demonstrated with a value of $2H^+/2e^-$ using sulfite pulses (Kobayashi et al. 1982).

Perhaps the most unlikely respiration in thermodynamic terms is that of *Desulfuromonas acetoxidans* which survives by oxidising acetate to CO_2 (E_m -290 mV) using sulfur as electron acceptor (E_m -270 mV) (Pfennig and Biebl 1976). Not only is the energy span very small but sulfur reduction is expected to be periplasmic both on the basis of the c-type cytochrome location and due to the impermeability of the membrane to colloidal sulfur. Thus a vectorial electron transfer mechanism could not give rise to a proton gradient of the correct sign. No proposals have been made which overcome this anomaly.

References

Aitken A (1976) Protein evolution in the cyanobacteria. Nature (London) 263:793–796

Akey CW, Moffat K, Wharton DC, Edelstein SJ (1980) Characterisation of crystals of a cytochrome oxidase from *Pseudomonas aeruginosa* by X-ray diffraction and electron microscopy. J Mol Biol 136:19–43

Akimenko VK, Trutko SM (1984) On the absence of correlation between CN^- resistant respiration and cytochrome d content in bacteria. Arch Microbiol 138:58–63

Aleem MIH (1968) Mechanism of oxidative phosphorylation in the chemiautotroph – *Nitrobacter agilis*. Biochim Biophys Acta 162:338–347

Aleem MIH (1977) Coupling of energy with electron transfer reactions in chemolithotrophic bacteria. Symp Soc Gen Microbiol 27:351–381

Alefounder PR, Ferguson SJ (1980) The location of the dissimilatory nitrite reductase and the control of dissimilatory nitrate reduction in *Paracoccus denitrificans*. Biochem J 192:231–240

Alefounder PR, Ferguson SJ (1981) A periplasmic location for methanol dehydrogenase from *Paracoccus denitrificans* – implications for proton pumping by cytochrome aa_3. Biochem Biophys Res Commun 98:778–784

Ambler RP (1971) The amino acid sequence of cytochrome c-551.5 (cytochrome c_7) from the green photosynthetic bacterium *Chloropseudomonas ethylica*. FEBS Lett 18:351–353

Ambler RP (1977) Structurally similar cytochromes from diverse bacterial electron transport chains. In: Senez J, Le Gall J, Peck H (eds) CNRS Int Symp Electron Transport in Microorganisms, Marseilles

Ambler RP, Taylor E (1973) Amino acid sequence of cytochrome c_5 from *Pseudomonas mendocina*. Biochem Soc Trans 1:166–168

Ambler RP, Daniel M, Hermoso J, Meyer TE, Bartsch RG, Kamen MD (1979) Cytochrome c_2 sequence variation among the recognised species of purple non-sulfur photosynthetic bacteria. Nature (London) 278:659–660

Ambler RP, Bartsch RG, Daniel M, Kamen MD, McClellan L, Meyer TE, Beeumen J van (1981) Amino acid sequences of bacterial cytochrome c′ and c-556. Proc Natl Acad Sci USA 78:6854–6857

Amminuddin M, Nicholas DJD (1973) Electron transfer during sulfide and sulfite oxidation in *Thiobacillus denitrificans*. J Gen Microbiol 82:115–123

Anderson JR, Strumeyer DH, Pramer D (1968) Purification and properties of peroxidase from *Nitrosomonas europaea*. J Bacteriol 96:93–97

Anthony C (1975) The microbial metabolism of c-1 compounds – the cytochromes of *Pseudomonas* AMl. Biochem J 146:289–298

Antonini E, Finazzi-Agro A, Avigliano L, Guerrieri P, Rotilio G, Mondovi B (1970) Kinetics of electron transfer between azurin and cytochrome c-551 from *Pseudomonas*. J Biol Chem. 245:4847–4856

Araiso T, Ronnberg M, Dunford HB, Ellfolk N (1980) The formation of the primary compound from H_2O_2 and *Pseudomonas* cytochrome c peroxidase. FEBS Lett 118:99–102

Baccarini-Melandri A, Zannoni D, Melandri BA (1973) Respiratory sites and energy conservation in membranes of dark grown cells of *Rhodopseudomonas capsulata*. Biochem Biophys Acta 314:298–311

Baccarini-Melandri A, Jones OTG, Hauska G (1978) Cytochrome c_2, an electron carrier shared by the respiratory and photosynthetic electron transport chain of *Rhodopseudomonas capsulata*. FEBS Lett 86:151–153

Badziong W, Thauer RK (1978) Growth yields and growth rates of *Desulfovibrio vulgaris* (Marburg) growing on hydrogen plus sulfate and hydrogen plus thiosulfate as the sole energy sources. Arch Microbiol 117:209–214

Badziong W, Thauer RK (1980) Vectorial electron transport in *Desulfovibrio vulgaris* (Marburg) growing on hydrogen plus sulfate as sole energy source. Arch Microbiol 125:167–174

Baines BS, Hubbard JAM, Poole RK (1984) Purification and characterisation of two cytochrome oxidases (caa_3 and o) from the thermophilic bacterium PS3. Biochem Biophys Acta 766:438–445

Bamforth CW, Quayle JR (1978) Aerobic and anaerobic growth of *Paracoccus denitrificans* on methanol. Arch Microbiol 119:91–97

Barber D, Parr SR, Greenwood C (1976) Some spectral and steady state kinetic properties of *Pseudomonas* cytochrome oxidase. Biochem J 157:431–438

Barber D, Parr SR, Greenwood C (1977) The reduction of *Pseudomonas* cytochrome c-551 oxidase by chromous ions. Biochem J 163:629–632

Barber D, Parr SR, Greenwood C (1978) The reactions of *Pseudomonas* cytochrome c-551 oxidase with potassium cyanide. Biochem J 175:239–249

Barrett J (1956) The prosthetic group of cytochrome a_2. Biochem J 64:626–639

Bartsch RG (1978) Cytochromes. In: Clayton RK, Sistrom WR (eds) The photosynthetic bacteria. Plenum, New York, pp 249–279

Bartsch RG, Meyer TE, Robinson AB (1968) Complex c-type cytochromes with bound flavin. In: Okunuki K, Kamen MD, Sekuzu I (eds) The structure and function of cytochromes. Univ Park Press, Baltimore, pp 443–451

Beardmore-Gray M, O'Keefe DT, Anthony C (1982) The auto-reducible cytochromes c of the methylotrophs *Methylophilus methylotrophus* and *Pseudomonas* AMl. Biochem J 207:161–165

Bell GR, LeGall J, Peck HD (1974) Evidence for the periplasmic location of hydrogenase in *Desulfovibrio gigas*. J Bacteriol 120:994–997

Bell GR, Lee JP, Peck HD, LeGall J (1978) Reactivity of *Desulfovibrio gigas* hydrogenase towards artifical and natural electron donors or acceptors. Biochemie 60:315–320

Bendall DS, Davenport HE, Hill R (1971) Cytochrome components in chloroplasts of the higher plants. Meth Enzymol 23:327–344

Bengis C, Nelson N (1977) Subunit structure of chloroplast photosystem I Reaction Centre. J Biol Chem 252:4564–4569

Bennoun P (1982) Evidence for a respiratory chain in the chloroplast. Proc Natl Acad Sci USA 79:4352–4356

Berger H, Wharton DC (1980) Small angle X-ray scattering studies of oxidised and reduced cytochrome c oxidase from *Pseudomonas aeruginosa*. Biochim Biophys Acta 622:355–359

Berry EA, Trumpower BL (1985) Isolation of ubiquinol oxidase from *Paracoccus denitrificans* and resolution into cytochrome bc_1 and cytochrome caa_3 complexes. J Biol Chem 260:2458–2467

Bertrand P, Bruschi M, Denis M, Gayda JP, Manca F (1982) Cytochrome c-553 from *Desulfovibrio vulgaris*: potentiometric characterisation by optical and epr studies. Biochem Biophys Res Commun 106:756–760

Beeumen J van (1980) On the evolution of bacterial cytochromes c. In: Peeters H (ed) Protides of the biological fluids. Pergamon, New York, pp 69–74

Beeumen J van, Branden B van den, Tempst P, DeLey J (1980) Cytochromes c-556 from three genetic races of *Agrobacterium*. Eur J Biochem 107:475–483

Bianco P, Haladjian J (1981) Current potential responses for a tetrahemic protein – method for determining the individual half wave potentials of cytochrome c_3 from *Desulfovibrio desulfuricans* Norway. Electrochem Acta 26:1001–1004

Biebl H, Pfennig N (1977) Growth of sulfate reducing bacteria with sulfur as electron acceptor. Arch Microbiol 112:115–117

Biggins J (1967) Photosynthetic reactions by lysed protoplasts and particle preparations from the blue green alga *Phormidium luridum*. Plant Physiol 42:1447–1456

Blankenship RE, Parson WW (1978) The photochemical electron transfer reactions of photosynthetic bacteria and plants. Annu Rev Biochem 47:635–653

Blatt Y, Pecht I (1979) Allosteric cooperative interactions among redox sites of *Pseudomonas* cytochrome oxidase. Biochemistry 18:2917–2922

Bohme H, Brutsch S, Weithman G, Boger P (1980a) Isolation and characterisation of soluble c-553 and membrane-bound f-553 from thylakoids of the green alga *Scenedesmus acutus*. Biochim Biophys Acta 590:248–260

Bohme H, Pelzer B, Boger P (1980b) Purification and characterisation of cytochrome f-556.5 from the blue green alga *Spirulina platensis*. Biochim Biophys Acta 592:528–535

Bohner H, Boger P (1978) Reciprocal formation of a cytochrome c-553 and plastocyanin in *Scenedesmus*. FEBS Lett 85:337–340

Bohner H, Bohme H, Boger P (1980) Reciprocal formation of plastocyanin and cytochrome c-553 and the influence of cupric ions on photosynthetic electron transport. Biochim Biophys Acta 592:103–112

Boogerd FC, Versefeld HW van, Stouthamer AH (1981) Respiration-driven proton translocation with nitrite and nitrous oxide in *Paracoccus denitrificans*. Biochim Biophys Acta 638:181–191

Bosshard HR, Davidson MW, Knaff DB, Millet F (1986) Complex formation and electron transfer between mitochondrial cytochrome c and flavocytochrome c-552 of *Chromatium vinosum*. J Biol Chem 261:190–193

Bouges-Bouquet B (1977) Cytochrome f and plastocyanin kinetics in *Chlorella pyrenoidosa*. Biochim Biophys Acta 462:362–370

Bowyer JR, Crofts AR (1980) The photosynthetic electron transport chain of *Chromatium vinosum* chromatophores. Biochim Biophys Acta 591:298–311

Bowyer JR, Crofts AR (1981) On the mechanism of photosynthetic electron transfer in *Rhodopseudomonas capsulata* and *sphaeroides*. Biochim Biophys Acta 636:218–233

Bowyer JR, Tierney GV, Crofts AR (1979) Cytochrome c_2 – Reaction centre coupling in chromatophores of *Rhodopseudomonas sphaeroides* and *capsulata*. FEBS Lett 101:207–212

Bowyer JR, Meinhardt SV, Tierney GV, Crofts AR (1981) Resolved difference spectra of redox centres involved in photosynthetic electron flow in *Rhodopseudomonas capsulata* and *sphaeroides*. Biochim Biophys Acta 635:167–186

Bruschi M, LeGall J (1972) C-type cytochromes of *Desulfovibrio vulgaris* – the primary structure of cytochrome c-553. Biochim Biophys Acta 271:48–60

Bruschi M, LeGall J, Hatchikian CE, Dubourdieu M (1969) Cristallisation et propriétés d'un cytochrome intervenant dans la réduction du thiosulfate par *Desulfovibrio gigas*. Bull Soc Fr Physiol Veg 15:381–386

Bruschi M, Hatchikian EC, LeGall J, Moura JJG, Xavier AV (1976) Purification, characterisation and biological activity of three forms of ferredoxin from the sulfate reducing bacterium *D. gigas*. Biochim Biophys Acta 449:275–284

Bruschi M, Hatchikian CE, Golovleva LA, LeGall J (1977) Purification and characterisation of cytochrome c_3, ferredoxin and rubredoxin isolated from *D. desulfuricans* Norway. J Bacteriol 129:30–38

Campbell LL, Postgate JR (1965) Classification of the spore-forming sulfate reducing bacteria. Bacteriol Rev 29:359–363

Campbell WH, Orme-Johnson WH, Burris RH (1973) A comparison of the physical and chemical properties of four cytochromes c from *Azotobacter vinelandii*. Biochem J 135:617–630

Carver MA, Jones CW (1983) The terminal respiratory chain of the methylotrophic bacterium *Methylophilus methylotrophus*. FEBS Lett 155:187–191

Case GD, Leigh JS (1974) Intramembrane location of photosynthetic electron carriers in *Chromatium vinosum* revealed by epr interactions with Gadolinium. Proc 11th Rare Earth Res Conf, Traverse City, Michigan

Case GD, Parson WW (1971) Thermodynamics of the primary and secondary photochemical reactions in *Chromatium vinosum*. Biochim Biophys Acta 253:187–202

Case GD, Parson WW (1973) Shifts of bacteriochlorophyll and carotenoid absorption bands linked to cytochrome c-555 photooxidation in *Chromatium vinosum*. Biochim Biophys Acta 325:441–453

Case GD, Parson WW, Thornber JP (1970) Photooxidation of cytochromes in reaction centre preparations from *Chromatium vinosum* and *Rhodopseudomonas viridis*. Biochim Biophys Acta 223:122–128

Castor LN, Chance B (1955) Photochemical action spectra of carbon monoxide inhibited respiration. J Biol Chem 217:453–465

Chambers LA, Trudinger PA (1975) Are thiosulfate and trithionate intermediates in dissimilatory sulfate reduction? J Bacteriol 123:36–40

Chance B, Smith L, Castor LN (1953) New methods for the study of the carbon monoxide compounds of respiratory enzymes. Biochim Biophys Acta 12:289–298

Chang CK (1985) On the structure of cytochrome cd_1. J Biol Chem 260:9520–9522

Chaudhry GR, Suzuki I, Lees H (1980) Cytochrome oxidase of *Nitrobacter agilis*: isolation by hydrophobic interaction chromatography. Can J Biochem 26:1270–1274

Chaudhry GR, Suzuki I, Duckworth NW, Lees H (1981) Isolation and properties of cytochrome c-553, cytochrome c-550 and cytochrome c-549, 554 from *Nitrobacter agilis*. Biochim Biophys Acta 637:18–27

Ching Y, Ondrias MR, Rousseau DL, Muhoberac BB, Wharton DC (1982) Resonance Raman spectra of haem c and d_1 in *Pseudomonas* cytochrome oxidase. FEBS Lett 138:239-244

Clark RD, Hind G (1983) Isolation of a five polypeptide cytochrome bf complex from spinach chloroplasts. J Biol Chem 258:10348–10354

Clayton RK, Wang RT (1971) Photochemical reaction centres from *Rhodopseudomonas sphaeroides*. Meth Enzymol 23A:696–704

Cobley JG (1976a) Energy conserving reactions on phosphorylating electron transport particles from *Nitrobacter winogradskyi*. Biochem J 156:481–491

Cobley JG (1976b) Reduction of cytochromes by nitrite in electron transport particles from *Nitrobacter* – proposal for a mechanism of H^+ translocation. Biochem J 156:493–498

Cobley JG, Haddock BA (1975) The respiratory chain of *Thiobacillus ferrooxidans*: the reduction of cytochromes by Fe^{2+} and the preliminary characterisation of rusticyanin, a novel "blue" copper protein. FEBS Lett 60:29–33

Cogdell RJ, Crofts AR (1972) Some observations on the primary acceptor of *Rhodopseudomonas viridis*. FEBS Lett 27:176–178

Colby J, Dalton H (1978) Resolution of the methane monooxygenase of *Methylococcus capsulatus* (Bath) into three components. Biochem J 171:461–468

Cole JA, Ward FB (1973) Nitrite reductase-deficient mutants of *E. coli* K12. J Gen Microbiol 76:21–29

Cole JA, Coleman KJ, Compton BE, Kavanagh BM, Keevil CW (1974) Nitrite and NH_3 assimilation by anaerobic *E. coli*. J Gen Microbiol 85:11–22

Coleman KJ, Cornish-Bowden A, Cole JA (1978) Purification and properties of nitrite reductase from *E. coli* K12. Biochem J 175:483–493

Colman PM, Freeman HC, Guss JM, Murata M, Norris VA, Ramshaw JAM, Venkatappa MP (1978) X-ray crystal structure analysis of plastocyanin at 2.7 Å resolution. Nature (London) 272:319–324

Cooper RG (1970) Cytochemical detection of cytochrome oxidase activity in bacteria. J Med Microbiol 3:243–250

Corker GA, Sharpe SA (1975) Influence of light on the epr detectable electron transport components of *Rhodospirillum rubrum.* Photochem Photobiol 21:49–61

Coulson AFW, Oliver RIC (1979) Isolation and properties of cytochrome c peroxidase from *Pseudomonas denitrificans.* Biochem J 181:159–169

Cox, JC, Boxer DH (1978) The purification and some properties of rusticyanin, a blue copper protein involved in iron II oxidation from *Thiobacillus ferrooxidans.* Biochem J 174:497–502

Cox JC, Ingledew WJ, Haddock BA, Lawford HG (1978) The variable cytochrome content of *Paracoccus denitrificans* grown aerobically under different conditions. FEBS Lett 93:261–265

Crofts AR, Wraight CA (1983) The electrochemical domain of photosynthesis. Biochim Biophys Acta 726:149–185

Cross AR, Anthony C (1980a) The purification and properties of the soluble cytochromes c of the obligate methylotroph – *Methylophilus methylotrophus.* Biochem J 192: 421–427

Cross AR, Anthony C (1980b) The electron transport chain of the obligate methylotroph *Methylophilus methylotrophus.* Biochem J 192:429–439

Cusanovich MA, Bartsch RG, Kamen MD (1968) Light induced absorbance changes in *Chromatium* chromatophores. Biochim Biophys Acta 153:397–417

Cypionka H, Meyer O (1983) CO insensitive respiratory chain of *Pseudomonas carboxydovorans.* J Bacteriol 156:1178–1187

Daldal F, Cheng S, Applebaum J, Davidson E, Prince RC (1986) Cytochrome c_2 is not essential for photosynthetic growth of *Rhodopseudomonas capsulata.* Proc Natl Acad Sci USA 83:2012–2016

Davies HC, Smith L, Nava ME (1983) Reaction of cytochrome c in the electron transport chain of *Paracoccus denitrificans.* Biochim Biophys Acta 725:238–245

Davis DJ, Krogmann DW, San Pietro A (1980) Electron donation to photosystem I. Plant Physiol 65:697–702

Dawson MJ, Jones CW (1981) Energy conservation in the terminal region of the respiratory chain of the methylotrophic bacterium *Methylophilus methylotrophus.* Eur J Biochem 118:113–118

Deisenhofer J, Epp O, Miki K, Huber R, Michel H (1984) X-ray structure analysis of a membrane protein complex. J Mol Biol 180:385–398

DeKlerk H, Kamen MD (1966) A high potential non-heme iron protein from the facultative photoheterotroph *Rhodopseudomonas gelatinosa.* Biochim Biophys Acta 112:175–178

Delwiche CC, Bryan BA (1976) Denitrification. Annu Rev Microbiol 30:241–262

DerVartanian DV, Xavier AV, LeGall J (1978) Epr determination of the oxidation reduction potentials of the hemes in cytochrome c_3 from *Desulfovibrio vulgaris.* Biochimie 60:321–325

Dhesi R, Timkovich R (1984) Patterns of product inhibition for bacterial nitrite reductase. Biochem Biophys Res Commun 123:966–972

Douglas MW, Ward FB, Cole JA (1974) Formate hydrogen lyase activity of c-552 deficient mutants of *E. coli* K12. J Gen Microbiol 80:557–560

Drozd JW (1974) Respiration-driven proton translocation in *Thiobacillus neapolitanus* c. FEBS Lett 49:103–105

Drozd JW (1977) Energy conservation in *Thiobacillus neapolitanus* c: sulfide and sulfite oxidation. J Gen Microbiol 98:309–312

Drozd JW (1980) Respiration in the ammonia oxidising chemoautotrophic bacteria. In: Knowles CJ (ed) Diversity in bacterial respiratory systems. CRC Press, Boca Raton, pp 87–111

Drucker H, Campbell LL (1969) Electrophoretic and immunological differences between the cytochrome c_3 from *Desulfovibrio desulfuricans* and that from *D. vulgaris*. J Bacteriol 100:358–364

Drucker H, Trousil EB, Campbell LL, Barlow E, Margoliash E (1970) Amino acid composition, heme content and molecular weight of cytochrome c_3 from *Desulfovibrio desulfuricans* and *D. vulgaris*. Biochemistry 9:1515–1518

Dubourdieu M, LeGall J, Favaudon V (1975) Physicochemical properties of flavodoxin from *Desulfovibrio vulgaris*. Biochim Biophys Acta 376:519–532

Dugan PR, Lundgren DG (1965) Energy supply for the chemiautotroph *Ferrobacillus ferrooxidans*. J Bacteriol 89:825–834

Duine JA, Frank J (1980) Studies on methanol dehydrogenase from *Hyphomicrobium* x. Biochem J 187:213–219

Duine JA, Frank J, Ruiter LG (1979) Isolation of a methanol dehydrogenase with functional coupling to cytochrome c. J Gen Microbiol 115:523–526

Dutton PL (1971) Oxidation reduction potential dependence of the interactions of cytochromes, bacteriochlorophyll and carotenoids at 77 °K in chromatophores of *Chromatium vinosum* and *Rhodopseudomonas gelatinosa*. Biochim Biophys Acta 226:63–80

Dutton PL, Leigh JS (1973) Epr characterisation of *Chromatium* D hemes, non-heme irons and the components involved in primary photochemistry. Biochim Biophys Acta 314:178–190

Dutton PL, Prince RC (1978) Reaction centre-driven cytochrome interactions in electron and proton translocation and energy coupling. In: Clayton RK, Sistrom WR (eds) The photosynthetic bacteria. Plenum, New York, pp 525–570

Dutton PL, Petty KM, Bonner HS, Morse SD (1975) Cytochrome c_2 and reaction centre of *Rhodopseudomonas sphaeroides* membranes – extinction coefficients, content, half reduction potentials, kinetics and electric field alterations. Biochim Biophys Acta 387:536–556

Duysens LNM (1955) Role of cytochrome and pyridine nucleotide in algal photosynthesis. Science 121:210–211

Ellfolk N, Soininen R (1970) *Pseudomonas* cytochrome c peroxidase – purification procedure. Acta Chem Scand 24:2126–2136

Ellfolk N, Soininen R (1971) *Pseudomonas* cytochrome c peroxidase – the size and shape of the enzyme molecule. Acta Chem Scand 25:1535–1540

Ellfolk N, Ronnberg M, Soininen R (1973) *Pseudomonas* cytochrome c peroxidase – kinetics of the peroxidatic reaction mechanism. Acta Chem Scand 27:2171–2178

Ellfolk N, Ronnberg M, Aasa R, Andreasson L, Vanngard T (1983) Properties and function of the two hemes in *Pseudomonas* cytochrome c peroxidase. Biochim Biophys Acta 743:23–30

Ellfolk N, Ronnberg M, Aasa R, Andreasson L, Vanngard T (1984) Anion binding to resting and half reduced *Pseudomonas* cytochrome c peroxidase. Biochim Biophys Acta 784:62–67

Erickson RH, Hooper AB (1972) Preliminary characterisation of a variant CO binding protein from *Nitrosomonas*. Biochim Biophys Acta 275:231–244

Errede BJ, Kamen MD (1978) Comparative kinetic studies of cytochrome c in reactions with mitochondrial cytochrome oxidase and reductase. Biochemistry 17:1015–1027

Evans MCW, Crofts AR (1974a) A thermodynamic characterisation of the cytochromes of chromatophores of *Rhodopseudomonas capsulata*. Biochim Biophys Acta 357:78–88

Evans MCW, Crofts AR (1974b) In situ characterisation of photosynthetic electron transport in *Rhodopseudomonas capsulata*. Biochim Biophys Acta 357:89–102

Fauque GD, Herve D, LeGall J (1979a) Structure function relationships in hemoproteins – the role of cytochrome c_3 in the reduction of colloidal sulfur by sulfate reducing bacteria. Arch Microbiol 121:261–264

Fauque GD, Bruschi M, LeGall J (1979b) Purification and some properties of cytochrome c-553(550) isolated from *Desulfovibrio desulfuricans* (Norway). Biochem Biophys Res Commun 86:1020–1029

Fee JA, Choc MG, Findling KL, Lorence R, Yoshida T (1980) Properties of a copper containing cytochrome c_1aa_3 complex: a terminal oxidase of the extreme thermophile – *Thermus thermophilus*. Proc Natl Acad Sci USA 77:147–151

Ferenci T (1974) Carbon monoxide stimulated respiration in methane utilising bacteria. FEBS Lett 41:94–98

Ferguson SJ (1982a) Is a proton pumping cytochrome oxidase essential for energy conservation in *Nitrobacter*? FEMS Lett 146:239–243

Ferguson SJ (1982b) Aspects of the control and organisation of bacterial electron transport. Biochem Soc Trans 10:198–200

Ferguson SJ (1984) A novel bacterial dehydrogenase cofactor turns up in a serum amine oxidase. Trends Biochem Sci 9:367–368

Fiechtner MD, Kassner RJ (1979) Redox properties and haem environment of cytochrome c-551.5 of *Desulfuromonas acetoxidans*. Biochim Biophys Acta 579:269–278

Findlay JE, Akagi JM (1970) Role of thiosulfate in bisulfite reduction as catalysed by *Desulfovibrio vulgaris*. J Bacteriol 103:741–744

Fischer U, Truper HG (1977) Cytochrome c-550 of *Thiocapsa roseopercicina* – properties and reduction by sulfide. FEMS Lett 1:87–90

Foote N, Thompson AC, Barber D, Greenwood C (1983) *Pseudomonas* cytochrome c-551 peroxidase – a purification procedure and study of CO binding kinetics. Biochem J 209:701–707

Foote N, Peterson J, Gadsby PMA, Greenwood C, Thomson AJ (1984) A study of the oxidised form of *Pseudomonas* cytochrome c-551 peroxidase using magnetic circular dichroism. Biochem J 223:369–378

Foote N, Peterson J, Gadsby PMA, Greenwood C, Thomson AJ (1985) Redox-linked spin state changes in the diheme cytochrome c-551 peroxidase from *Pseudomonas aeruginosa*. Biochem J 230:227–237

Forti G, Bertole ML, Zanetti G (1965) Purification and properties of cytochrome f from parsley leaves. Biochim Biophys Acta 109:33–40

Fowler CF, Nugent NA, Fuller RC (1971) The isolation and characterisation of a photochemically active complex from *Chloropseudomonas ethylica*. Proc Natl Acad Sci USA 68:2278–2282

Froud SJ, Anthony C (1984) The purification and characterisation of the o-type cytochrome oxidase from *Methylophilus methylotrophus* and its reconstitution into a methanol oxidase electron transport chain. J Gen Microbiol 130:2201–2212

Fujita T (1966a) Studies on soluble cytochromes from Enterobacteriaceae – cytochromes b-562 and c-550. J Biochem 60:329–334

Fujita T (1966b) Studies on soluble cytochromes from Enterobacteriaceae – detection, purification and properties of cytochrome c-552 in anaerobic *E. coli*. J Biochem 60:204–215

Fujita T, Sato R (1966) Studies on soluble cytochromes in Enterobacteriaceae – involvement of cytochrome c-552 in anaerobic nitrite metabolism. J Biochem 60:691–700

Fujita T, Sato R (1967) Physiological function of cytochrome c-552. J Biochem 62:230–238

Fukumori Y, Yamanaka T (1979) Flavocytochrome c of *Chromatium vinosum*. J Biochem 85:1405–1414

Fukumori Y, Nakayama K, Yamanaka T (1985) Cytochrome c oxidase of *Pseudomonas* AMI: purification and molecular and enzymatic properties. J Biochem 98:493–499

Gabellini N, Hauska G (1983) Isolation of cytochrome b from the cytochrome bc_1 complex of *Rhodopseudomonas sphaeroides* GA. FEBS Lett 154:171–174

Gabellini N, Bowyer JR, Hurt E, Melandri BA, Hauska G (1982) A cytochrome bc_1 complex with ubiquinol cytochrome c_2 oxidoreductase activity from *Rhodopseudomonas sphaeroides*. Eur J Biochem 126:105–111

Gennis RB, Casey RP, Azzi A, Ludwig B (1982) Purification and characterisation of the cytochrome oxidase from *Rhodopseudomonas sphaeroides*. Eur J Biochem 125:189–195

Gibson J (1961) Cytochrome pigments from the green photosynthetic bacterium *Chlorobium thiosulfatophilum*. Biochem J 79:151–158

Goodhew C, Brown K, Pettigrew GW (1986) Haem staining in gels, a useful tool in the study of bacterial c-type cytochromes. Biochim Biophys Acta 852:288–294

Gool A van, Laudelout H (1966) The mechanism of nitrite oxidation in *Nitrobacter winogradskyi*. Biochim Biophys Acta 113:41–50

Gray GO, Knaff DB (1982) The role of cytochrome c-552-cytochrome c complex in the oxidation of sulfide in *Chromatium vinosum*. Biochim Biophys Acta 680:290–296

Gray GO, Gaul DF, Knaff DB (1983) Partial purification and characterisation of two soluble c-types cytochromes from *Chromatium vinosum*. Arch Biochem Biophys 222:78–86

Gray JC (1978) Purification and properties of monomeric cytochrome f from Charlock, *Sinapsis arvensis*. Eur J Biochem 82:133–144

Gray JC (1980) Maternal inheritance of cytochrome f in interspecific *Nicotiana* hybrids. Eur J Biochem 112:39–46

Greenwood C, Barber D, Parr SR, Antonini E, Brunori M, Colosimo A (1978) Reaction of *Pseudomonas aeruginosa* cytochrome c-551 oxidase with oxygen. Biochem J 173: 11–17

Greenwood C, Foote N, Peterson J, Thomson AJ (1984) The nature of species prepared by photolysis of half reduced, fully reduced and fully reduced carbon monoxy c-551 peroxidase from *Pseudomonas aeruginosa*. Biochem J 223:379–391

Groen B, Frank J, Duine JA (1984) Quinoprotein alcohol dehydrogenase from ethanol-grown *Pseudomonas aeruginosa*. Biochem J 223:921–924

Groen B, Kleff MAG van, Duine JA (1986) Quinohemeprotein alcohol dehydrogenase apoenzyme from *Pseudomonas testosteronii*. Biochem J 234:611–615

Grondelle R van, Duysens LNM, Wal HN van der (1976) Functions of three cytochromes in photosynthesis of whole cells of *Rhodospirillum rubrum* as studied by flash spectroscopy. Biochim Biophys Acta 449:169–187

Grondelle R van, Duysens LNM, Wel JA van der, Wal HN van der (1977) Function and properties of a soluble c-type cytochrome c-551 in secondary photosynthetic electron transport in whole cells of *Chromatium vinosum* as studied with flash spectroscopy. Biochim Biophys Acta 461:188–201

Gross EL (1979) Cation induced increases in the rate of P700 recovery in photosystem I particles. Arch Biochem Biophys 195:198–204

Gudat JC, Singh J, Wharton DC (1973) Cytochrome oxidase from *Pseudomonas aeruginosa* – purification and some properties. Biochim Biophys Acta 292:376–390

Guerlesquin F, Lapierre GB, Bruschi M (1982) Purification and characterisation of cytochrome c_3 (26 K) from *Desulfovibrio desulfuricans* (Norway). Biochem Biophys Res Commun 105:530–538

Guikema JA, Sherman LA (1980) Electrophoretic profiles of cyanobacterial membrane polypeptides showing heme-dependent peroxidase activity. Biochim Biophys Acta 637:189–201

Haddock BA, Jones CW (1977) Bacterial respiration. Bacteriol Rev 41:47–99

Haehnel W (1977) Electron transport between plastocyanin and chlorophyll AI in chloroplasts. Biochim Biophys Acta 459:418–441

Haehnel W, Propper A, Krause H (1980) Evidence for complexed plastocyanin as the immediate electron donor of P700. Biochim Biophys Acta 593:384–399

Haehnel W, Berzborn RJ, Andersson B (1981) Localisation of the reaction side of plastocyanin from immunological and kinetic studies with inside out thylakoid vesicles. Biochim Biophys Acta 637:389–399

Halsey YD, Byers B (1975) A large photoreactive particle from *Chromatium vinosum* chromatophores. Biochim Biophys Acta 387:349–367

Hartingsveldt JV, Stouthamer AH (1973) Mapping and characterisation of mutants of *Pseudomonas aeruginosa* affected in nitrate respiration in anaerobic or aerobic growth. J Gen Microbiol 74:97–106

Hatchikian EC, LeGall J, Bruschi M, Dubourdieu M (1972) Regulation of the reduction of sulfite and thiosulfate by ferredoxin, flavodoxin and cytochrome cc_3 in extracts of the sulfate reducer *Desulfovibrio gigas*. Biochim Biophys Acta 258:701–708

Hatchikian EC, Papavassiliou P, Bianco P, Haladjian J (1983) Characterisation of cytochrome c_3 from the thermophilic sulfate reducer – *Thermodesulfobacter commune*. J Bacteriol 159:1040–1046

Hauska G, McCarty RE, Berzborn RJ, Racker E (1971) Partial resolution of the enzymes catalysing photophosphorylation. J Biol Chem 246:3524–3531

Hauska G, Hurt E, Gabellini N, Lockau W (1983) Comparative aspects of quinol cytochrome c plastocyanin oxidoreductases. Biochim Biophys Acta 726:97–133

Henry Y, Bessieres P (1984) Denitrification and nitrite reduction – *Pseudomonas aeruginosa* nitrite reductase. Biochimie 66:259–289

Higgins IJ, Taylor SC, Tonge GM (1976) The respiratory system of *P. extorquens*. Proc Soc Gen Microbiol 3:179.

Hill KE, Wharton DC (1978) Reconstitution of the apoenzyme of cytochrome oxidase from *Pseudomonas aeruginosa* with heme d_1 and other heme groups. J Biol Chem 253: 489–495

Hind G, Olson JM (1968) Electron transport pathways in photosynthesis. Annu Rev Plant Physiol 19:249–282

Ho KK, Krogman DW (1980) Cytochrome f from Spinach and cyanobacteria. J Biol Chem 255:3855–3861

Ho KK, Ulrich EL, Krogman DW, Gomez-Lojero C (1979) Isolation of photosynthetic catalysts from cyanobacteria. Biochim Biophys Acta 545:236–248

Hochman A, Carmeli C (1977) Reconstitution of photosynthetic electron transport and photophosphorylation in cytochrome c_2-deficient membrane preparations of *Rhodopseudomonas capsulata*. Arch Biochem Biophys 179:349–359

Hochman A, Fridberg I, Carmeli C (1975) The location and function of cytochrome c_2 in *Rhodopseudomonas capsulata* membranes. Eur J Biochem 58:65–72

Hochman A, Berman T, Plotkin B, Schejter A (1985) Isolation and properties of the soluble c-type cytochromes of the dinoflagellate *Peridinium cinctum*. Arch Biochem Biophys 243:161–167

Hoffman PS, Morgan TV, DerVartanian DV (1980) Respiratory properties of cytochrome c-deficient mutants of *Azotobacter vinelandi*. Eur J Biochem 110:349–354

Hollocher TC, Tate ME, Nicholas DJD (1981) Oxidation of ammonia by *Nitrosomonas europaea* – definitive ^{18}O tracer evidence that hydroxylamine formation involves a monooxygenase. J Biol Chem 256:10834–10836

Holton RW, Myers J (1967a) Extraction, purification and spectral properties of cytochromes c(549, 552 and 554) of *Anacystis nidulans*. Biochim Biophys Acta 131:362–374

Holton RW, Myers (1967b) Physicochemical properties and quantitative relationships of cytochromes c(549, 552 and 554) of *Anacystis nidulans*. Biochim Biophys Acta 131: 375–384

Hon-nami K, Oshima T (1977) Purification and some properties of cytochrome c-552 from an extreme thermophile *Thermus thermophilus* HB8. J Biochem 82:769–776

Hon-nami K, Oshima T (1980) Cytochrome oxidase from an extreme thermophile – *Thermus thermophilus* HB8. Biochem Biophys Res Commun 92:1023–1029

Hooper AB (1968) A nitrite reducing enzyme from *Nitrosomonas europaea* – preliminary characterisation with hydroxylamine as electron donor. Biochim Biophys Acta 162:49–65

Hooper AB (1969) Lag phase of ammonia oxidation by resting cells of *Nitrosomonas europaea*. J Bacteriol 97:968–969

Hooper AB, Nason A (1965) Characterisation of hydroxylamine cytochrome c reductase from the chemoautotrophs *Nitrosomonas europaea* and *Nitrocystis oceanus*. J Biol Chem 240:4044–4057

Hooper AB, Terry KR (1977) Hydroxylamine oxidase from *Nitrosomonas*: inactivation by hydrogen peroxide. Biochemistry 16:455–459

Hooper AB, Maxwell DC, Terry KR (1978) Hydroxylamine oxidoreductase from *Nitrosomonas*: absorption spectra and content of heme and metal. Biochemistry 17:2984–2989

Hopper DJ (1983) Redox potential of the cytochrome c in the flavocytochrome p-cresol methyl hydroxylase. FEBS Lett 161:100–102

Hopper DJ, Taylor DG (1977) The purification and properties of p-cresol acceptor oxidoreductase (hydroxylating), a flavocytochrome from *Pseudomonas putida*. Biochem J 167:155–162

Hori K (1961) Electron transporting components participating in nitrate and oxygen respirations from halotolerant *Micrococcus*. J Biochem 50:440–449

Horio T, Higashi T, Sasagawa M, Kusai K, Nakai M, Okunuki K (1960) Preparation of crystalline *Pseudomonas* cytochrome c-551 and its general properties. Biochem J 77:194–201

Horio T, Kamen MD, DeKlerk H (1961 a) Relative oxidation reduction potentials of heme groups in two soluble double heme proteins. J Biol Chem 236:2783–2787

Horio T, Higashi T, Yamanaka T, Matsubara H, Okunuki K (1961 b) Purification and properties of cytochrome oxidase from *Pseudomonas aeruginosa*. J Biol Chem 236:944–951

Horowitz PM, Falksen K, Muhoberac BB, Wharton DC (1982) Activation of *Pseudomonas* cytochrome oxidase by limited proteolysis with subtilisin. J Biol Chem 257:9258–9260

Hudig H, Drews G (1982) Characterisation of a b-type cytochrome oxidase of *Rhodopseudomonas capsulata*. FEBS Lett 146:389–392

Hudig H, Drews G (1983) Characterisation of a new membrane bound cytochrome of *Rhodopseudomonas capsulata*. FEBS Lett 152:251–255

Hudig H, Drews G (1984) Reconstitution of b-type cytochrome oxidase from *Rhodopseudomonas capsulata* in liposomes and turnover studies of proton translocation. Biochim Biophys Acta 765:171–177

Hurt E, Hauska G (1981) A cytochrome f/b6 complex of five polypeptides with plastoquinol plastocyanin oxidoreductase activity from spinach chloroplasts. Eur J Biochem 117:591–599

Husain M, Sadana JC (1974) Nitrite reductase from *Achromobacter fischeri* – molecular weight and subunit structure. Eur J Biochem 42:283–289

Huynh BH, Moura JJG, Moura I, Kent TA, LeGall J (1980) Evidence for a three iron centre in a ferredoxin from *Desulfovibrio gigas*. J Biol Chem 255:3242–3244

Huynh BH, Liu MC, Moura JJG, Moura I, Ljungdahl PO, Munck E, Payne WJ, Peck HD, DerVartanian DV, LeGall J (1982) Mossbauer and epr studies on nitrite reductase from *Thiobacillus denitrificans*. J Biol Chem 257:9576–9581

Ingledew WJ (1982) *Thiobacillus ferrooxidans* – the bioenergetics of an acidophilic chemolithotroph. Biochim Biophys Acta 683:89–117

Ingledew WJ, Chappell JB (1975) ATP induced changes in the midpoint potential of *Nitrobacter* cytochromes – an indication of anisotropy. Fed Proc 34:488

Ingledew WJ, Cobley JG (1980) A potentiometric and kinetic study on the respiratory chain of ferrous iron grown *Thiobacillus ferrooxidans*. Biochim Biophys Acta 590:141–158

Ingledew WJ, Cox JC, Jones RW, Garland PB (1978) Vectorial oxidoreductions – ferrous iron oxidase complex of *Thiobacillus ferroxidans* and the nitrate reductase of *Escherichia coli*. In: Scarpa A, Dutton PL, Leigh JS (eds) Frontiers of biological energetics, vol 1. Academic Press, London New York, pp 334–341

Iwasaki H (1960) Participation of cytochromes in denitrification. J Biochem 47:174–184

Iwasaki H, Matsubara T (1971) Cytochrome c-557(551) and cytochrome cd_1 in *Alcaligenes faecalis*. J Biochem 69:847–857

Iwasaki H, Matusbara T (1972) A nitrite reductase from *Achromobacter cycloclastes*. J Biochem 71:645–652

John P, Whatley FR (1975) *Paracoccus denitrificans* and the evolutionary origin of the mitochondrion. Nature (London) 254:495–498

Johnson MK, Thomson AJ, Walsh TA, Barber D, Greenwood C (1980) Epr studies on *Pseudomonas* nitrosyl nitrite reductase. Biochem J 189:285–294

Johnson PA, Quayle JR (1964) Oxidation of methanol, formaldehyde and formate by methanol-grown *Pseudomonas* AMl. Biochem J 93:281–290

Jones CW (1977) Aerobic respiratory systems in bacteria. In: Haddock BA, Hamilton WA (eds) 27th Symp Soc for General Microbiology

Jones CW, Redfearn ER (1967) The cytochrome system of *Azotobacter vinelandii*. Biochim Biophys Acta 143:340–353

Jones CW, Brice JM, Downs AJ, Drozd JW (1975) Bacterial respiration linked proton translocation and its relationship to respiratory chain composition. Eur J Biochem 52:265–271

Jurtshuk P, Mueller TJ, Wang TY (1981) Isolation and purification of the cytochrome oxidase of *Azotobacter vinelandii*. Biochim Biophys Acta 637:374–382

Kakuno T, Bartsch RG, Nishikawa K, Hori T (1971) Redox components associated with chromatophores from *Rhodospirillum rubrum*. J Biochem 70:79–94

Kalkinnen N, Ellfolk N (1978) IUPAC 11th Int Symp Chemistry of Natural products, pp 79–82

Katan MB (1976) Detection of cytochromes on SDS gels by their intrinsic fluorescence. Anal Biochem 74:132–137

Ke B, Chaney TH, Reed DW (1970) The electrostatic interaction between the reaction centre BChl derived from *Rhodospseudomonas sphaeroides* and horse cytochrome c and its effects on light activated electron transport. Biochim Biophys Acta 216:373–383

Kennel SJ, Kamen MD (1971) Iron containing proteins in *Chromatium vinosum*. Purification and properties of cholate-solubilised cytochrome complex. Biochim Biophys Acta 253:153–156

Kenney WC, Singer TP (1977) Evidence for a thioether linkage between the flavin and polypeptide chain of *Chromatium* cytochrome c-552. J Biol Chem 252:4767–4772

Kenney WC, McIntyre W, Yamanaka T (1977) Structure of the covalently bound flavin of *Chlorobium* cytochrome c-553. Biochim Biophys Acta 483:467–474

Kienzl PF, Peschek GA (1982) Oxidation of c-type cytochromes by the membrane bound cytochrome oxidase of blue green algae. Plant Physiol 69:580–584

Kihara T, Chance B (1969) Cytochromes photooxidation at liquid nitrogen temperatures in photosynthetic bacteria. Biochim Biophys Acta 189:116–124

Kikuchi G, Saito Y, Motokawa Y (1965) On cytochrome oxidase as the terminal oxidase of dark respiration of non-sulfur purple bacteria. Biochim Biophys Acta 94:1–14

King M-T, Drews G (1976) Isolation and partial characterisation of the cytochrome oxidase from *Rhodopseudomonas palustris*. Eur J Biochem 68:5–12

Kita K, Konishi K, Anraku Y (1984) Terminal oxidases of *E. coli* aerobic respiratory chain. J Biol Chem 259:3368–3374

Kitada M, Krulwich JA (1984) Purification and characterisation of the cytochrome oxidase from alkalophilic *Bacillus firmus* RAB. J Bacteriol 158:963–966

Knaff DB, Buchanan BB (1975) Cytochrome b and photosynthetic sulfur bacteria. Biochim Biophys Acta 376:549–560

Knaff DB, Kraichoke S (1983) Oxidation reduction and electron paramagnetic resonance properties of a cytochrome complex from *Chromatium vinosum*. Photochem Photobiol 37:243–246

Knaff DB, Buchanan BB, Malkin R (1973) Effect of redox potential on light induced cytochrome and bacteriochlorophyll reactions in chromatophores from the photosynthetic green bacterium *Chlorobium*. Biochim Biophys Acta 325:94–101

Knaff DB, Worthington TM, White CC, Malkin R (1979) A partial purification of membrane-bound b and c-type cytochromes from *Chromatium vinosum*. Arch Biochem Biophys 192:158–163

Knaff DB, Whetstone R, Carr JW (1980) The role of soluble cytochrome c-551 in cyclic electron flow driven active transport in *Chromatium vinosum*. Biochim Biophys Acta 590:50–58

Kobayashi K, Hasegewa H, Takagi M, Ishimoto M (1982) Proton translocation associated with sulfite reduction in a sulfate reducing bacterium, *Desulfovibrio vulgaris*. FEBS Lett 142:235–237

Kodama T (1970) Effects of growth conditions on formation of a cytochrome system of a denitrifying bacterium *Pseudomonas stutzeri.* Plant Cell Physiol 11:231–239

Kodama T, Mori T (1969) A double peak c-type cytochrome c-552, 558 of a denitrifying bacterium *Pseudomonas stutzeri.* J Biochem 65:621–628

Koike J, Hattori A (1975) Growth yield of a denitrifying bacterium *Pseudomonas denitrificans* under aerobic and denitrifying conditions. J Gen Microbiol 88:1–19

Krinner M, Hauska G, Hurt E, Lockau W (1982) A cytochrome f-b6 complex with plastoquinol cytochrome c oxidoreductase activity from *Anabaena variabilis.* Biochim Biophys Acta 681:110–117

Kroger A (1980) Bacterial electron transport to fumarate. In: Knowles CJ (ed) Diversity of bacterial respiratory systems, vol 2. CRC Press, Boca Raton, pp 1–17

Kuo LM, Davies HC, Smith L (1985) Monoclonal antibodies to cytochrome c from *Paracoccus denitrificans:* effects on electron transfer reactions. Biochim Biophys Acta 809: 388–395

Kuronen T, Ellfolk N (1972) A new purification procedure and molecular properties of *Pseudomonas* cytochrome c oxidase. Biochim Biophys Acta 275:308–318

Kuronen T, Saraste M, Ellfolk N (1975) The subunit structure of *Pseudomonas* cytochrome oxidase. Biochim Biophys Acta 393:48–54

Kusai A, Yamamaka T (1973a) Cytochrome c-553 of *Chlorobium thiosulfatophilum* is a sulfide cytochrome c reductase. FEBS Lett 34:235–237

Kusai A, Yamanaka T (1973b) The oxidation mechanism of thiosulfate and sulfide in *Chlorobium thiosulfatophilum.* Biochim Biophys Acta 325:304–314

La Monica RF, Marrs B (1976) The branched respiratory system of photosynthetically grown *Rhodopseudomonas capsulata.* Biochim Biophys Acta 423:431–439

Lach HJ, Ruppel HG, Boger P (1973) Cytochrome c-553 from the alga *Bumilleriopsis filiformis.* Pflanzenphysiologie 70:432–451

Lam Y, Nicholas DJD (1969) A nitrite reductase with cytochrome oxidase activity from *Micrococcus denitrificans.* Biochim Biophys Acta 180:459–472

Laycock MV, Craigie JS (1971) Purification and characterisation of cytochrome c-553 from the chrysophycean alga *Monochrysis lutherii.* Can J Biochem 49:641–646

Le Gall J, Forget N (1978) Purification of electron transfer components from sulfate reducing bacteria. Meth Enzymol 53:613–633

Le Gall J, Postgate JR (1973) The physiology of sulfate reducing bacteria. Adv Microb Physiol 10:81–133

Le Gall J, Mazza G, Dragoni N (1965) Le cytochrome c_3 des *Desulfovibrio gigas.* Biochim Biophys Acta 90:385–387

Le Gall J, Bruschi-Heriaud M, DerVartanian DV (1971) Epr and light absorption studies on c-type cytochromes of the anaerobic sulfate reducer – *Desulfovibrio.* Biochim Biophys Acta 234:499–522

Lees H (1960) Energy metabolism in the chemolithotrophic bacteria. Annu Rev Microbiol 14:83–98

Lemberg R, Barrett J (1973) Cytochromes. Academic Press, London New York

Lenhoff HM, Kaplan NO (1956) A cytochrome c peroxidase from *Pseudomonas fluorescens.* J Biol Chem 220:967–982

Liu CY, Webster DA (1974) Spectral characteristics and interconversions of the reduced, oxidised and oxygenated forms of purified cytochrome o. J Biol Chem 249:4261–4266

Liu MC, Peck HD (1981) Isolation of a hexaheme cytochrome from *Desulfovibrio desulfuricans* and its identification as a new type of nitrite reductase. J Biol Chem 256:13159–13164

Liu MC, DerVartanian DV, Peck HD (1980) On the nature of oxidation reduction properties of nitrite reductase from *Desulfovibrio desulfuricans.* Biochem Biophys Res Commun 96:278–285

Liu MC, Peck HD, Payne W, Anderson JL, DerVartanian DV, Le Gall J (1981a) Purification and properties of the diheme cytochrome c-552 from *Pseudomonas perfectomarinus.* FEBS Lett 129:155–160

Liu MC, Peck HD, Abou-Jaoude A, Chippaux M, Le Gall J (1981 b) A reappraisal of the role of the low potential c-type cytochrome in NADH dependent nitrite reduction and its relationship with a copurified NADH oxidase in *E. coli* K12. FEMS Lett 10:333–337

Liu MC, Payne WJ, Peck HD, Le Gall J (1983) Comparison of the cytochromes from aerobically and anaerobically grown cells of *Pseudomonas perfectomarinus.* J Bacteriol 154:278–286

Lockau W (1979) The inhibition of photosynthetic electron transport in Spinach chloroplasts by low osmolarity. Eur J Biochem 94:365–373

Lockau W (1981) Evidence for a dual role of cytochrome c-553 and plastocyanin in photosynthesis and respiration of the cyanobacterium *Anabaena variabilis.* Arch Microbiol 128:336–340

Lorence RM, Yoshida T, Findling KL, Fee JA (1981) Observations on the c-type cytochromes of the extreme thermophile *Thermus thermophilus* HB8: cytochrome c-552 is located in the periplasmic space. Biochem Biophys Res Commun 99:591–599

Lu WP (1986) A periplasmic location for the thiosulfate oxidising multi-enzyme system from *Thiobacillus versutus.* FEMS Lett 34:313–317

Lu WP, Kelly DP (1984) Purification and characterisation of two essential cytochromes of the thiosulfate oxidising multienzyme system from *Thiobacillus* A2 (*versutus*). Biochim Biophys Acta 765:106–117

Lu WP, Poole RK, Kelly DP (1984) Oxidation reduction potentials of and spectral properties of some cytochromes from *Thiobacillus versutus* A2. Biochim Biophys Acta 767: 326–334

Ludwig B, Schatz G (1980) A two subunit cytochrome c oxidase from *Paracoccus denitrificans.* Proc Natl Acat Sci USA 77:196–200

Ludwig B, Grabo M, Gregor I, Lustig A, Regenass M, Rosenbusch JP (1982) Solubilised cytochrome oxidase from *Paracoccus denitrificans* is a monomer. J Biol Chem 257: 5576–5578

Ludwig B, Suda K, Cerletti N (1983) Cytochrome c_1 from *Paracoccus denitrificans.* Eur J Biochem 137:597–602

Makinen MW, Schichman SA, Hill SG, Gray HB (1983) Heme-heme orientation and electron transfer kinetic behaviour of multisite oxidation reduction enzymes. Science 222:929–931

Malkin R, Bearden AJ (1973) Light induced changes of bound chloroplast plastocyanin as studied by epr spectroscopy – the role of plastocyanin in non-cyclic photosynthetic electron transport. Biochim Biophys Acta 292:169–185

Marrs BL, Gest H (1973) Genetic mutations affecting the respiratory electron transport system of the photosynthetic bacterium *Rhodopseudomonas capsulata.* J Bacteriol 114:1052–1057

Matsubara T, Iwasaki H (1972) NO-reducing activity of *Alcaligenes faecalis* cytochrome cd_1. J Biochem 72:57–64

Matsubara T, Mori T (1968) N_2O – its production and reduction to N_2. J Biochem 64:863–871

Matsuda H, Nishi N, Tsuji K, Tanaka K, Kakuno T, Yanashita J, Horio T (1984) Reconstitution of photosynthetic cyclic electron transport system from Photoreaction Unit, Ubiquinone 10 protein, cytochrome c_2 and polar lipids purified from *Rhodospirillum rubrum.* J Biochem 95:431–442

Matsushita K, Shinagawa E, Adachi O, Ameyama M (1982 a) o-type cytochrome oxidase in the membrane of aerobically grown *Pseudomonas aeruginosa.* FEBS Lett 139:255–258

Matsushita K, Shinagawa E, Adachi O, Ameyama M (1982 b) Membrane bound cytochromes c of *Pseudomonas aeruginosa* grown aerobically. J Biochem 92:1607–1613

Matsuzaki E, Kamimura Y, Yamasaki T, Yakushiji (1975) Purification and properties of cytochrome f from *Brassica komatsuna* leaves. Plant Cell Physiol 16:237–246

McEwan AG, Greenfield AJ, Wetzstein HG, Jackson JB, Ferguson SJ (1985) Nitrous oxide reduction by *Rhodospirillaceae* and the nitrous oxide reductase of *Rhodopseudomonas capsulata.* J Bacteriol 164:823–830

McIntyre W, Edmondson DE, Hopper DJ, Singer TP (1981) 8-O-tyrosyl flavin adenine nucleotide. The prosthetic group of bacterial p-cresol methyl hydroxylase. Biochemistry 20:3068–3075

Mehard CW, Prezelin BL, Haxo FT (1975) Isolation and characterisation of dinoflagellate and chrysophyte cytochrome f (553-4). Phytochemistry 14:2379–2389

Meijer EM, Zwaan JW van der, Stouthamer AH (1979) Location of the proton consuming site in nitrite reduction in *Paracoccus denitrificans.* FEMS Lett 5:369–372

Meyer TE, Bartsch RG (1976) The reaction of flavocytochromes c of the phototrophic sulfur bacteria with thiosulfate, sulfite, cyanide and mercaptans. In: Singer TP (ed) Flavins and flavoproteins. Elsevier, Amsterdam, pp 312–317

Meyer TE, Kamen MD (1982) New perspectives on c-type cytochromes. Adv Protein Chem 35:105–212

Meyer TE, Bartsch RG, Cusanovich MA, Mathewson JH (1968) The cytochromes of *Chlorobium limicola* f. thiosulfatophilum. Biochim Biophys Acta 153:854–861

Meyer TE, Bartsch RG, Kamen MD (1971) Cytochrome c_3 – a class of electron transfer heme proteins found in both photosynthetic and sulfate reducing bacteria. Biochim Biophys Acta 245:453–464

Meyer TE, Kennel SJ, Tedro SM, Kamen MD (1973) Iron protein content of *Thiocapsa pfennigii* – a purple sulfur bacterium of atypical chlorophyll composition. Biochim Biophys Acta 292:634–643

Meyer TE, Vorkink WP, Tollin G, Cusanovich MA (1985) *Chromatium* flavocytochrome c – kinetics of reduction of the heme subunit and the flavocytochrome mitochondrial cytochrome c complex. Arch Biochem Biophys 236:52–58

Michels PAM, Haddock BA (1980) Cytochrome c-deficient mutants of *Rhodopseudomonas capsulata.* FEMS Lett 7:327–331

Miki K, Okunuki K (1969a) Cytochromes of *Bacillus subtilis* – purification and spectral properties of cytochrome c-550 and cytochrome c-554. J Biochem 66:831–843

Miki K, Okunuki K (1969b) Cytochromes of *Bacillus subtilis* – physicochemical and enzymatic properties of cytochromes c-550 and c-554. J Biochem 66:845–854

Miller DJ, Wood PM (1982) Characterisation of the c-type cytochromes of *Nitrosomonas europaea* with the aid of fluorescent gels. Biochem J 207:511–517

Miller DJ, Wood PM (1983a) CO binding c-type cytochrome and a high potential cytochrome c in *Nitrosomonas europaea.* Biochem J 211:503–506

Miller J, Wood PM (1983b) The soluble cytochrome oxidase from *Nitrosomonas europaea.* J Gen Microbiol 129:1645–1650

Mitchell P (1975) The proton motive Q-cycle – a general formulation. FEBS Lett 59:137–139

Mitra S, Bersohn R (1980) Location of the heme groups in cytochrome cd_1 oxidase of *Pseudomonas aeruginosa.* Biochemistry 19:3200–3203

Mitsui A, Tsushima K (1986) Crystallisation and some properties of *Euglena* cytochrome c-552. In: Okunuki K, Kamen MD, Sezuku I (eds) Structure and function of cytochromes. Univ Park Press, Baltimore, pp 459–466

Miyata M, Mori T (1969) The denitrifying enzyme as a nitrite reductase and the electron donating system for denitrification. J Biochem 66:463–471

Morita S (1968) Evidence for three photochemical systems in *Chromatium* D. Biochim Biophys Acta 153:241–247

Morita S, Conti SF (1963) Localisation and nature of the cytochromes of *Rhodomicrobium vanniellii.* Arch Biochem Biophys 100:302–307

Moser CC, Pachence JM, Blaise JK, Dutton PL (1982) Submicromolar binding and biphasic electron transfer kinetics of cytochrome c and vesicular reaction centres. Biophys J 37:225a

Moura JJG, Xavier AV, Cookson DJ, Moore GR, Williams RJP (1977) Redox states of cytochrome c in the absence and presence of ferredoxin. FEBS Lett 81:275–280

Moura JJG, Santos H, Moura I, Le Gall J, Moore GR, Williams RJP, Xavier AV (1982) NMR redox studies of *Desulfovibrio vulgaris* cytochrome c_3. Eur J Biochem 127:151–155

Muhoberac BB, Wharton DC (1983) Electron paramagnetic resonance study of the interaction of some anionic ligands with oxidised *Pseudomonas* cytochrome c oxidase. J Biol Chem 258:3019–3027

Muscatello U, Carafoli E (1969) The oxidation of exogenous and endogenous cytochrome c in mitochondria – a biochemical and ultrastructural study. J Cell Biol 40:602–621

Nagata T, Yamanaka T, Okunuki K (1970) Amino acid composition and N-terminus of *Pseudomonas* cytochrome oxidase. Biochim Biophys Acta 221:668–671

Nakano K, Kikimoto Y, Yagi T (1983) Amino acid sequence of cytochrome c-553 from *Desulfovibrio vulgaris* Miyazaki. J Biol Chem 258:12409–12412

Nelson BD, Gellerfors P (1976) Alkali induced reduction of b-cytochromes in purified complex III of beef heart mitochondria. Biochim Biophys Acta 396:202–209

Nelson N, Neumann J (1972) Isolation of a cytochrome b6-f particle from chloroplasts. J Biol Chem 247:1917–1924

Nelson N, Racker E (1972) Purification of spinach cytochrome f and its photooxidation by photosystem I particles. J Biol Chem 247:3848–3853

Newton N (1969) The two heme nitrite reductase of *Micrococcus denitrificans*. Biochim Biophys Acta 185:316–331

Nicholls P, Sone N (1984) Kinetics of cytochrome c and TMPD oxidation by cytochrome c oxidase from the thermophilic bacterium PS3. Biochim Biophys Acta 767:240–247

Niwa S, Ishikawa H, Nikai S, Takabe T (1980) Electron transfer reactions between cytochrome f and plastocyanin from *Brassica komatsuna*. J Biochem 88:1177–1183

Odom JM, Peck HD (1981a) Hydrogen cycling as a general mechanism for energy coupling in the sulfate-reducing bacteria *Desulfovibrio* sp. FEMS Lett 12:47–50

Odom JM, Peck HD (1981b) Localisation of dehydrogenases, reductases and electron transfer components in the sulfate-reducing bacterium *Desulfovibrio gigas*. J Bacteriol 147:161–169

Odom JM, Peck HD (1984) Hydrogenases, electron transfer proteins and energy coupling in the sulfate-reducing bacteria *Desulfovibrio*. Annu Rev Microbiol 38:551–592

Ohta S, Tobari J (1981) Two cytochromes of *Methylomonas* J. J Biochem 90:215–224

Okamura MY, Steiner LA, Feher G (1974) Characterisation of reaction centres from photosynthetic bacteria – subunit structure of the protein mediating the primary photochemistry in *Rhodopseudomonas sphaeroides* R26. Biochemistry 13:1394–1403

O'Keefe DT, Anthony C (1980a) The two cytochromes c in the facultative methylotroph *Pseudomonas* AM1. Biochem J 192:411–419

O'Keefe DT, Anthony C (1980b) The interaction betwen methanol dehydrogenase and the autoreducible cytochromes c of the facultative methylotroph *Pseudomonas* AM1. Biochem J 190:481–484

Olson JM, Chance B (1960) Oxidation reduction reactions in the photosynthetic bacterium *Chromatium* – absorption spectrum changes in whole cells. Arch Biochem Biophys 88:26–53

Olson JM, Nadler KD (1965) Energy transfer and cytochrome function in a new type of photosynthetic bacterium. Photochem Photobiol 4:783–791

Olson JM, Shaw EK (1969) Cytochromes from the green photosynthetic bacterium *Chloropseudomonas ethylica*. Photosynthetica 3:288–290

Olson JM, Sybesma C (1963) In: Gest H, San Pietro A, Vernon LP (eds) Bacterial photosynthesis. Antioch Press, Yellow Springs, Ohio, pp 413–422

Olson JM, Carroll JW, Clayton ML, Gardner GM, Linkins AE, Moreth CMC (1969) Light induced absorption changes of cytochromes and carotenoids in a sulfur bacterium containing bacteriochlorophyll b. Biochim Biophys Acta 172:338–339

O'Neill JG, Wilkinson JF (1977) Oxidation of ammonia by methane oxidising bacteria and the effects of ammonia on methane oxidation. J Gen Microbiol 100:407–412

Orii Y, Shimada H (1978) Reaction of cytochrome c with nitrite and nitrous oxide. J Biochem 84:1543–1552

Overfield RE, Wraight CA (1980a) Oxidation of cytochromes c and c_2 by bacterial photosynthetic reaction centres in phospholipid vesicles – studies with neutral membranes. Biochemistry 19:3322–3327

Overfield RE, Wraight CA (1980b) Oxidation of cytochrome c and c_2 by bacterial photosynthetic centres in phospholipid vesicles – studies with negative membranes. Biochemistry 19:3328–3334

Overfield RE, Wraight CA, DeVault D (1979) Microsecond photooxidation kinetics of cytochrome c_2 from *Rhodopseudomonas sphaeroides* – in vivo and solution studies. FEBS Lett 105:137–142

Pachence JM, Dutton PL, Blaisie JK (1981) The reaction centre profile structure derived from neutron diffraction. Biochim Biophys Acta 635:267–283

Pachence JM, Dutton PL, Blaisie JK (1983) A structural investigation of cytochrome c binding to photosynthetic reaction centres in reconstituted membranes. Biochim Biophys Acta 724:6–19

Packham NK, Tiede DM, Mueller P, Dutton PL (1980) Construction of a flash activated cyclic electron transport system by using bacterial reaction centres and the ubiquinone-cytochrome bc_1/c segment of mitochondria. Proc Natl Acad Sci USA 77:6339–6343

Palmer G, Babcock GT, Vickery LE (1976) A model for cytochrome oxidase. Proc Natl Acad Sci USA 73:2206–2210

Parr SR, Wilson MT, Greenwood C (1974) The reaction of *Pseudomonas aeruginosa* cytochrome c oxidase with sodium metabisulfite. Biochem J 139:273–276

Parr SR, Wilson MT, Greenwood C (1975) Reaction of *Pseudomonas aeruginosa* cytochrome c oxidase with CO. Biochem J 151:51–59

Parr SR, Barber D, Greenwood C, Phillips BW, Melling J (1976) Purification procedure for the soluble cytochrome oxidase and some other respiratory proteins of *Pseudomonas aeruginosa*. Biochem J 157:423–430

Parr SR, Barber D, Greenwood C, Brunori M (1977) The electron transfer reaction between azurin and the cytochrome oxidase from *Pseudomonas aeruginosa*. Biochem J 167:447–455

Parson WW (1969) Cytochrome photooxidation in *Chromatium vinosum* chromatophores – each P870 oxidises two cytochrome c-422 hemes. Biochim Biophys Acta 189:397–403

Parson WW, Case GD (1970) In *Chromatium vinosum* a single photochemical reaction centre oxidises both cytochrome c-552 and c-555. Biochim Biophys Acta 205:232–245

Payne WJ (1981) Denitrification. Wiley Interscience, New York

Peck HD, Deacon TE, Davidson JT (1965) Studies on adenosine 5 phosphosulfate reductase from *Desulfovibrio desulfuricans* and *Thiobacillus thioparus* – assay and purification. Biochim Biophys Acta 96:429–446

Perini F, Kamen MD, Schiff JA (1964) Iron containing proteins in *Euglena* – detection and characterisation. Biochim Biophys Acta 88:74–90

Peschek GA (1981) Spectral properties of a cyanobacterial cytochrome oxidase : evidence for a cytochrome aa_3. Biochem Biophys Res Commun 98:72–79

Peschek GA (1982) Proton pump coupled to cytochrome c oxidase in the cyanobacterium *Anacystis nidulans*. J Bacteriol 153:539–542

Peschek GA, Schmetterer G (1982) Evidence for plastoquinol cytochrome f/b563 reductase as a common electron donor to P700 and cytochrome oxidase in cyanobacteria. Biochem Biophys Res Commun 108:1188–1195

Pettigrew GW, Meyer TE, Bartsch RG, Kamen MD (1975) pH dependence of the oxidation reduction potential of cytochrome c_2. Biochim Biophys Acta 430:197–208

Pettigrew GW, Bartsch RG, Meyer TE, Kamen MD (1978) Redox potentials of the photosynthetic bacterial cytochromes c_2 and the structural basis for variability. Biochim Biophys Acta 503:509–523

Pettigrew GW, Leitch FA, Moore GR (1983) The effect of iron hexacyanide binding on the determination of redox potentials of cytochromes and copper proteins. Biochim Biophys Acta 725:409–416

Pfennig N, Biebl H (1976) *Desulfuromonas acetoxidans* gen. nov. and sp. nov., a new anaerobic sulfur-reducing acetate-oxidising bacterium. Arch Microbiol 110:3–12

Phillips AL, Gray JC (1983) Isolation and characterisation of the cytochrome bf complex from pea. Eur J Biochem 137:553–560

Poole RK (1983) Bacterial cytochrome oxidases – a structurally and functionally diverse group of electron transfer proteins. Biochim Biophys Acta 726:205–243

Postgate JR (1956) Cytochrome c_3 and desulfoviridin – pigments of the anaerobic *Desulfovibrio desulfuricans*. J Gen Microbiol 14:545–572

Postgate JR (1979) The sulfate reducing bacteria. Cambridge Univ Press, Cambridge

Prakash O, Sadana JC (1972) Purification, characterisation and properties of nitrite reductase of *Achromobacter fischeri*. Arch Biochem Biophys 148:614–632

Prince RC (1978) The reaction centre and associated cytochromes of *Thiocapsa pfennigii*. Biochim Biophys Acta 501:195–207

Prince RC, Bashford CL (1979) Equilibrium and kinetic measurements of the redox potentials of cytochromes c_2 in vitro and in vivo. Biochim Biophys Acta 547:447–454

Prince RC, Dutton PL (1977a) Single and multiple turnover reactions in the ubiquinone cytochrome bc_2 oxidoreductase of *Rhodopseudomonas sphaeroides* – the physical chemistry of the major electron donor to cytochrome c_2 and its coupled reactions. Biochim Biophys Acta 462:731–747

Prince RC, Dutton PL (1977b) pH dependence of the oxidation reduction potential of cytochromes c_2 in vivo. Biochim Biophys Acta 459:573–577

Prince RC, Olson JM (1976) Some thermodynamic and kinetic properties of the primary photochemical reactants in a complex from a green photosynthetic bacterium. Biochim Biophys Acta 423:357–362

Prince RC, Leigh JS, Dutton PL (1974a) An electron spin resonance characterisation of *Rhodopseudomonas capsulata*. Biochem Soc Trans 2:950–953

Prince RC, Cogdell RJ, Crofts AR (1974b) The photooxidation of horse heart cytochrome c and native cytochrome c_2 by reaction centres from *Rhodopseudomonas sphaeroides*. Biochim Biophys Acta 347:1–13

Prince RC, Baccarini-Melandri A, Hauska GA, Melandri BA, Crofts AR (1975) Asymmetry of an energy transducing membrane and the location of cytochrome c_2 in *Rhodopseudomonas sphaeroides* and *Rhodopseudomonas capsulata*. Biochim Biophys Acta 387:212–227

Prince RC, Leigh JS, Dutton PL (1976) Thermodynamic properties of the reaction centre of *Rhodopseudomonas viridis*. Biochim Biophys Acta 440:622–636

Prince RC, Bashford CL, Takamiya K, Berg WH van den, Dutton PL (1978) Second-order kinetics of the reduction of cytochrome c_2 by the ubiquinol cytochrome bc_2 oxidoreductase of *Rhodopseudomonas sphaeroides*. J Biol Chem 253:4137–4142

Prince RC, Matsuura K, Hurt E, Hauska G, Dutton PL (1982) Reduction of cytochromes b_6 and f in isolated plastoquinol plastocyanin oxidoreductase driven by photochemical reaction centres from *Rhodopseudomonas sphaeroides*. J Biol Chem 257:3379–3381

Probst I, Bruschi M, Pfennig N, LeGall JL (1977) Cytochrome c-551.5 (c_7) from *Desulfuromonas acetoxidans*. Biochim Biophys Acta 460:58–64

Pucheu NL, Kerber NL, Garcia AF (1976) Isolation and purification of reaction centre from *Rhodopseudomonas viridis* NHTC133 by means of LDAO. Arch Microbiol 109:301–305

Puttner I, Solioz M, Carafoli E, Ludwig B (1983) DCCD does not inhibit proton pumping by cytochrome c oxidase of *Paracoccus denitrificans*. Eur J Biochem 134:33–37

Rees MK (1968) Effect of chloride on the oxidation of hydroxylamine by *Nitrosomonas europaea* cells. J Bacteriol 95:243–244

Rees MK, Nason A (1965) A P450 like cytochrome and a soluble terminal oxidase identified as cytochrome o from *Nitrosomonas europaea*. Biochem Biophys Res Commun 21: 248–250

Renner ED, Becker GE (1970) Production of NO and N_2O during denitrification by *Corynebacterium nephridii*. J Bacteriol 101:821–826

Rickle GK, Cusanovich MA (1979) The kinetics of photooxidation of c-type cytochromes by *Rhodospirillum rubrum* reaction centres. Arch Biochem Biophys 197:589–598

Rieder R, Wiemken V, Bachofen R, Bosshard HR (1985) Binding of cytochrome c_2 to the isolated reaction centre of *Rhodospirillum rubrum* involves the backside of cytochrome c_2. Biochem Biophys Res Commun 128:120–126

Riederer-Henderson MA, Peck HD (1970) Formic dehydrogenase of *Desulfovibrio gigas*. Bacteriol Proc 70:134
Riegler J, Peschke J, Mohwald H (1984) 2-dimensional electron transfer from cytochrome c to photosynthetic reaction centres. Biochem Biophys Res Commun 125:592–599
Ritchie GAF, Nicholas DJD (1974) The partial characterisation of purified nitrite reductase and hydroxylamine oxidase from *Nitrosomonas europaea*. Biochem J 138:471–480
Robinson MK, Martinkus K, Kennelly PJ, Timkovich R (1979) Implications of the integrated rate law for the reactions of *Paracoccus* nitrite reductase. Biochemistry 18:3921–3926
Ronnberg M (1976) *Pseudomonas* cytochrome c peroxidase – product inhibition studies. Acta Chem Scand B30:721–726
Ronnberg M, Ellfolk N (1975) Kinetics of peroxidatic oxidation of *Pseudomonas* respiratory chain components. Acta Chem Scand B29:719–729
Ronnberg M, Ellfolk N (1979) Heme-linked properties of *Pseudomonas* cytochrome c peroxidase – evidence for non-equivalence of hemes. Biochim Biophys Acta 581:325–333
Ronnberg M, Osterlund K, Ellfolk N (1980) Resonance Raman spectra of *Pseudomonas* cytochrome c peroxidase. Biochim Biophys Acta 623:23–30
Ronnberg M, Araiso T, Ellfolk N, Dunford HB (1981 a) The reaction between reduced azurin and oxidised cytochrome c peroxidase from *Pseudomonas aeruginosa*. J Biol Chem 256:2471–2474
Ronnberg M, Araiso T, Ellfolk N, Dunford HB (1981 b) The catalytic mechanism of *Pseudomonas* cytochrome c peroxidase. Arch Biochem Biophys 207:197–204
Rosen D, Okamura MY, Feher G (1980) Interaction of cytochrome c with reaction centres of *Rhodopseudomonas sphaeroides* R-26; determination of number of binding sites and dissociation constants by equilibrium dialysis. Biochemistry 19:5687–5692
Rosen D, Okamura MY, Abresch EC, Valkirs GE, Feher G (1983) Interaction of cytochrome c with reaction centres of *Rhodopseudomonas sphaeroides* R26: localisaton of the binding site by chemical cross-linking and immunochemical studies. Biochemistry 22:335–341
Rowe JJ, Sherr BF, Payne WJ, Eagon RG (1977) Unique NO binding complex formed by denitrifying *Pseudomonas aeruginosa*. Biochem Biophys Res Commun 77:253–258
Santos M, Moura JJG, Moura I, Le Gall J, Xavier AV (1984) NMR studies of electron transfer mechanisms in a protein with interacting redox centres – *Desulfovibrio gigas* cytochrome c_3. Eur J Biochem 141:283–296
Sapshead LM, Wimpenny JWT (1972) The influence of O_2 and NO_3^- on the formation of the cytochrome pigments of the aerobic and anaerobic respiratory chain of *Micrococcus denitrificans*. Biochim Biophys Acta 267:388–397
Saraste M, Virtanen I, Kuronen T (1977) The quaternary structure of *Pseudomonas* cytochrome oxidase studied by electron microscopy. Biochim Biophys Acta 492:156–162
Sawada E, Satoh T, Kitamura H (1978) Purification and properties of a dissimilatory nitrite reductase of a denitrifying phototrophic bacterium. Plant Cell Physiol 19:1339–1351
Sawhney V, Nicholas DJD (1977) Sulfite and NADH dependent nitrate reductase from *Thiobacillus denitrificans*. J Gen Microbiol 100:49–58
Sawhney V, Nicholas DJD (1978) Sulfide linked nitrite reductase from *Thiobacillus denitrificans* with cytochrome oxidase activity. J Gen Microbiol 106:119–128
Schmid GH, Radunz A, Menke W (1975) The effect of an antiserum to plastocyanin on various chloroplast preparations. Z Naturforsch 30c:201–212
Scholes PB, Smith L (1968) Composition and properties of the membrane-bound respiratory chain system of *Micrococcus denitrificans*. Biochim Biophys Acta 153:363–375
Scholes PB, McLain G, Smith L (1971) Purification and properties of a c-type cytochrome from *Micrococcus denitrificans*. Biochemistry 10:2072–2075
Seibert M, DeVault D (1970) Relationship between the laser-induced oxidation of the high and low potential cytochromes of *Chromatium vinosum*. Biochim Biophys Acta 205:222–231
Sewell DL, Aleem MJH (1969) The mechanism of pyridine nucleotide reduction by nitrite in *Nitrobacter agilis*. Biochim Biophys Acta 172:467–475

Sewell DL, Aleem MIH, Wilson DF (1972) The oxidation reduction potentials and rates of oxidation of the cytochromes of *Nitrobacter agilis*. Arch Biochem Biophys 153:312–319
Shapleigh JP, Payne WJ (1985) Nitric oxide dependent proton translocation in various denitrifiers. J Bacteriol 163:837–840
Shidara S (1980) Components of the cytochrome system of *Alcaligenes* sp.NCIB 11015 with special reference to particulate bound c-type cytochromes. J Biochem 87:1177–1184
Shill DA, Wood PM (1984) A role for cytochrome c_2 in *Rhodopseudomonas viridis*. Biochim Biophys Acta 764:1–7
Shimada H, Orii Y (1975) The NO compounds of *Pseudomonas aeruginosa* nitrite reductase and their probable participation in the nitrite reduction. FEBS Lett 54:237–240
Shimada H, Orii Y (1976) Oxidation reduction behaviour of the heme c and heme d moieties of *Pseudomonas aeruginosa* nitrite reductase and the formation of an oxygenated intermediate at heme d_1. J Biochem 80:135–140
Shimada H, Orii Y (1978) The pH dependent reactions of *Pseudomonas aeruginosa* nitrite reductase with nitric oxide and nitrite. J Biochem 84:1553–1558
Shimazaki K, Takamiya K, Nishimura M (1978) Studies on electron transfer systems in the marine diatom *Phaeodactylum tricormitum*. J Biochem 83:1631–1638
Shioi Y, Takamiya K, Nishimura M (1972) Studies on energy and electron transfer systems in the green photosynthetic bacterium *Chloropseudomonas* 2K. J Biochem 71:285–291
Shioi Y, Takamiya K, Nishimura M (1976) Light induced oxidation reduction reactions of cytochromes in the green sulfur photosynthetic bacterium *Prosthecochloris aestuarii*. J Biochem 80:811–820
Shiozawa JA, Alberte RS, Thornber JP (1974) The P700 chlorophyll *a* protein – isolation and some characteristics of the complex in higher plants. Arch Biochem Biophys 165:388–397
Silvestrini MC, Colosimo A, Brunori M, Walsh TA, Barber D, Greenwood C (1979) A reevaluation of some basic structural and functional properties of *Pseudomonas* cytochrome oxidase. Biochem J 183:701–709
Silvestrini MC, Brunori M, Wilson MT, Darley-Usmar VM (1981) The electron transfer system of *Pseudomonas aeruginosa* – a study of the pH dependent transitions between redox forms of azurin and cytochrome c-551. J Inorg Biochem 14:327–338
Silvestrini MC, Tordi MG, Colosimo A, Antonini E, Brunori M (1982) The kinetics of electron transfer between *Pseudomonas aeruginosa* cytochrome c-551 and its oxidase. Biochem J 203:445–451
Silvestrini MC, Citro G, Colosimo A, Chersi A, Zito R, Brunori M (1983) Purification of *Pseudomonas* cytochrome oxidase by immunological methods. Anal Biochem 129: 318–325
Singh J, Wasserman AR (1971) The use of disc gel electrophoresis with non-ionic detergent in the purification of cytochrome f from spinach grana membranes. J Biol Chem 246:3532–3541
Singh J, Wharton DC (1973) Cytochrome c-556, a diheme protein from *Pseudomonas aeruginosa*. Biochim Biophys Acta 292:391–401
Singleton R, Deins J, Campbell LL (1982) Cytochrome c_3 from the sulfate reducing anaerobe *Desulfovibrio africanus* Benghazi – antigenic properties. J Bacteriol 152: 527–529
Smith DD, Selman BR, Voegeli KK, Johnson G, Willey RA (1977) Chloroplast membrane sidedness – location of plastocyanin determined by chemical modifiers. Biochim Biophys Acta 459:468–482
Snyder SW, Hollocher TC (1984) N_2O reductase and the 120 K Cu protein of N_2-producing denitrifying bacteria are different entities. Biochem Biophys Res Commun 119:588–592
Soininen R, Ellfolk N (1972) *Pseudomonas* cytochrome c peroxidase – some kinetic properties of the peroxidation reaction and enzymatic determination of the extinction coefficients of *Pseudomonas* cytochrome c-551 and azurin. Acta Chem Scand 26:861–872
Soininen R, Ellfolk N (1975) *Pseudomonas* cytochrome c peroxidase – effect of *Pseudomonas* neutral proteinase on the enzyme molecule. Acta Chem Scand B29:134–136

Solioz M, Carafoli E, Ludwig B (1982) The cytochrome c oxidase of *Paracoccus denitrificans* pumps protons in a reconstituted system. J Biol Chem 257:1579–1582

Sone N, Hinkle PC (1982) Proton transport by cytochrome oxidase from the thermophilic bacterium PS3 reconstituted by liposomes. J Biol Chem 257:12600–12604

Sone N, Yanagita Y (1982) A cytochrome aa_3 type terminal oxidase of a thermophilic bacterium – purification, properties and proton-pumping. Biochim Biophys Acta 682:216–226

Sone N, Yanagita Y (1984) High vectorial proton stoichiometry by cytochrome oxidase from the thermophilic bacterium PS3 reconstituted in liposomes. J Biol Chem 259:1405–1408

Sone N, Ohyama T, Kagawa Y (1979) Thermostable single band cytochrome oxidase. FEBS Lett 106:39–42

Sone N, Yanagita Y, Hon-nami K, Fukumori Y, Yamanaka T (1983) Proton-pumping activity of *Nitrobacter agilis* and *Thermus thermophilus* cytochrome c oxidases. FEBS Lett 155:150–154

Sperl GT, Forrest HS, Gibson DT (1974) Substrate specificity of the purified primary alcohol dehydrogenases from methanol oxidising bacteria. J Bacteriol 118:541–550

Stanier RY, Palleroni NJ, Douderoff M (1966) The aerobic Pseudomonads – a taxonomic study. J Gen Microbiol 43:159–272

Steenkamp DJ, Peck HD (1981) Proton translocation associated with nitrite respiration in *Desulfovibrio desulfuricans*. J Biol Chem 256:5450–5458

Steffens GCM, Buse G, Oppliger W, Ludwig B (1983) Sequence homology of bacterial and mitochondrial cytochrome c oxidases. Biochem Biophys Res Commun 116:335–340

Steinmetz MA, Fischer U (1981) Cytochromes of the non-sulfate-utilising bacterium, *Chlorobium limicola*. Arch Microbiol 130:31–37

Steinmetz MA, Fischer U (1982) Cytochromes of the green sulfur bacterium *Chlorobium vibrioforme* f. thiosulfatophilum – purification, characterisation and sulfur metabolism. Arch Microbiol 131:19–26

Steinmetz MA, Truper HG, Fischer U (1983) Cytochrome c-555 and iron sulfur proteins of the non-thiosulfate utilising green sulfur bacterium *Chlorobium vibrioforme*. Arch Microbiol 135:186–190

Stewart AC, Bendall DS (1980) Photosynthetic electron transport in a cell-free preparation from the thermophilic blue green alga *Phormidium laminosum*. Biochem J 188:351–361

Stirling DI, Dalton H (1978) Purification and properties of NADP-linked formaldehyde dehydrogenase from *Methylococcus capsulatus* (Bath). J Gen Microbiol 107:19–29

St. John RT, Hollocher TC (1977) Nitrogen 15 tracer studies on the pathway of denitrification in *Pseudomonas aeruginosa*. J Biol Chem 252:212–218

Stouthamer AH (1980) Bioenergetic studies on *Paracoccus denitrificans*. Trends Biochem Sci 5:164–166

Stouthamer AH, Riet J vant, Oltman LF (1980) Respiration with nitrate as acceptor. In: Knowles CJ (ed) Diversity of bacterial respiratory systems. CRC Press, Boca Raton, Florida, pp 19–48

Sugimura Y, Hase T, Matsubara H (1981) Amino acid sequence of cytochrome c-553 from a brown alga *Petalonia fascia*. J Biochem 90:1213–1219

Sugimura Y, Hosoya K, Yoshizaki F, Shimokoryama M (1984) Purification and characterisation of cytochrome c-552 from a red alga *Polysiphona urceolata*. J Biochem 96: 1681–1687

Susor WA, Krogman DW (1966) Triphosphopyridine nucleotide reduction with cell-free preparations of *Anabaena variabilis*. Biochim Biophys Acta 120:65–72

Suzuki H, Iwasaki H (1962) Azurin and cytochrome c in a denitrifying bacterium. J Biochem 52:193–199

Suzuki I (1974) Mechanisms of inorganic oxidation and energy coupling. Annu Rev Microbiol 28:85–101

Suzuki I, Kwok S-C (1981) A partial resolution and reconstitution of the ammonia oxidising system of *Nitrosomonas europaea* – the role of cytochrome c-554. Can J Biochem 59:484–488

Swank RT, Burris RH (1969) Purification and properties of cytochromes c of *Azotobacter vinelandii*. Biochim Biophys Acta 180:473–489

Swarthoff T, Veek-Horsley KM van der, Amesz J (1981) The primary charge separation, cytochrome oxidation and triplet formation in preparations from the green photosynthetic bacterium *Prosthecochloris aestuarii*. Biochim Biophys Acta 635:1–12

Sybesma C (1967) Light-induced cytochrome reactions in the green photosynthetic bacterium *Chloropseudomonas ethylicum*. Photochem Photobiol 6:261–267

Takabe T, Ishikawa H, Niwa S, Tanaka Y (1984) Electron transfer reactions of chemically modified plastocyanin with P700 and cytochrome f – importance of local charges. J Biochem 96:385–393

Takakuwa S (1975) Purification and some properties of a cytochrome c-552 from *Thiobacillus thiooxidans*. J Biochem 78:181–185

Takano T, Dickerson RE, Schichman SA, Meyer TE (1979) Crystal data, molecular dimensions and molecular symmetry in cytochrome oxidase from *Pseudomonas aeruginosa*. J Mol Biol 133:185–188

Takenaka K, Takabe T (1984) Importance of local positive charges on cytochrome f for electron transfer to plastocyanin and potassium ferricanide. J Biochem 96:1813–1821

Tanaka K, Takahashi M, Asada K (1978) Isolation of monomeric cytochrome f from Japanese radish and a mechanism for autoreduction. J Biol Chem 253:7397–7403

Tanaka Y, Fukumori Y, Yamanaka T (1982) The complete amino acid sequence of *Nitrobacter agilis* cytochrome c-550. Biochim Biophys Acta 707:14–20

Tanaka Y, Fukumori Y, Yamanaka T (1983) Purification of cytochrome a_1c_1 from *Nitrobacter agilis* and characterisation of the nitrite oxidation system. Arch Microbiol 135:265–271

Taniguchi S, Kamen MD (1965) The oxidase system of heterotrophically grown *Rhodospirillum rubrum*. Biochim Biophys Acta 96:395–428

Terry KR, Hooper AB (1981) Hydroxylamine oxidoreductase – a 20 heme, 200000 molecular weight cytochrome c with unusual denaturation properties which forms a 63000 monomer after heme removal. Biochemistry 20:7026–7032

Thauer RK, Badziong W (1980) Respiration with sulfate as electron acceptor. In: Knowles CJ (ed) Diversity of bacterial respiratory systems, CRC Press Boca Raton Florida, pp 65–85

Thauer RK, Jungerman K, Decker K (1977) Energy conservation in chemotrophic anaerobic bacteria. Bacteriol Revs 41:100–180

Then J, Truper HG (1983) Sulfide oxidation in *Ectothiorhodospora abdelmalekii* – evidence for a catalytic role of cytochrome c-551. Arch Microbiol 135:254–258

Thomas PE, Ryan D, Levin W (1976) An improved staining procedure for the detection of the peroxidase activity of cytochrome P450 on sodium dodecyl sulfate polyacrylamide gels. Anal Biochem 75:168–176

Thornber JP (1969) Comparison of a chlorophyll a-protein complex isolated from a blue green alga with chlorophyll-protein complexes obtained from green bacteria and higher plants. Biochim Biophys Acta 172:230–241

Thornber JP, Olson JM (1971) Chlorophyll proteins and reaction centre preparations from photosynthetic bacteria, algae and higher plants. Photochem Photobiol 14:329–341

Thornber JP, Cogdell RJ, Selfor REB, Webster GD (1980) Further studies of the composition and spectral properties of the photochemical reaction centres of bacteriochlorophyll b containing bacteria. Biochim Biophys Acta 593:60–75

Tiede DM, Leigh JS, Dutton PL (1978) Structural organisation of the *Chromatium vinosum* reaction centre associated c-type cytochromes. Biochim Biophys Acta 503:524–544

Tikhonova GV, Lisenkova LL, Doman NG, Skulachev VP (1967) Puti perenosa elektronov u zhelezookisliaushchikh bakteri *Thiobacillus ferrooxidans*. Biokhimiya 32:725–734

Timkovich R, Cork MS (1983) Magnetic susceptibility measurements on *Pseudomonas* cytochrome cd_1. Biochim Biophys Acta 742:162–168

Timkovich R, Robinson MK (1979) Evidence for water as the product of oxygen reduction in cytochrome cd_1. Biochem Biophys Res Commun 88:649–655

Timkovich R, Dickerson RE, Margoliash E (1976) Amino acid sequence of *Paracoccus denitrificans* cytochrome c-550. J Biol Chem 251:2197–2206

Timkovich R, Dhesi R, Martinkus KJ, Robinson MK, Rea TM (1982) Isolation of *Paracoccus denitrificans* cytochrome cd_1 – comparative kinetics with other nitrite reductases. Arch Biochem Biophys 215:47–58

Timkovich R, Cork MS, Taylor PV (1984) Proposed structure for the non-covalently associated heme prosthetic group of dissimilatory nitrite reductases. J Biol Chem 259: 15089–15093

Titani K, Ericsson LH, Hon-nami K, Miyazawa T (1985) Amino acid sequence of cytochrome c-552 from *Thermus thermophilus* HB8. Biochem Biophys Res Commun 128:781–787

Tollin G, Meyer TE, Cusanovich MA (1982) Intramolecular electron transport in *Chlorobium thiosulfatophilum* flavocytochrome c. Biochemistry 21:3849–3856

Tonge GM, Harrison DEF, Knowles CJ, Higgins IJ (1975) Properties and partial purification of the methane oxidising enzyme system from *Methylosinus trichosporium.* FEBS Lett 58:293–299

Tonge GM, Harrison DEF, Higgins IJ (1977) Purification and properties of the methane monooxygenase enzyme system from *Methylosinus trichosporium.* Biochem J 161: 333–344

Tordi MG, Silvestrini MC, Colosimo A, Tuttobello L, Brunori M (1985) Cytochrome c-551 and azurin oxidation catalysed by *Pseudomonas aeruginosa* cytochrome oxidase. Biochem J 230:797–805

Tronson DA, Ritchie GAF, Nicholas DJD (1973) Purification of c-type cytochromes from *Nitrosomonas europaea.* Biochim Biophys Acta 310:331–343

Trosper TL, Benson DL, Thornber JP (1977) Isolation and spectral characteristics of the photochemical reaction centre of *Rhodopseudomonas viridis.* Biochim Biophys Acta 460:318–330

Trumpower B (1981) Function of the iron sulfur protein of the cytochrome bc_1 segment in electron transfer and energy conserving reactions of the mitochondrial electron transport chain. Biochim Biophys Acta 639:129–155

Truper HG, Fischer U (1982) Anaerobic oxidation of sulfur compounds as electron donors for bacterial photosynthesis. Philos Trans R Soc London Ser B 298:529–542

Truper HG, Rogers LA (1971) Purification and properties of adenylyl sulfate reductase from a phototrophic sulfur bacterium *Thiocapsa roseopersicina.* J Bacteriol 108:1112–1121

Tsang DCY, Suzuki I (1982) Cytochrome c-554 as a possible electron donor in the hydroxylation of ammonia and CO in *Nitrosomonas europaea.* Can J Biochem 60:1018–1024

Tsuji T, Fujita Y (1972) Electron donor specificity observed in photosystem 1 reactions of membrane fragments of the blue green alga *Anabaena variabilis* and the higher plant *Spinacea oleracea.* Plant Cell Physiol 13:93–99

Tsuji K, Yagi T (1980) Significance of the hydrogen burst from growing cultures of *Desulfovibrio vulgaris* Miyazaki and the role of hydrogenase and cytochrome c_3 in the energy production system. Arch Microbiol 125:35–42

Valkirs GE, Feher G (1982) Topography of reaction centre subunits in the membrane of the photosynthetic bacterium *Rhodopseudomonas sphaeroides.* J Cell Biol 95:179–188

Versefeld HW van, Stouthamer AH (1978) Electron transport chain and coupled oxidative phosphorylation in methanol-grown *Paracoccus denitrificans.* Arch Microbiol 118:13–20

Vickery LE, Palmer G, Wharton DC (1978) Heme c-heme d_1 interaction in *Pseudomonas* cytochrome oxidase – a reappraisal of spectroscopic evidence. Biochem Biophys Res Commun 80:458–463

Villalain J, Moura I, Liu MC, Payne WJ, LeGall J, Xavier AV, Moura JG (1984) NMR and epr studies of a dihaem cytochrome from *Pseudomonas stutzeri* (ATCC 11607). Eur J Biochem 141:305–312

Visser JW, Rijgersberg KP, Amesz J (1974) Light induced reactions of ferredoxin and P700 at low temperatures. Biochim Biophys Acta 368:235–246

Vrij W, Azzi A, Konings WN (1983) Structural and functional properties of cytochrome c oxidase from *Bacillus subtilis* W23. Eur J Biochem 131:97–103

Wakabayashi S, Matsubara H, Kim CH, King TE (1982) Structural study on bovine cytochrome c_1. J Biol Chem 257:9335–9344

Walsh TA, Johnson MK, Greenwood C, Barber D, Springall JP, Thomson AJ (1979) Some magnetic properties of *Pseudomonas* cytochrome oxidase. Biochem J 177:29–39

Wasserman AR (1980) Chloroplast cytochrome f. Meth Enzymol 69:181–202

Webster DA (1975) The formation of hydrogen peroxide during the oxidation of reduced nicotinamide adenine dinucleotide by cytochrome o of *Vitreoscilla*. J Biol Chem 250:4955–4958

Westen HM van der, Mayhew SG, Veeger C (1978) Separation of hydrogenase from intact cells of *Desulfovibrio vulgaris*. FEBS Lett 86:122–126

Wharton DC, Gibson QH (1976) Cytochrome oxidase from *Pseudomonas aeruginosa* – reaction with oxygen and carbon monoxide. Biochim Biophys Acta 430:445–453

Wharton DC, Weintraub ST (1980) Identification of nitric oxide and nitrous oxide as products of nitrite reduction by *Pseudomonas* nitrite reductase. Biochem Biophys Res Commun 97:236–242

Wharton DC, Gudat JC, Gibson QH (1973) Cytochrome oxidase from *Pseudomonas aeruginosa* – reaction with copper protein. Biochim Biophys Acta 292:611–620

Widdowson D, Anthony C (1975) Microbial metabolism of c-1 compounds – the electron transport chain of *Pseudomonas* AMl. Biochem J 152:349–356

Wildner GF, Hauska G (1974) Localisation of the reaction site of cytochrome c-552 in chloroplasts of *Euglena gracilis*. Arch Biochem Biophys 164:136–144

Willey DL, Aufret AD, Gray JC (1984) Structure and topology of cytochrome f in pea chloroplast membranes. Cell 36:555–562

Williams JC, Steiner LA, Ogden RC, Simon MI, Feher G (1983) Primary structure of the M-subunit of the reaction centre from *Rhodopseudomonas sphaeroides*. Proc Natl Acad Sci USA 80:6505–6509

Witt HT (1979) Energy conversion in the functional membrane of photosynthesis – analysis by light pulse and electric pulse methods. Biochim Biophys Acta 505:355–427

Wood PM (1974) Rate of electron transfer between plastocyanin, cytochrome f, related proteins and artificial redox reagents in solution. Biochim Biophys Acta 357:370–379

Wood PM (1977) The roles of c-type cytochromes in algal photosynthesis. Eur J Biochem 72:605–612

Wood PM (1978a) Periplasmic location of the terminal reductase in nitrite respiration. FEBS Lett 92:214–219

Wood PM (1978b) Interchangeable copper and iron proteins in algal photosynthesis. Eur J Biochem 87:9–19

Wood PM (1978c) A chemiosmotic model for sulfate respiration. FEBS Lett 95:12–18

Wood PM (1980a) Interrelationships of the two c-type cytochromes in *Rhodopseudomonas sphaeroides* photosynthesis. Biochem J 192:761–764

Wood PM (1980b) Do photosynthetic bacteria contain cytochrome c_1? Biochem J 189:385–391

Wood PM (1981) Fluorescent gels as a general technique for characterising bacterial c-type cytochromes. Anal Biochem 111:235–239

Wood PM (1983) Why do c-type cytochromes exist? FEBS Lett 164:223–226

Wood PM (1984) Bacterial proteins with CO binding b or c-type haem – functions and absorption spectroscopy. Biochim Biophys Acta 768:293–317

Wood PM, Bendall DS (1975) The kinetics and specificity of electron transfer from cytochromes and copper proteins to P700. Biochim Biophys Acta 387:115–128

Wood PM, Bendall DS (1976) The reduction of plastocyanin by plastoquinol-1 in the presence of chloroplasts. Eur J Biochem 61:337–344

Wood PM, Willey DL (1980) Use of fluorescent gels to characterise a membrane cytochrome c of *Pseudomonas aeruginosa*. FEMS Lett 7:273–277

Wynn RM, Gaul DF, Shaw RW, Knaff DB (1985) Identification of the components of a putative cytochrome bc_1 complex in *Rhodopseudomonas viridis*. Arch Biochem Biophys 238:373–377
Yagi T (1970) Solubilisation, purification and properties of particulate hydrogenase from *Desulfovibrio vulgaris*. J Biochem 68:649–657
Yagi T (1979) Purification and properties of cytochrome c-553, an electron transport acceptor for formate dehydrogenase of *Desulfovibrio vulgaris* Miyazaki. Biochim Biophys Acta 548:96–105
Yagi T (1984) Spectral and kinetic abnormality during the reduction of cytochrome c_3 catalysed by hydrogenase with hydrogen. Biochim Biophys Acta 767:288–294
Yagi T, Honya M, Tamiya N (1968) Purification and properties of hydrogenases of different origins. Biochim Biophys Acta 153:699–705
Yagi T, Maruyama K (1971) Purification and properties of cytochrome c_3 of *D. vulgaris* (Miyazaki). Biochim Biophys Acta 243:214–224
Yakushiji E (1971) Algal cytochromes. Meth Enzymol 22:364–368
Yamanaka T (1972) A cytochrome c peroxidase isolated from *Thiobacillus novellus*. Biochim Biophys Acta 275:74–82
Yamanaka T (1975) A comparative study of the redox reactions of cytochromes c with certain enzymes. J Biochem 77:493–499
Yamanaka T, Fujii K (1980) Cytochrome a-type terminal oxidase derived from *Thiobacillus novellus*. Biochem Biophys Acta 591:53–62
Yamanaka T, Fukumori Y (1977) *Thiobacillus novellus* cytochrome oxidase can separate some eukaryotic cytochromes. FEBS Lett 77:155–158
Yamanaka T, Fukumori Y (1981) Functional and structural comparisons between prokaryotic and eukaryotic aa_3-type cytochrome c oxidases from an evolutionary point of view. Plant Cell Physiol 22:1223–1230
Yamanaka T, Okunuki K (1963a) Reconstitution of *Pseudomonas* cytochrome oxidase from haem a_2 and its protein moiety and some properties of the reconstituted enzyme. Biochem Z 338:62–72
Yamanaka T, Okunuki K (1963b) Crystalline *Pseudomonas* cytochrome oxidase – the spectral properties of the enzyme. Biochim Biophys Acta 67:394–406
Yamanaka T, Okunuki K (1968) Comparative studies of the reactivities of cytochrome c with cytochrome c oxidases. In: Okunuki K, Sekuzu I, Kamen MD (eds) Structure and function of cytochromes. Academic Press, London New York, pp 390–403
Yamanaka T, Okunuki K (1974) Cytochromes. In: Microbial iron metabolism. Academic Press, London New York, pp 349–400
Yamanaka T, Shinra M (1974) Cytochrome c-552 and cytochrome c-554 derived from *Nitrosomonas europaea*. J Biochem 75:1265–1273
Yamanaka T, Ota A, Okunuki K (1961) A nitrite reducing system reconstructed with purified cytochrome components of *Pseudomonas aeruginosa*. Biochim Biophys Acta 53:294–308
Yamanaka T, Takenami S, Akijama N, Okunuki K (1971) Purification and properties of cytochrome c-550 and cytochrome c-551 derived from the facultative chemoautotroph – *Thiobacillus novellus*. J Biochem 70:349–358
Yamanaka T, Fukumori Y, Wada K (1978) Cytochrome c-554 derived from the blue green alga *Spirulina platensis*. Plant Cell Physiol 19:117–126
Yamanaka T, Shinra M, Takahashi K, Shibasaka M (1979a) Highly purified hydroxylamine oxidoreductase derived from *Nitrosomonas europaea*. J Biochem 86:1101–1108
Yamanaka T, Fujii K, Kamita Y (1979b) Subunits of cytochrome a-type terminal oxidase derived from *Thiobacillus novellus* and *Nitrobacter agilis*. J Biochem 86:821–824
Yamanaka T, Yoshioka T, Kimura K (1981a) Purification of sulfite cytochrome c reductase of *Thiobacillus novellus* and reconstitution of its sulfite oxidase system with purified components. Plant Cell Physiol 22:613–622
Yamanaka T, Kamita Y, Fukumori Y (1981b) Molecular and enzymatic properties of cytochrome aa_3 type terminal oxidase derived from *Nitrobacter agilis*. J Biochem 89:265–273

Yamanaka T, Tanaka Y, Fukumori Y (1982) *Nitrobacter agilis* cytochrome c-550 – isolation, physicochemical and enzymatic properties and primary structure. Plant Cell Physiol 23:441–449

Yamanaka T, Fukumori Y, Yamazaki T, Kato H, Nakayama K (1985) A comparative survey of several bacterial aa_3 type cytochrome c oxidase. J Inorg Biochem 23:273–277

Yanagita Y, Sone N, Kagawa Y (1983) Proton pumping and oxidase activity of thermophilic cytochrome oxidase remain after extensive proteolysis. Biochem Biophys Res Commun 113:575–580

Yang T (1982) Tetramethylphenylenediamine oxidase of *Pseudomonas aeruginosa.* Eur J Biochem 121:335–341

Yang T, O'Keefe D, Chance B (1979) The oxidation reduction potentials of cytochrome $o+c_4$ and cytochrome o purified from *Azotobacter vinelandii.* Biochem J 181:763–766

Yoshida T, Fee TA (1984) Studies of cytochrome c oxidase activity of the cytochrome c_1 aa_3 complex of *Thermus thermophilus.* J Biol Chem 259:1031–1036

Yoshida T, Lorence RM, Choc MG, Tarr GE, Findling KL, Fee JA (1984) Respiratory proteins from the extremely thermophilic aerobe, *Thermus thermophilus.* J Biol Chem 259:112–123

Yu L, Yu CA (1982) Isolation and properties of the cytochrome bc_1 complex from *Rhodopseudomonas sphaeroides.* Biochem Biophys Res Commun 108:1285–1292

Yu K, Mei QC, Yu CA (1984) Characterisation of purified cytochrome bc_1 complex from *Rhodopseudomonas sphaeroides.* J Biol Chem 259:5752–5760

Yun L, Leonard K, Weiss H (1981) Membrane bound and water soluble cytochrome c_1 from *Neurospora* mitochondria. Eur J Biochem 116:199–205

Zannoni D, Baccarini-Melandri A, Melandri BA, Evans EH, Prince RC, Crofts AR (1974) Energy transduction in photosynthetic bacteria – the nature of cytochrome c oxidase in the respiratory chain of *Rhodopseudomonas sphaeroides.* FEBS Lett 48:152–155

Zannoni D, Melandri BA, Baccarini-Melandri A (1976) Composition and function of the branched oxidase system in wild type and respiration-deficient mutants of *Rhodopseudomonas capsulata.* Biochim Biophys Acta 423:413–430

Zannoni D, Prince RC, Dutton PL, Marrs BL (1980) Isolation and characterisation of a cytochrome c_2-deficient mutant of *Rhodopseudomonas capsulata.* FEBS Lett 113: 289–293

Zhu QS, Wal HN van der, Grondelle R van, Berden JA (1983) Kinetics of flash induced electron transfer between bacterial reaction centres, mitochondrial ubiquinol cytochrome c reductase and cytochrome c. Biochim Biophys Acta 725:121–130

Zhu QS, Wal HN van der, Grondelle R van, Berden JA (1984) Flash-induced electron transfer through mitochondrial QH_2-cytochrome c oxidoreductase in the presence of bacterial reaction centres and cytochrome c. Biochim Biophys Acta 765:48–57

Zumft WG, Matsubara T (1982) A novel kind of multicopper protein as terminal oxidoreductase of N_2O respiration in *Pseudomonas perfectomarinus.* FEBS Lett 148:107–112

Zumft WG, Sherr BF, Payne WJ (1979) A reappraisal of the NO binding protein of denitrifying Pseudomonads. Biochem Biophys Res Commun 88:1230–1236

Zumft WG, Coyle CL, Frunzke K (1985) The effect of oxygen on the chromatographic behaviour and properties of nitrous oxide reductase. FEBS Lett 183:240–244

[illegible]

Chapter 4 The Biosynthesis of Cytochrome c

4.1 Gene Structure and the Control of Gene Expression

It is ironic that, in spite of the wealth of information on gene structure and expression in bacteria, very little is known about the bacterial cytochromes. Only recently has the gene structure of cytochrome c_2 been determined and there is no genetic information on the controls that must operate to allow the striking metabolic versatility of the bacteria. In contrast, the yeast isocytochrome c system has been extensively characterised by classical genetic methods and more recently using the powerful approaches of recombinant DNA technology. We divide consideration of the subject into the structure of cytochrome c genes and the control of their expression.

4.1.1 The Structure of the Genes for Mitochondrial Cytochrome c

Bakers' yeast (*Saccharomyces cerevisiae*) synthesises two isocytochromes c (iso-1 and iso-2) encoded at two unlinked nuclear loci, CYC1 and CYC7 (Sherman et al. 1966; Downie et al. 1977). A large number of mutations have been analysed which include changes within the translated part of the gene, changes in the untranslated regions and changes unlinked to the structural genes. The latter two are of interest in the control of gene expression and will be discussed in the following section. The analysis of frameshift mutations in the coding region of the CYC1 gene (Stewart and Sherman 1974) allowed prediction of the nucleotide sequence of the N-terminal region. On this basis a complementary oligonucleotide probe was synthesised and used to select restriction fragments of the yeast genome which contained the iso-1-cytochrome c gene (Smith et al. 1979). The cloned CYC1 gene has since been used as a hybridisation probe to isolate other cytochrome c genes including the yeast iso-2-gene (CYC7) (Montgomery et al. 1980) and the related genes from *Schizosaccharomyces pombe* (Russell and Hall 1982), rat (Scarpulla et al. 1981) and chick (Limbach and Wu 1983). The rat gene was then used as a probe in the genomes of the mouse (Limbach and Wu 1985a) and Drosophila (Limbach and Wu 1985b).

Coding Region and Introns. Of the cytochrome c gene sequences determined, only those of the rat and mouse have an intervening sequence in the coding region. This is a 105 nucleotide insert within the codon for Gly 56, which is excised in the messenger via donor and acceptor splice sequences that obey consensus rules (Scarpulla et al. 1981; Limbach and Wu 1985a).

For genes that evolved as single units, different coding segments should incorporate silent third base substitutions at the same rate. However, if the N-terminal halves of the bakers' yeast iso-1 and -2 cytochromes are compared, it is found that 65% of the codons have incorporated such changes while only 32% have in the C-terminal region of the molecule. One possible reason for this C-terminal conservatism is that this region has a dual function like, for example, the viral genes which also contain origins of DNA replication (Godson et al. 1978). A more likely explanation, however, is that the segments have a different evolutionary background. The possibility of recombination or gene conversion within a cytochrome c gene has been demonstrated for the CYC1−11 mutation in yeast iso-1-cytochrome c. This mutation contains a nonsense codon at the position of Pro 71 in iso-1-cytochrome c. It can revert by recombining with the non-allelic iso-2-cytochrome c gene so that the segment for (Thr 69−Ala 83) in the latter replaces the defective segment in the mutant (Ernst et al. 1981). These authors suggest that the perfect homology in DNA sequence of the iso-1- and iso-2-cytochrome c genes in the codon region (75−85) arises from a recent recombinational event of this sort.

Indeed, the possibility of shuffling segments of genes by recombination has been proposed as the raison d'etre for intervening sequences (Gilbert 1978). Thus the differences in silent third base substitutions in exons I and II of the rat gene (42% and 74% respectively) compared to those regions of yeast iso-1-cytochrome c may also reflect separate evolutionary history (Scarpulla et al. 1981).

Pseudogenes. Pseudogenes are related to functional genes in base sequence but cannot be expressed. They can be divided into three groups depending on their origin. The first group contains genes which have arisen through DNA duplication and sequence divergence but which have incorporated a mutation which prevents their expression. For example, there are seven β-globin genes in mouse which form a linkage group on the chromosome but only three are active genes (Jahn et al. 1980). The second group arises during retroviral replication and is flanked by retrovirus repeat elements (Leuders et al. 1982). An example is the α-globin ψ 3 gene. The third group, called "processed retropseudogenes" by Weiner et al. (1986), is formed by reverse transcription of mRNA and insertion of the copy into the chromosomes. This group has 3′ poly-A regions and no intervening sequences. Pseudogenes are a peculiarity of the mammalian genome. It is not known whether they have arisen in this group as a consequence of retrovirus action or due to normal cellular activities.

The cloned structural gene for rat cytochrome c (RC4) hybridises with ~30 different restriction fragments of the rat genome (Scarpulla et al. 1982). Similar studies have been performed with the mouse (Limbach and Wu 1985a). Most of these fragments also hybridise with the 5′ non-coding region

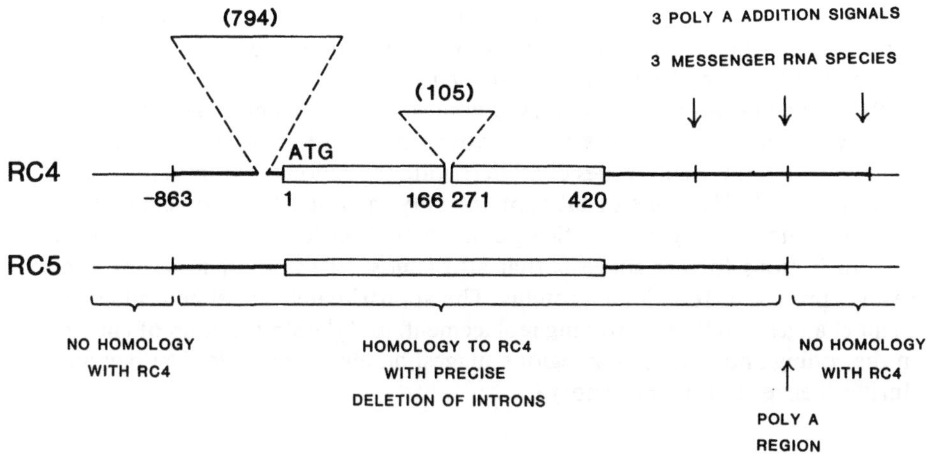

Fig. 4.1. The structure of the rat cytochrome c gene (RC4) and the pseudogene clone RC5. The coding region of the functional gene RC4 is *boxed*. One intervening sequence is situated at position 166 and a second is situated just upstream from the translation start. Three mRNAs have been identified which differ in 3′ non-coding length. The pseudogene RC5 is proposed to derive from the messenger of intermediate length. It lacks the two introns of RC4 and contains a poly-A stretch downstream from the region of homology with RC4. Other pseudogenes also lack the introns of RC4 and differ in the extent of homology with RC4 in the 5′ and 3′ non-coding regions indicative of their derivation from transcripts of different lengths

but only two hybridise with the coding region intron segment, one of which is the cloned gene itself.

Several of the hybridising restriction fragment clones (RC5, 6, 8, 9, 10 and 13) have been sequenced (Scarpulla and Wu 1983; Scarpulla 1984). The sequences share features which are consistent with their being pseudogenes originating by reverse transcription of mRNA (Fig. 4.1). Thus they all lack the two introns of RC4 and all have 3′ poly-A tracts. These tracts occur at different distances downstream from the coding region and, beyond the tracts, there is no sequence similarity to RC4. Such a pattern would arise if mRNA was heterogeneous at the 3′ end. Hybridisation of mRNA to the RC4 probe showed that such heterogeneity indeed exists and three mRNAs are present which differ in length at the 3′ end. Pseudogenes have been found which correspond to each of these messengers but most of the group of 30 seem to derive from the messenger of intermediate length. This is not due to a greater concentration of this particular transcript and Scarpulla (1984) suggested that the tendency to pseudogene formation arises from a strong secondary structure upstream from the 3′ end. This may facilitate binding of enzymes involved in reverse transcription or integration.

Thus there is convincing evidence that these genes arose from mRNA. Several factors indicate that they cannot be expressed as functional proteins. Since RNA polymerase II promoters are believed to reside upstream from transcription starts, a pseudogene derived from a correctly transcribed mRNA should always be inactive unless fortuitously transposed to a new promoter. Although RC 5, 6 and 8 retain 98% sequence homology in the coding region to RC 4, the changes that are observed almost all encode amino acid substitutions which have never been seen in cytochromes c. More strikingly, some of the mouse and rat pseudogenes contain terminator codons or frameshift mutations and the RC 12 clone of rat contains only the last 19 amino acid codons and the 3′ non-coding region (Scarpulla 1985). Consideration of the pattern of substitutions further suggests that all or most of these pseudogenes were never expressed as functional proteins. The approximately equal percentage of silent changes and those involving replacement, and the similar rates of change in the coding and non-coding regions suggest an absence of selection pressure during their evolutionary history.

4.1.2 The Control of Expression of Mitochondrial Cytochrome c

In bacteria, the regulation of gene expression is mainly by the action of specific DNA sequences upstream from the coding regions of genes. Analysis of the genes of eukaryotic cytochromes c has revealed that similar control features may be present.

The Promoter and Transcription Start. The TATA or Hogness box is a consensus sequence (TATAA$^{G}_{A}$) upstream from several eukaryotic genes and is likely to be an element of the RNA polymerase II promoter (Goldberg 1979). Close relatives are found in the bakers' yeast iso-1 (−121), bakers' yeast iso-2 (−94), drosophila DC 3 (−81) and 4 (−100) and chick (−160) cytochrome c genes (Fig. 4.2). These promoter sequences are all situated within 170 nucleotides of the coding region, whereas in rat no such sequence was identified out to −257 (Scarpulla et al. 1981). This is due to the presence of a 794 nucleotide intron at −8 in the rat gene (Scarpulla 1984). Transcription starts 61 base pairs upstream from the intron boundary (Fig. 4.1).

By analogy with other eukaryotic genes, the Hogness box is expected to lie approximately 30 nucleotides upstream from the transcription start point. The yeast iso-1-cytochrome c gene is the only one in which the transcription start has been well characterised. This was done by hybridisation of cellular mRNA with the cloned gene and digestion of single-stranded DNA by S 1 nuclease. The protected DNA was then sequenced (Faye et al. 1981). As suspected from related work (Boss et al. 1981) there are several mRNA species differing at the 5′ end, indicative of multiple transcription starts in the region −29 to −90 (Fig. 4.2). If the Hogness sequence at −121 directs transcription start at −90, it is possible that the two other Hogness-like sequences (Fig. 4.2) direct transcription of the shorter messengers. Indeed, if the region (−139 to −99) is deleted, transcription is re-directed to −29 perhaps under the influence of

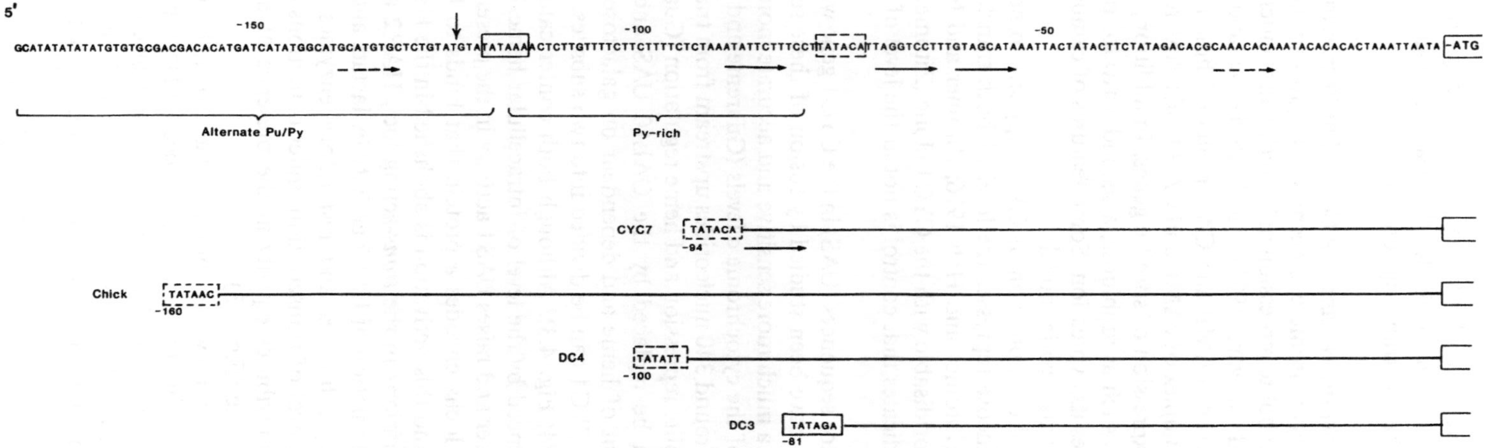

Fig. 4.2. The 5′ non-coding region of yeast iso-l-cytochrome c. The sequence of the 5′ non-coding region of yeast iso-l-cytochrome c is shown (Smith et al. 1979). It is *numbered* from the translation initiation codon ATG. The ↓ indicates the nearest ATG to the coding region. *Solid arrows* indicate transcription starts observed by Faye et al. (1981) and Boss et al. (1981). Nucleotide −29 becomes a major transcription start (*broken arrow*) when nucleotides −99 to −139 are deleted (Faye et al. 1981). In vitro transcription of the CYC1 gene starts at a position upstream from the Hogness sequence at −116 to −121 (*broken arrow* at −140; Arcangioli and Lescure 1985). The relative positions of the Hogness sequences for CYC7, Drosophila DC3 and 4 and chick are shown. The *broken lines* around the proposed Hogness sequences indicate that they do not fully match the consensus sequence of TATA$^{A}_{G}$A. CYC7 does however encode a messenger with a single 5′ end at −77 (Montgomery et al. 1982). Transcription starts are not known for Drosophila or chick genes (Limbach and Wu 1983, 1985b)

CATAAA (−57) (Faye et al. 1981; Fig. 4.2). Although detailed analysis of other messengers is not yet available, the yeast iso-1-gene seems unusually complex in the number of transcription starts.

Upstream Activator Sequences and the Control of Transcription. The biosynthesis of yeast iso-1-cytochrome c is decreased in glucose media (a phenomenon analogous to catabolite repression in bacteria) and under anaerobic conditions (Sherman and Stewart 1971). In bacteria, catabolite repression is mediated by the levels of cAMP and CAP protein. These rise in the absence of glucose and the complex of cAMP and CAP binds close to appropriate promoters to enhance expression of specific genes. In addition, in the presence of some carbon sources such as arabinose, a second activator protein, specific to that operon, enhances its expression. Some features of catabolite repression in yeast resemble this prokaryotic model.

The major control of expression of CYC1 appears to be at the transcriptional level. Thus glucose represses the level of both translatable mRNA for iso-1-cytochrome c (Zitomer and Hall 1976; Zitomer and Nichols 1978) and the level of RNA hybridisable with the CYC1 clone (Zitomer et al. 1979). The latter experiment indicates that control is not at the level of processing of the primary transcript.

Upstream activator sequences (UAS) in the CYC1 gene which influence the rate of transcription have been studied by fusion of these segments to *E. coli* lac Z thus providing a much more sensitive and accurate monitor of expression than measurement of the cytochrome c levels (Guarente and Ptashne 1981). A UAS site is located around 340 nucleotides upstream from translation start and is sensitive to catabolite repression and heme regulation (Guarente and Mason 1983). The site can be replaced by the GAL10 UAS and expression then becomes independent of heme and dependent on galactose.

The UAS site of CYC1 can be divided into two subsites, UAS1 and UAS2 (Guarente et al. 1984; Fig. 4.3). Although both confer catabolite repression, only UAS1 is influenced by the level of intracellular heme. Cobalt-heme acts as a gratuitous inducer and raises UAS1 activity in the presence of lactate. The *trans*-acting HAP1 locus encodes a protein that binds to UAS1 in a heme-responsive fashion and this activation is abolished in the Hap1-1 mutant. On the other hand, mutations in the *trans*-acting loci, HAP2 and HAP3, selectively abolish the stimulation of UAS2 activity by lactate and influence the expression of other cytochromes and respiratory enzymes (Guarente 1986). UAS1 dominates control of transcription under conditions of glucose repression while UAS2 contributes equally in the presence of a non-fermentable substrate such as lactate (Fig. 4.3).

The HAP-1 protein may be among those isolated by Arcangioli and Lescure (1985) by studying the tendency of specific DNA-binding proteins to retard the electrophoretic mobility of a DNA fragment containing UAS. The isolated proteins protect the UAS site from DNase I digestion, are lowered under anaerobic conditions and are absent in mutants defective in heme syn-

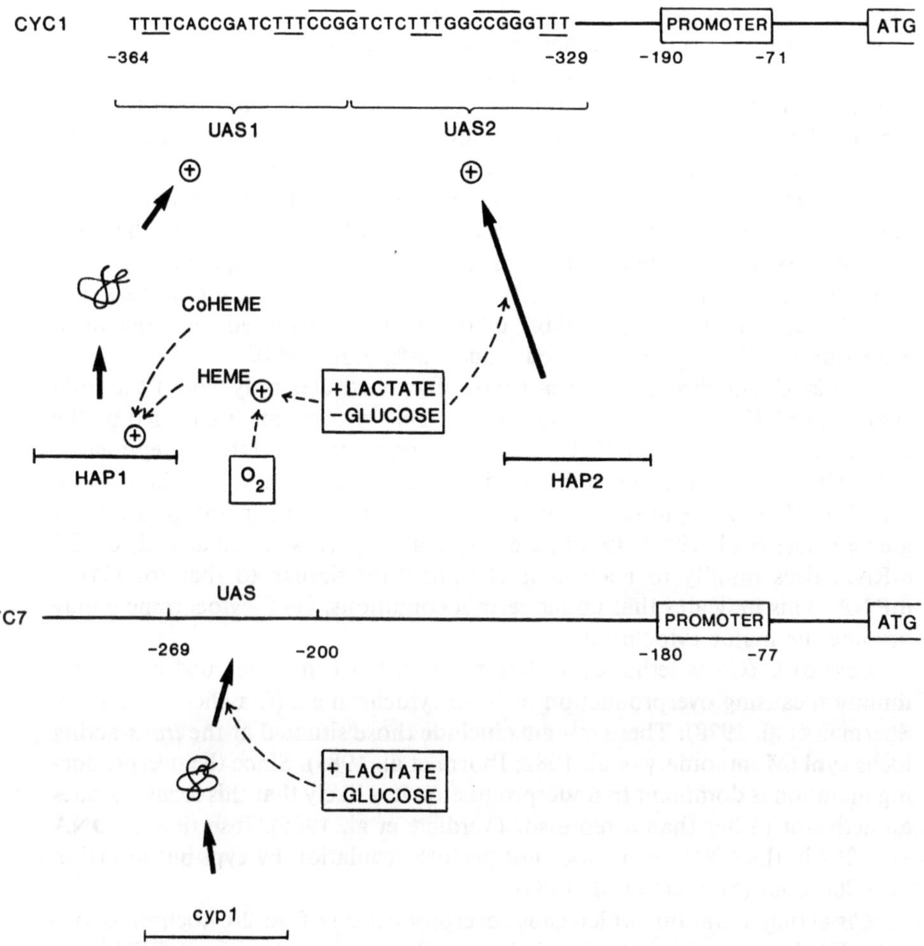

Fig. 4.3. The control of expression of bakers' yeast isocytochromes. The nucleotide sequence of the UAS site of CYC1 was determined by Guarente and Mason (1983) and is numbered from the coding sequence. Repeated segments are *under-* or *over-lined.* The promoter region contains the stretch of alternating purines and pyrimidines (−120 to −190), the Hogness sequence TATAAA at −117 to −121 and the major transcription start sites of −90 and −71 (see Fig. 4.2). HAP1 and HAP2 are *trans*-acting gene loci. HAP1 expression is probably influenced by heme and via heme, by oxygen and catabolite repression. The mechanism of activation of UAS2 is less well characterised but it is independently influenced by catabolite repression perhaps via the protein encoded by HAP2. Little is yet known of the regulation by a third *trans*-acting locus, HAP3, which also influences UAS2 (Guarente 1986).

	Relative transcription activity		
	(+glucose)	(+lactate)	heme deficient
UAS1	1	10	0.005
UAS2	0.1	10	

CYC7 expression is regulated by UAS (−200 to −269). This region binds the product of cyp1 under the control of carbon source. Coheme is cobalt substituted heme; ⊕ stimulation thesis. Heme is not required for binding which may indicate that it is involved in synthesis of the protein rather than as a cofactor (Fig. 4.3).

Thus synthesis of iso-1 cytochrome c is controlled by the binding of specific activator proteins to adjacent UAS regions, upstream from the promoter. These regions are homologous and this may reflect homology in the binding proteins themselves. In some respects this type of regulation resembles that conferred by other eukaryotic enhancers (Yaniv 1984). Such enhancer proteins may regulate entry of polymerase II to the DNA by altering the local chromatin structure. One possibility is that they encourage supercoiling of the DNA in the region (−120 to −190) which contains alternating purines and pyrimidines (Fig. 4.2). Supercoiling of this region is required for efficient in vitro transcription to occur (Lescure and Arcangioli 1984).

The iso-2-cytochrome c of yeast typically constitutes only 5% of the total cytochrome c (Sherman and Stewart 1971). This cytochrome is encoded by the CYC7 gene which, like the CYC1 locus, is susceptible to catabolite repression, but, unlike the latter, is not influenced by heme (Matner and Sherman 1982). CYC7 mRNA is depressed in anaerobic conditions and in the presence of glucose (Laz et al. 1984). It is interesting that as glucose is exhausted, CYC7 mRNA rises rapidly to reach a level transiently similar to that for CYC1 mRNA. This indicates that under certain conditions, iso-2-cytochrome c may become the major cytochrome c.

Loss of CYC1 is lethal on lactate medium but can be rescued by further mutation causing overproduction of iso-2-cytochrome c (Clavilier et al. 1976; Sherman et al. 1978). These mutants include those situated at the *trans*-acting locus cypl (Montgomery et al. 1982; Iborra et al. 1985). Since the overproducing mutation is dominant to underproduction it is likely that this locus encodes an activator rather than a repressor (Verdiere et al. 1985). Insertion of DNA at −269 in the CYC7 gene does not perturb regulation by cypl but insertion at −200 does (Verdiere et al. 1985).

Cis-acting mutations which cause overproduction of iso-2-cytochrome c involve DNA rearrangements or deletions in the upstream region of CYC7. Insertion of the transposon Tyl brings CYC7 under the control of the inserted element (Montgomery et al. 1982) while a similar effect is seen for a 5 kb deletion which results in the attachment of a new regulatory region to the CYC7 locus (McKnight et al. 1981). These studies indicate that the stimulation of transcription of CYC7 in the absence of glucose may be mediated by the binding of an activator protein encoded by cypl to a UAS in the region −200 to −269 of the CYC7 gene (Fig. 4.3).

Isocytochromes and Alleles. The two yeast isocytochromes considered above are the best studied examples of two mitochondrial cytochromes c occurring in the same organism. Prior to DNA hybridisation studies the only other examples of isocytochromes c were those of the mouse and the housefly. The mouse testis contains a distinct cytochrome c which differs in 13/104 residues from that present in other tissues (Hennig 1975). It is found in cells of the germinal epithelium which give rise to the spermatozoa (Goldberg et al. 1977) and thus may be associated specifically with the haploid state containing the Y-chromosome. The 13 substitutions have a strikingly local distribution at the "back" of the molecule far from the exposed heme edge. This suggests some specific role for this region in the testis isocytochrome but no discriminating functional tests have yet been performed.

Kim (1980) found a testis-specific rat cytochrome c, the composition of which suggested about 20 differences from that found in other tissues. However, neither the rat nor the mouse testis-specific cytochrome c gene has yet been identified among the hybridising DNA fragments and thus the controls which govern the developmental expression of the isocytochromes are not known.

The two isocytochromes c of the housefly (*Musca domestica*) (Yamanaka et al. 1980) are also differentially expressed during development. One form predominates in larvae and in testis and ovaries, while the other is the major form in adult tissues (Inoue et al. 1984). The two differ in six positions in the sequence and these again are situated at the "back" of the molecule.

The same authors could not detect a comparable larval-type isocytochrome in the fruit fly (*Drosophila melanogaster*) in spite of the fact that this species contains two functional genes, DC3 and DC4 (Limbach and Wu 1985b). The latter encodes the known protein sequence while the former differs from this in 32/104 amino acid positions. This is a large divergence in terms of cytochrome c evolution and indeed 13 of the substitutions have not previously been found in the composite eukaryotic sequence (although only Asp 70 occurs at a previously unvaried position). The divergence time implied by this degree of difference is similar to the time of divergence of the insects and vertebrates (600 million years; Vol. 2, Chap. 6).

DC4 hybridises to a single messenger of 900 nucleotides which occurs at a relatively high concentration but varies with the tissue and with the stage of development. In contrast, the messenger encoded by DC3 occurs at a constant but low level throughout development. Also its size of 2100 nucleotides is much longer than expected from consideration of probable transcription start and stop sequences and the reason for this is not known.

Two genes were isolated from commercial chick DNA which differ by 20/1600 nucleotides (Limbach and Wu 1983) none of which are in the coding region. However, an individual chick contains only one gene and the two therefore represent alleles in the chick population.

The Control of Translation. Sherman et al. (1980) found that normal amounts of iso-1-cytochrome c were produced regardless of whether the initiation codon was at −3, −2, 3 or 5 rather than the normal −1. Indeed, the rat gene retains codons for Glu-Phe-Lys corresponding to the N-terminal region of yeast iso-1-cytochrome c but these are now upstream from a new initiation codon corresponding to position 5 of the yeast sequence (Scarpulla et al. 1981). Thus initiation occurs successfully within a span of 25 nucleotides. This implies that there is no requirement for a 5′ sequence at a set distance from the translation start. Translation appears to start at the AUG closest to the 5′ end of the message and this is consistent with the long 5′ stretches which lack ATG triplets in eukaryotic genes (Fig. 4.2). Kozak (1978) has proposed that the 40S ribosomal subunit binds to the 5′ end of the message and migrates until the first AUG is encountered. This is in contrast to translation start in prokaryotes which is directed by 5′ non-coding sequences complementary to 16S rRNA (Steitz 1979).

Although we have emphasised transcriptional control of gene expression there is evidence that CYC1 is also controlled at the translational level. Thus mutations at the CYC3 locus are thought to affect the enzyme involved in attachment of heme to iso-1-cytochrome c (Matner and Sherman 1982). Mutants contain normal levels of CYC1 mRNA yet very low levels of the apo-iso-1-cytochrome c suggesting that the apoprotein regulates its own translation.

4.1.3 The Control of Gene Expression in Bacterial Respiration and Photosynthesis

In Chapter 3 we have outlined the flexibility that characterises bacterial respiration and photosynthesis. For example, in the denitrifying organisms the presence of NO_3^- induces formation of the enzymes of denitrification which are repressed in aerobic growth; the bacterial oxidases aa_3 and o may appear at different ambient oxygen tensions; the Rhodospirillaceae can suppress pigment formation in the dark in the presence of oxygen and an oxidisable substrate.

In many of these adaptations there are associated changes in the c-type cytochromes (see for example Fig. 1.2). In some, the adaptation is a phenotypically simple process which should be amenable to genetic analysis. An example is the effect of copper and iron on the synthesis of plastocyanin and cytochrome c in algae. In *Chlamydomonas reinhardii* (Wood 1978), *Scenedesmus* (Bohner and Boger 1978; Bohner et al. 1980), *Anabaena variabilis* and *Plectonema boryanum* (Sandman and Boger 1980), copper depletion results in a fall in plastocyanin and a rise in cytochrome c such that their combined concentration stays roughly constant and in excess of P700 (Chap. 3, Sect. 3.6.5). The replacement of one by the other does not affect growth rate or oxygen utilisation and in some instances the flexibility conferred by the presence of both genes has been lost by mutation. Thus a mutant of *Chlamydomonas* (Gorman and Levine 1965) lacks plastocyanin but did not synthesise cytochrome c, probably because of regulation by Cu in the medium (Wood 1978). The naturally occurring *Chlamydomonas mundana* also lacks plastocyanin but appears to have overcome the regulatory control by a further mutation so that the cytochrome c is synthesised constitutively (Wood 1978). *Dunaliella parva* and higher plants synthesise only plastocyanin (although this does not prove the absence of the cytochrome c gene) while *Euglena gracilis* and *Bumilleriopsis filiformis* (Kunert and Boger 1975) resemble *C. mundana* in lacking plastocyanin. In an investigation of the molecular basis for the influence of Cu^{2+} on gene expression in *Chlamydomonas reinhardii*, Merchant and Bogorad (1986) found that no immunoreactive plastocyanin or cytochrome c precursor is detectable under conditions where the mature protein is not present. This is consistent with the finding of Wood (1978) that addition of Cu^{2+} to extracts of cells grown in the absence of Cu^{2+} did not produce the plastocyanin chromophore. Using in vitro transcription of cellular RNA and immunological detection of the products, Merchant and Bogorad (1986) showed that translatable mRNA for pre-apoplastocyanin was detectable during both Cu^{2+}-deficient and Cu^{2+}-supplemented growth, whereas mRNA for pre-apocytochrome c was found only in Cu^{2+}-deficient cells. Thus the positive regulation by Cu^{2+} on plastocyanin expression may be at the translational level while the negative regulation on cytochrome c expression appears to be at the transcriptional level. This is just one example where the techniques of molecular biology are being used to dissect the regulation of gene expression. Applications of these techniques promise a rapid expansion on our knowledge of bioenergetic adaptation in bacteria.

4.2 Posttranslational Processing and Modification

4.2.1 General Aspects of Posttranslational Transport and Processing

Almost all proteins are synthesised on cytoplasmic ribosomes or ribosomes bound to the cytoplasmic side of membranes. For proteins that are destined to be secreted, to be taken up by cellular organelles or to be specifically incorporated in membranes, there must therefore be sorting and transport mechanisms which determine their destination. Of these, the processes which govern protein secretion by the rough endoplasmic reticulum (RER) are the most thoroughly characterised and have influenced studies in other systems. Our main concern in this section is to discuss the sorting and transport of redox proteins associated with the mitochondria, the chloroplast and the bacterial plasma membrane.

As we have seen in Chapters 2 and 3, the bioenergetic processes of respiration and photosynthesis are intimately associated with cell membranes. Electron transport in these systems is a vectorial reaction involving a unique arrangement of redox centres relative to the membrane. The correct positioning of the polypeptides that carry the redox centres is therefore crucial to the integrity of the bioenergetic process.

Protein translocation across a membrane may be cotranslational or posttranslational (Wickner and Lodish 1985). These two possibilities are described by the signal hypothesis and the membrane trigger hypothesis respectively.

According to the signal hypothesis, the nascent N-terminal region of the polypeptide chain is a signal or leader sequence which directs translocation across a membrane. Machinery exists to ensure that further elongation of the native chain does not take place until the leader sequence penetrates the membrane, probably via a protein pore (Walter and Blobel 1982). Elongation then proceeds so that the nascent chain appears on the opposite side of the membrane where the signal peptide is removed by a specific protease. Integral membrane proteins differ from secreted proteins in having a "stop transfer" sequence which arrests translocation before synthesis is complete (Blobel 1980). Thus a distinguishing feature of the signal hypothesis is the independent action of discrete short sequences of the growing polypeptide chain.

In contrast, the membrane trigger hypothesis predicts that the nascent polypeptide chain folds into domains before interaction with the membrane (Wickner 1980). This interaction triggers a conformational change which results in spontaneous penetration of the protein into or through the membrane. After translocation, cleavage of the N-terminal region traps secreted protein on one side of the membrane.

There is considerable debate as to the relative merits of the two theories. In general, it appears that uptake of proteins by the RER is cotranslational while uptake by mitochondria and chloroplasts is posttranslational. In bacteria the timing of translocation relative to translation varies with the particular protein studied. In some, the nascent chains appear immediately on the periplasmic side of the cell membrane. In others, translocation only occurs after most of the protein has been synthesised or after protein synthesis is complete (Randall 1983; Michaelis and Beckwith 1982).

There are two general approaches to the study of co- and posttranslational transport of proteins. A combination of radioactive pulse labelling and immunological identification allows the study of specific newly synthesised polypeptide chains in vivo. If translocation is posttranslational, a cytoplasmic pool of precursor polypeptide will be detectable which can be chased across the membrane and appear as mature protein after a time lag in a process that is insensitive to inhibitors of protein synthesis. For most cases studied, the precursor is larger than the mature protein by the size of the N-terminal presequence and the two can be distinguished by SDS gel electrophoresis. If transport is cotranslational, no cytoplasmic pool will be detected. This may also be so for systems proposed to operate posttranslationally if the pool is small and turnover rapid. In these cases, precursor accumulation may be detected if transport is made defective, for example by uncouplers of oxidative phosphorylation.

The second approach involves cell-free protein synthesis programmed with mRNA from the source under study. This can now be done by in vitro transcription of cloned genes followed by in vitro translation to give a single labelled polypeptide. The transport and processing of the newly synthesised precursors is then initiated by addition of membranes, for example whole mitochondria, everted vesicles of RER or bacterial cell membranes. Identifica-

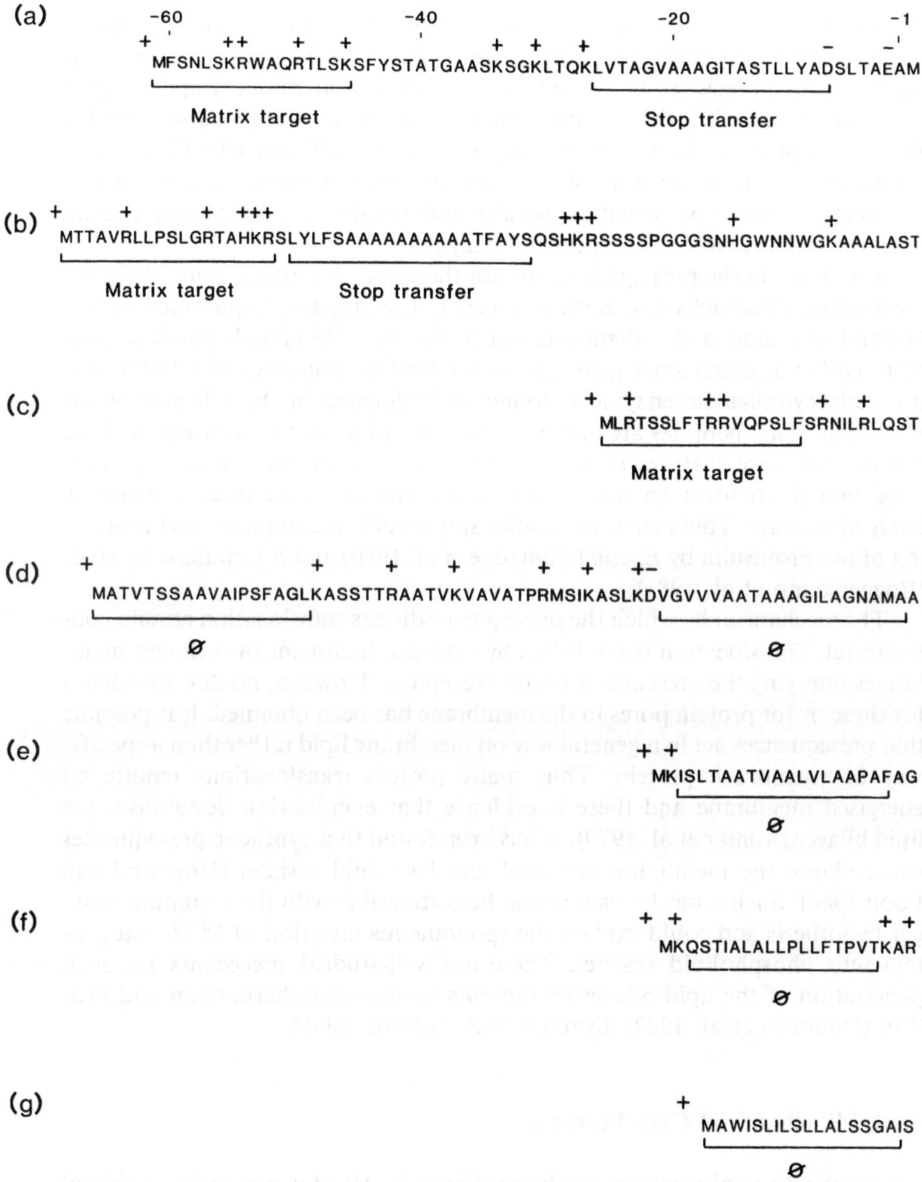

Fig. 4.4 a–g. The presequences of proteins. Positively (+) and negatively (−) charged amino acids are indicated. ∅ = Non-polar regions. Proteins imported to the mitochondrion: **a** Cytochrome c_1 (inner membrane); Sadler et al. (1984). **b** Cytochrome c peroxidase (intermembrane space); Kaput et al. (1982). **c** Alcohol dehydrogenase (matrix). Protein imported to the chloroplast: **d** Plastocyanin (thylakoid space); Smeekens et al. (1985). Proteins transported across bacterial plasma membrane:* **e** Cytochrome c_2 (periplasmic space); Daldal et al. (1986); **f** Alkaline phosphatase (periplasmic space); Inouye et al. (1982). Protein transported across eukarotic plasma membrane: **g** Immunoglobulin λ-chain; Burstein and Shechter (1976)

* See appendix note 21

tion of the precursor and mature forms is as for the in vivo approach. The extension of this approach to its simplest form would be the uptake and processing of purified precursor by phospholipid vesicles containing entrapped signal peptidase. Indeed, it is that system with the phage M13 coat precursor that led Wickner to propose the membrane trigger hypothesis (Wickner 1980). The successful uptake and processing of the viral coat protein implied that the whole precursor is both competent in uptake and requires no additional protein machinery apart from the processing enzyme.

Mutations in the presequence, but not the rest of a protein, cause defective localisation (Michaelis and Beckwith 1982). The singular importance of this N-terminal region is also demonstrated by fusion of the β-galactosidase gene of *E. coli* to a presequence gene fragment (Bassford and Beckwith 1979). The normally cytoplasmic enzyme is found to be located in the cell membrane. Although signal peptides are not conserved in amino acid sequence they share certain structural features shown in Fig. 4.4. They are uncharged sequences long enough to form an α-helix across a membrane and often flanked by charged residues. This structural relationship is reflected in the correct processing of pre-proinsulin by *E. coli* (Talmadge et al. 1980) and β-lactamase by yeast (Roggenkamp et al. 1981).

The mechanism by which the presequence directs translocation remain controversial. Translocation is abolished by protease treatment of recipient membranes implying the presence of protein receptors. However, no direct evidence for these or for protein pores in the membrane has been obtained. It is possible that presequences act in a general way on membrane lipid rather than a specific way on membrane protein. Thus many protein translocations require an energised membrane and there is evidence that energisation destabilises the lipid bilayer (Komar et al. 1979). It has been found that synthetic presequences can collapse the membrane potential and lyse lipid vesicles (Hurt and van Loon 1986). Such a mechanism would be compatible with the membrane trigger hypothesis and could explain the spontaneous insertion of M13 coat protein into phospholipid vesicles. There are well-studied precedents for such penetration of the lipid bilayer by proteins such as diphtheria toxin and mellitin (Donovan et al. 1982; Tosteson and Tosteson 1981).

4.2.2 Mitochondrial Cytochrome c

Although a few polypeptides (such as subunits I–III of cytochrome c oxidase) are encoded on mitochondrial DNA, most mitochondrial proteins are encoded in the nuclear genome and synthesised on cytoplasmic ribosomes (Schatz and Mason 1974). Such proteins therefore require sorting which involves specific binding and translocation across one or both mitochondrial membranes (Neupert and Schatz 1981).

Cytochrome c is synthesised on cytoplasmic ribosomes and functions in the intermembrane space of the mitochondrion. The major questions regarding import of cytochrome c to the mitochondrion are the nature of the primary translation product, the mechanism of translocation across the membrane and the process of heme attachment.

Considerable progress has been made in all these areas.

Primary Translation Product. Several lines of evidence show that the product of translation is a polypeptide identical in sequence to the mature protein. This distinguishes cytochrome c from most other mitochondrial proteins which are synthesised as higher molecular weight precursors. Thus rabbit reticulocyte lysate programmed with *Neurospora* poly A mRNA will produce a polypeptide chain of identical size to the mature protein (Zimmerman et al. 1979). If ^{35}S formyl methionyl tRNA is added to the cell-free system this protein becomes selectively labelled at the N-terminus since the initiator formyl methionine is not removed by methionine amino peptidase (Palmiter et al. 1978). Edman degradation after deformylation is consistent with the presence of methionine followed by the mature protein sequence (Zimmerman et al. 1979). This proves that there is no presequence in agreement with genetic studies (Stewart et al. 1971) and now confirmed by the gene sequences of yeast isocytochromes (Smith et al. 1979; Montgomery et al. 1980). Similar studies have been performed with rat liver in which cytochrome c synthesis had been enhanced by treatment with thyroxine (T3) (Matsuura et al. 1981). Cell-free translation in the presence of an "acetyl trap" to block the normal N-terminal acetylation of mammalian cytochromes c (Fig. 4.11) and protease inhibitors to inhibit degradation of the unacetylated product, yields a polypeptide with an N-terminal methionine followed by the sequence of the mature protein.

Thus the primary translation product of cytochrome c contains no pre-sequence. By analogy with ovalbumin (Palmiter et al. 1978), removal of the initiator methionine probably takes place during early growth of the nascent chain on the ribosome. In the case of mammalian cytochrome c this is immediately followed by N-terminal acetylation catalysed by an acetyl CoA-dependent N-acetyl transferase probably bound to the ribosome (Tsunasawa and Sakujama 1984). There is no natural acetylation of yeast cytochrome c but acetylation in certain frameshift mutants is observed (Tsunasawa et al. 1985; Sherman et al. 1980).

Until recently there was conflicting evidence as to whether apo- or holocytochrome c was transferred into the mitochondrion (reviewed in Sherman and Stewart 1971). These early studies on rat liver faced the problem of poor expression and lacked a method for distinguishing the apo- and holocytochromes. The use of dual labelling to detect newly formed protein and the lack of immunological cross-reaction between apo- and holocytochromes c of *Neurospora* allowed Korb and Neupert (1978) to show the appearance of apocytochrome c in the soluble fraction of a cell-free translation system followed by its disappearance and parallel appearance of holocytochrome c within the mitochondrion (Fig. 4.5).

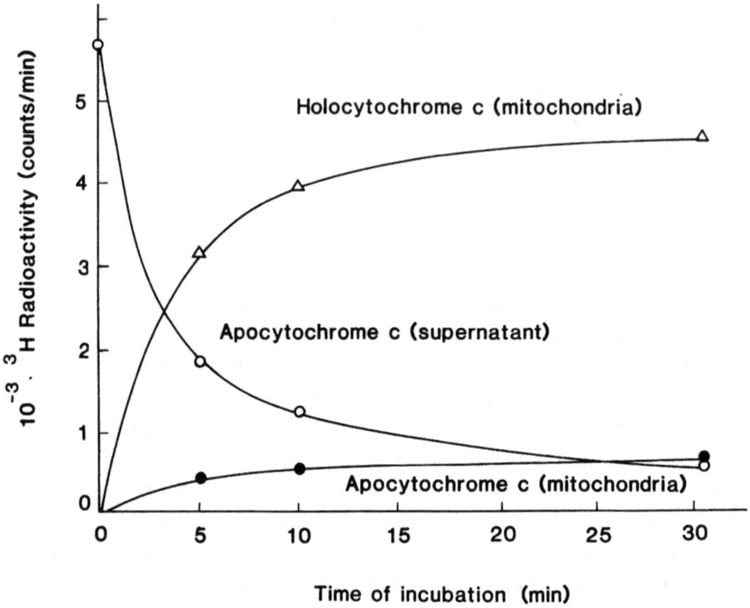

Fig. 4.5. Kinetics of formation of holocytochrome c from apocytochrome c in a reconstituted system. A postmitochondrial supernatant from *Neurospora* was incubated for 15 min with ^{3}H-leucine and then cycloheximide was added. Mitochondria were added and incubated for the times shown. Apocytochrome c was immune precipitated from the supernatant fraction and from the lysed mitochondria. Holocytochrome c was immune precipitated from the latter. The precipitates were analysed by SDS-PAGE and the radioactivities associated with cytochrome c were determined. ○ Apocytochrome c in supernatant; ● apocytochrome c in mitochondrion; △ holocytochrome c in mitochondrion (Hennig and Neupert 1981)

Thus apocytochrome c is the primary translation product. Because it can be prepared chemically in large amounts, studies on binding, translocation and maturation of cytochrome c by mitochondria are much easier than comparable studies on other mitochondrial precursors which are available only in the trace amounts synthesised by cell-free translation.

Import and Heme Attachment. When *Neurospora* mitochondria are treated with deuteroheme they will bind but not translocate apocytochrome c. This is due to inhibition of the heme attachment process which is required for translocation to occur. The binding is of high affinity (K_A 2×10^7 M^{-1}) and can be reversed by an excess of chemically prepared apocytochrome c from *Neurospora* and other eukaryotic species. In contrast, neither apocytochrome c-550 of the bacterium *Paracoccus denitrificans* nor *Neurospora* holocytochrome c are effective competitors (Hennig and Neupert 1981; Hennig et al. 1983). Matsuura et al. (1981) obtained similar results for rat liver and additionally showed that, of five CNBr fragments tested, only CNBr (66–104) prevented apocytochrome c binding. Hennig et al. (1983) suggested that the conserved sequence (70–80) may control binding and be exposed in the apo- but not the holocytochrome c. This sequence is not conserved in *Pa. denitrificans* cytochrome c-550.

The binding sites are trypsin-sensitive and are situated at the outer surface of the outer membrane. Uptake of other mitochondrial precursor proteins is not affected by apocytochrome c suggesting the presence of specific receptors for this protein. The mechanism of translocation is however poorly understood and this reflects the general uncertainty about protein translocation across membranes that was noted in the introduction. Rietveld et al. (1985) found that apocytochrome c can penetrate a lipid bilayer and cause extensive reorganisation of the lipid molecules. Using vesicles containing entrapped trypsin, Rietveld and Kruiff (1984) showed that apocytochrome c could cross the lipid barrier and became sensitive to proteolysis. Such a mechanism alone could not be expected to show saturation or specific inhibition. These characteristics could be incorporated however if the role of the receptor is to allow the apocytochrome c to bind specifically and adopt a conformation required for lipid penetration.

Any model of translocation must also explain the stringent coupling to heme attachment. As noted above, the uptake of apocytochrome c into mitochondria could be inhibited by deuteroheme which is believed to act as a competitive inhibitor of the heme attachment enzyme. Protoheme but not protoporphyrin could reverse this inhibition suggesting that it is heme and not porphyrin that is incorporated (Hennig and Neupert 1981). This is consistent with the enzymic studies discussed below and with the results of Colleran and Jones (1973) who found that *Physarum* cytochrome c contained heme almost solely derived from added ^{59}Fe hemin. No further experimental support has been obtained for the pathway proposed by Sano and Granick (1961) and Sano and Tanaka (1964) involving non-enzymic reaction of protoporphyrin and apoprotein followed by iron insertion by a putative ferrochelatase.

Horse or yeast iso-1-apocytochrome c can be converted into the holoprotein in the presence of an NADPH generating system, yeast mitochondria and ^{59}Fe hemin. The nature of the product was established by resistance to proteolysis in the ferrous state and spectroscopic examination confirmed the presence of two thioether bonds (Taniuchi et al. 1983). Attempts to solubilise and purify the enzyme revealed the presence of a Triton-soluble activity which was specific for yeast iso-1-apocytochrome c and did not react with horse apocytochrome c, and an insoluble activity which accepted either as a substrate (Visco et al. 1985). One possibility is that the true substrate for the latter is cytochrome c_1.*

The enzyme achieves not only the correct union of the two vinyls (2 and 4) with the two thiols (Cys 14 and 17) but also selects the unique chirality at each thioether out of the possible isomers. The process of heme attachment and subsequent protein folding is further discussed in Sect. 4.2.7.

The import of cytochrome c to the mitochondrion is summarised in Fig. 4.6.

* See appendix note 22

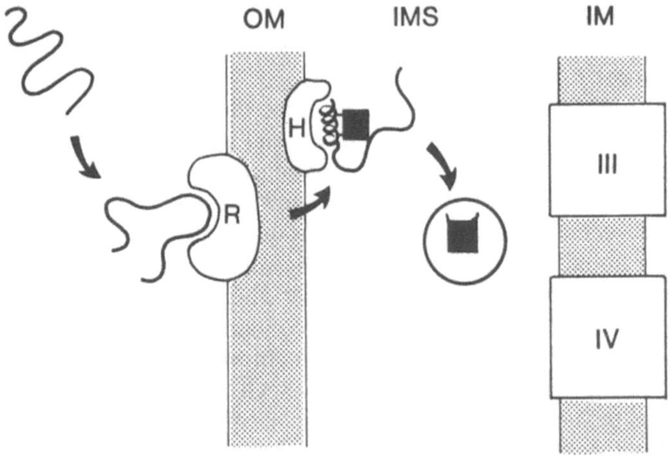

Fig. 4.6. The import of cytochrome c to mitochondria. The unstructured apoprotein binds to a receptor (*R*) on the outer membrane (*OM*) which facilitates translocation of the protein across the membrane. Translocation is tightly coupled to heme attachment (*H*). The mature protein is situated in the intermembrane space (*IMS*) mediating electron transfer between complexes *III* and *IV* on the inner membrane (*IM*)

4.2.3 Import of c_1 and Redox Enzymes of the Intermembrane Space

Cytochrome c_1 is synthesised as a larger precursor on cytoplasmic ribosomes (Nelson and Schatz 1979; Neupert and Schatz 1981). The precursor is processed in two stages after energy-dependent uptake into mitochondria. The first stage involves a matrix protease which cleaves the precursor to an intermediate form and the second involves a heme-dependent proteolysis probably at the outer surface of the inner membrane (Fig. 4.8).

These features are apparent from the results of Fig. 4.7. A heme-deficient yeast mutant lacking δ-aminolevulinic acid synthase accumulates the intermediate form in a pulse chase experiment. Addition of δ-aminolevulinic acid allows further processing to the mature form (Ohashi et al. 1982). The matrix protease catalysing the first stage of proteolysis is a common feature of mitochondrial protein processing. It is inhibited by chelating agents such as o-phenanthroline and is specific for mitochondrial precursor polypeptides (Boehni and Daum 1983). The intermediate form is probably situated on the outer surface of the inner membrane because it is sensitive to trypsin digestion in mitoplasts of the heme-deficient mutant (Ohashi et al. 1982). Little is known of the properties of the second protease.

This two-step processing is also a feature of cytochrome b_2 (Fig. 4.8), an enzyme present in the intermembrane space (Daum et al. 1982a). However, unlike cytochrome c_1, the second proteolysis does not require heme binding (Daum et al. 1982b). In the case of a second enzyme of the intermembrane space, cytochrome c peroxidase, no intermediate form could be detected either in pulse chase experiments with whole cells (Reid et al. 1982) or with in vitro

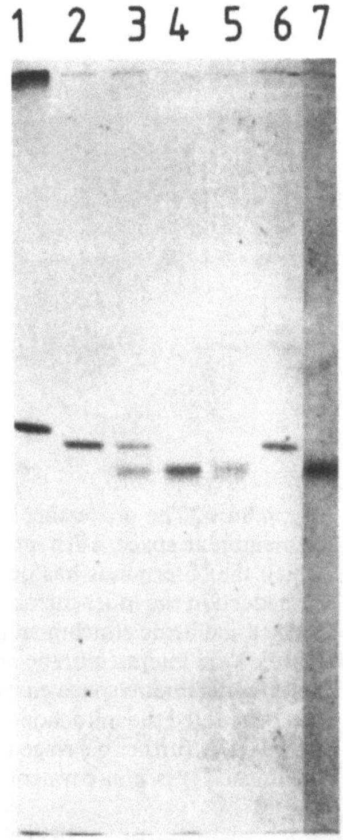

Fig. 4.7. The processing of cytochrome c_1 precursor. The heme-deficient yeast mutant GL-1 was pulse labelled for 1 h with ^{35}S-methionine. One-fifth of the cell suspension was mixed with trichloroacetic acid to stop the reaction (*lane 2*) and the remaining suspension was chased for 0.5 (*lane 3*), 1 (*lane 4*) or 2 h (*lane 5*) in the presence of δ-amino levulinic acid or for 2 h (*lane 6*) in the absence of any heme precursor. Reactions were stopped with trichloroacetic acid and cytochrome c_1 was immune precipitated and subjected to SDS-PAGE and fluorography. *Lane 1:* cytochrome c_1 synthesised by reticulocyte lysate programmed with yeast mRNA. *Lane 7:* cytochrome c_1 synthesised by wild-type cells grown in the presence of ^{35}S-sulfate. Note that an intermediate in processing is observed in *lanes 2* and *6*. Supply of the heme precursor causes disappearance of this intermediate and appearance of the mature protein (Ohashi et al. 1982)

studies. However, preparations of the matrix protease did yield an intermediate form (Gasser et al. 1982). It is proposed that the first proteolysis is rate limiting and no accumulation of the intermediate can be observed (Daum et al. 1982b).

The nucleotide sequences of the genes for cytochrome c_1 and cytochrome c peroxidase have been determined (Sadler et al. 1984; Kaput et al. 1982) and the amino acid sequences of the signal regions can be deduced. Each contains a long, uncharged region which may span the inner membrane as an α-helix and stop further transfer into the matrix (Fig. 4.4). This arrangement would leave the bulk of the protein in the intermembrane space and a highly basic N-terminal tail in the matrix to be cleaved by the matrix protease. The second cleavage would then occur to yield the mature proteins.*

Biosyntheses of cytochrome c and cytochrome c_1 are summarised in Fig. 4.9.

4.2.4 Posttranslational Transport in Bacteria

The gram-negative bacterium is surrounded by an outer membrane, an intermembrane region called the periplasm and the inner or plasma membrane (Chap. 3). Each of these locations contain a distinctive array of proteins, all originally synthesised on cytoplasmic ribosomes. Because of the ease of

* See appendix note 23

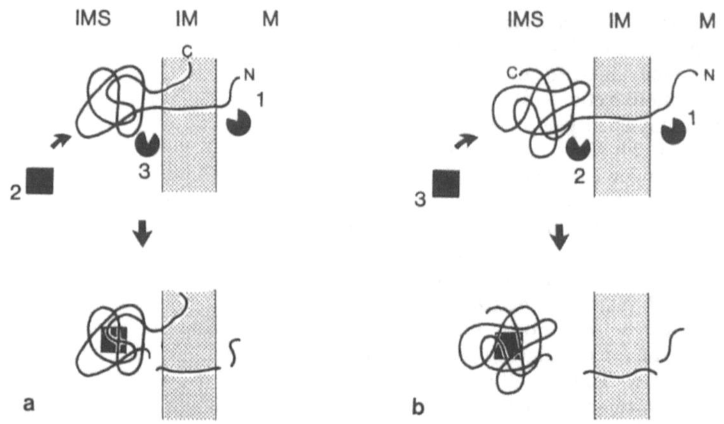

Fig. 4.8a, b. The processing of precursors of (**a**) cytochrome c_1 and (**b**) enzymes of the intermembrane space. **a** An intermediate stage in the processing of cytochrome c_1 is shown in which the N-terminus has penetrated the mitochondrial matrix (*M*) and the C-terminus is embedded in the inner membrane (*IM*). A matrix protease (*1*) cleaves the N-terminal presequence and heme attachment (*2*) is followed by further cleavage of the presequence by a protease of the intermembrane space (*IMS*) (*3*). **b** An intermediate stage in the processing of an intermembrane space enzyme such as cytochrome b_2 is shown in which the N-terminus has penetrated the mitochondrial matrix (*M*). A matrix protease (*1*) cleaves the presequence followed by further cleavage by a protease (*2*) of the intermembrane space (*IMS*). Heme attachment (*3*) is non-covalent and not required for the proteolytic processing to occur

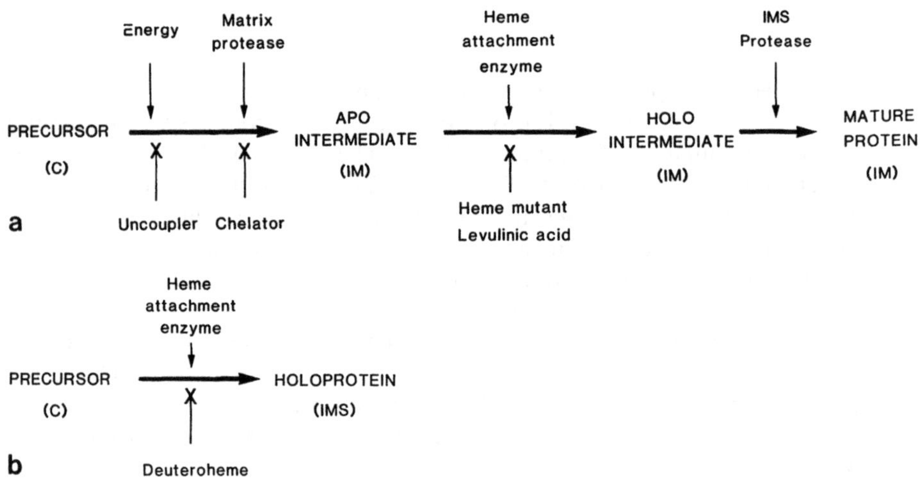

Fig. 4.9a, b. Stages in the processing of (**a**) cytochrome c_1 and (**b**) cytochrome c. In each case requirements for a stage to occur are indicated by *vertical arrows* above the process while antagonistic influences are shown by *crosses* below the process

genetic manipulation of bacteria, the transport of these proteins from site of synthesis to final destination is, in some cases, very well characterised. Mutations which influence transport fall into one of two general classes. There are those that alter the presequence, and therefore affect only the protein under study, and there are those that alter the common transport machinery and thus have pleiotropic effects. Prominent among the latter are the *sec* gene products which cause association of the nascent chain and ribosome with the cell membrane. The growing chain is translocated across the membrane directed by a presequence which contains a non-polar stretch of amino acids accompanied by several basic residues (Fig. 4.4). The presequences of all periplasmic proteins and many proteins of the inner membrane are then cleaved by the signal peptidase (Wolfe et al. 1983). This cotranslational process therefore closely resembles that of the RER. In such a tightly coupled process, precursors will not be detectable by pulse labelling of whole cells. They may accumulate, however, in "mini" cells, which have lost their chromosome, or in the presence of an uncoupler or an uncoupled mutant phenotype. In other cases elongation is not coupled to translocation and the latter may occur late or posttranslationally (Randall 1983; Wickner 1980).

There are few studies on the transport of c-type cytochromes in bacteria. According to the proposal of Wood (1983), extensively discussed in Chapter 3, all c-type cytochromes are predicted to be either periplasmic or bound to the periplasmic side of the plasma membrane. For both locations, transport across this membrane would be required. In a pioneering study, Garrard (1972) found no evidence for a cytoplasmic precursor pool for the periplasmic cytochrome c-550 of *Spirillum itersonii*. Heme was incorporated rapidly from a cytoplasmic pool of free heme and heme depletion by levulinic acid (an inhibitor of heme synthesis) caused parallel inhibition of synthesis of cytochrome c-550. These results indicate that protein synthesis, membrane translocation and heme attachment are tightly coupled for this protein.

Cytochrome c_2 from *Rhodopseudomonas capsulata* is a relative of cytochrome c-550 of *S. itersonii* (Chap. 3) and the gene sequence has been determined (Daldal et al. 1986). The coding region of the gene includes an N-terminal presequence of 21 residues with similar characteristics to those of other bacterial periplasmic proteins (Fig. 4.4). It is a reasonable hypothesis that all periplasmic c-type cytochromes will have such a presequence during synthesis and that this will be cleaved after translocation into the periplasm.

There is as yet no information concerning the transport and topology of c-type cytochromes associated with the bacterial plasma membrane. These include cytochrome c_1, cytochrome c_4 and the c-type components of certain terminal oxidases (Chap. 3). As indicated in Fig. 4.10, the topologies of membrane insertion of cytochromes c_1 of mitochondria and cytochrome f of chloroplasts are different and it is possible that bacterial cytochrome c_1 may resemble the latter rather than the mitochondrial counterpart in having a hydrophobic "stop transfer" sequence near its C-terminus (see following section). However, such sequences have not yet been described for prokaryotic proteins. From the amino acid sequence of cytochrome c_4 (Vol. 2, Chap. 3) it seems unlikely that it is a transmembrane protein. More probably cytochrome c_4 is associated with an integral membrane protein complex via hydrophobic

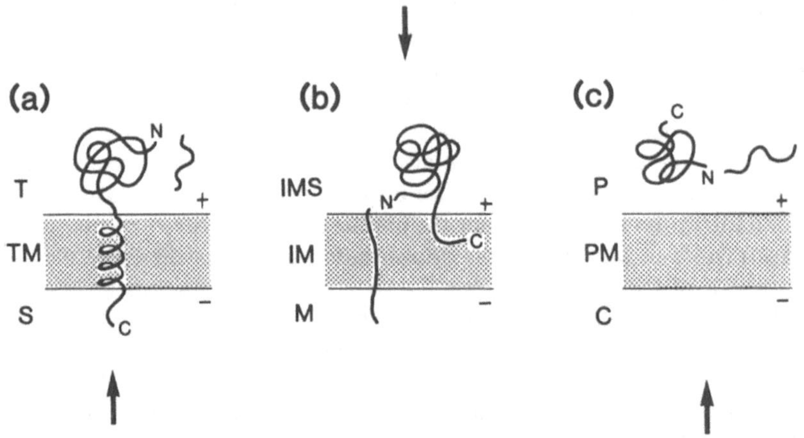

Fig. 4.10a–c. The topology of membrane insertion and translocation of c-type cytochromes. The *arrows* indicate the side of precursor synthesis. Positive (+) and negative (−) show the membrane potentials. **a** Translocation and insertion of cytochrome f in the thylakoid membranes (*TM*) with the bulk of the protein facing the thylakoid space (*T*) and a C-terminal tail (*c*) at the stromal side (*S*). Bacterial cytochromes c_1 may have a similar orientation. **b** Insertion of cytochrome c_1 in the mitochondrial inner membrane (*IM*) with the bulk of the protein facing the intermembrane space (*IMS*). Only a region of the presequence penetrates the matrix (*M*). **c** Translocation of cytochrome c_2 across the bacterial plasma membrane (*PM*) into the periplasmic space (*P*)

interactions that are broken by butanol treatment (Chap. 3.5.5). Such an association may occur after translation and translocation into the periplasmic space.

4.2.5 Posttranslational Transport in Chloroplasts

Proteins associated with the photosynthetic electron transport chain of the thylakoid membrane are mostly synthesised on cytoplasmic ribosomes (Ellis 1981) as larger precursors containing N-terminal presequences (Schmidt and Mishkind 1986). These precursors then must traverse the outer and inner envelope membranes of the chloroplasts and finally penetrate or pass across the thylakoid membrane. This complexity is unique to chloroplasts. Like mitochondria the uptake into the chloroplast is probably posttranslational and the uptake by the thylakoid membrane, which is not connected with the envelope membrane, certainly is. It is however difficult to exclude the possibility that traffic through the stroma is mediated by membrane vesicles rather than occurring in solution. If this were so then the crucial membrane insertion event would occur at the envelope rather than the thylakoid.

Plastocyanin and the small algal cytochrome c are functional homologues acting as diffusible links in the intrathylakoid space between the bf complex and the reaction centre of photosystem I (Chap. 3.6.5). Thus knowledge of the uptake of plastocyanin by the thylakoid may be applicable to the corresponding process for the small algal cytochrome c. In vitro translation of *Chlamydomonas* mRNA yields a plastocyanin precursor of M_r 17 K, much larger than the mature protein (Merchant and Bogorad 1986). The cytochrome c-553 precursor (14 K) induced in Cu-deficient cells (Sect. 4.1.3) is also considerably larger than the mature form of apparent M_r 6 K (Merchant and Bogorad 1986). The presequence of the plastocyanin from the white campion (*Silene pratevisis*) is known from analysis of the cDNA clone derived from mRNA (Fig. 4.4; Smeekens et al. 1985). The N-terminal 14 residues show some similarity to presequences of other proteins associated with the thylakoid and may represent a target sequence for this location. On the other hand, the rest of the presequence does not resemble those of other thylakoid proteins and probably contains a region directing the protein to the thylakoid lumen. A candidate for this function is the 20 residue hydrophobic segment proximal to the N-terminus of the mature protein (Fig. 4.4).

Cytochrome f is encoded within the chloroplast DNA (Gray 1980). It is therefore synthesised on stromal ribosomes before incorporation in the thylakoid membrane and in vitro translation studies suggest initial formation of a higher molecular weight precursor (Willey et al. 1984; Rothstein et al. 1985). The amino acid sequence of cytochrome f contains a 21 residue hydrophobic segment near the C-terminus which is flanked by basic residues. Willey et al. (1984) proposed that this is a "stop transfer" sequence which acts to anchor cytochrome f in the membrane. This model is supported by the removal of only a small C-terminal peptide (residues 271–285) by proteolysis of intact thylakoids.

4.2.6 Methylation of Cytochrome c

A wide range of chemical modifications can occur after synthesis of proteins. In the case of cytochrome c, only three have been observed: methylation, N-terminal acetylation and heme attachment.* The last two have been considered in Section 4.2.2 and methylation is considered below.

Nature and Specificity of Methylation. Methylation of mitochondrial cytochrome c has been observed only at certain specific lysines and, in one case, at the α-amino group. The pattern of methylation varies in different eukaryotic groupings (Fig. 4.11). In most cases trimethylation occurs although dimethyllysine is found, for example at position 72 in *Humicola lanuginosa* cytochrome c. The cytochrome c-557 of *Crithidia oncopelti* contains a novel N-terminal N,N-dimethyl proline. There is little evidence of methylation among cytochromes c outwith the mitochondrial group. A residue of monomethyllysine is found in the algal cytochrome c-553 of *Monochrysis lutheri* (Laycock 1972) but no lysine methylation has been observed in bacterial cytochromes c.

* See appendix note 24

Neurospora and yeast contain both methylated and unmethylated forms of cytochrome c and these can be separated by ion exchange chromatography (Scott and Mitchell 1969; Fouchere et al. 1972). The methylated form can be identified by a high $^{3}H/^{14}C$ ratio in double-labelling experiments with methyl ^{3}H-methionine and 2-^{14}C-methionine (Borun et al. 1972). Methyl lysines are stable to acid hydrolysis and can be separated by ion exchange chromatography (Paik and Kim 1967). NMR spectroscopy was required to identify the novel N,N-dimethyl proline of *Crithidia oncopelti* (Pettigrew and Smith 1977; Smith and Pettigrew 1980).

Effect of Methylation and Biological Significance. Methylation does not affect the conformation of cytochrome c (Looze et al. 1976), its susceptibility to denaturation (Polastro et al. 1976) or its redox properties (Polastro et al. 1977). However, the methylated form of yeast iso-1-cytochrome c binds two to three times more strongly to mitochondria than its unmethylated counterpart (Polastro et al. 1978). This is consistent with the finding of Scott and Mitchell (1969) that only methylated cytochrome c could be isolated from *Neurospora* mitochondria. The structural basis for the increased affinity is uncertain. Assuming no conformation changes, trimethylation of Lys 72 should increase the basicity of the front surface of cytochrome c involved in interaction with its inner membrane redox partners (Chap. 2.4). However, the methylated cytochrome c actually has a lower rather than higher pI (Kim et al. 1980; Paik et al. 1983). These authors suggested that methylation causes a subtle change in the hydrogen bonding around Lys-72 (see also Vol. 2, Chap. 4).

A long-standing proposal as to the biological significance of protein methylation is that it protects against proteolytic degradation by analogy with the protection against endonuclease action conferred on DNA by base

Fig. 4.11 a–g. Sites of methylation and acetylation in cytochrome c. The sequences are *numbered* from the conserved Gly 1 of vertebrate cytochrome c. **a** Vertebrate and invertebrate cytochromes c. Lamprey, dogfish, bullfrog, turtle, rattlesnake, turkey, mammals and *Samia cynthia* were specifically examined for methylated lysines and none were found (DeLange et al. 1970). **b** Wheat germ cytochrome c (DeLange et al. 1969). The same pattern of methylation and acetylation is found in other plants (e.g. Brown and Boulter 1973; Ramshaw and Boulter 1975; Brown et al. 1973) with the exception of *Enteromorpha intestinalis* which is trimethylated only at Lys 72 (Meatyard and Boulter 1974). **c** *Neurospora crassa* cytochrome c (DeLange et al. 1969). The same pattern of methylation is found in bakers' yeast (DeLange et al. 1970) and *Debaromyces kloeckeri* (Sugeno et al. 1971). Although *Candida krusei* cytochrome c contains one trimethyllysine (DeLange et al. 1970) its location in the sequence was not chemically determined. However, spectroscopic evidence (Vol. 2, Chap. 3) shows that Lys 72 is trimethylated. Surprisingly this cytochrome is still reactive with *Neurospora* SAM methyltransferase (Durban et al. 1978). Cytochrome c from *Ustilago sphaerogena* is unusual among the fungal cytochromes c in having no methyllysine (Bitar et al. 1972). **d** *Humicola lanuginosa* cytochrome c (Morgan et al. 1972). Two forms were isolated which differed only in methylation at residue 86. **e** *Hansenula anomala* cytochrome c (Becam and Lederer 1981). Residue 55 is partly monomethyl- and partly dimethyllysine. **f** *Crithidia oncopelti* cytochrome c-557 (Pettigrew et al. 1975; Smith and Pettigrew 1980). The N-terminal proline is dimethylated. This is probably also true for *Crithidia fasciculata* (Hill and Pettigrew 1975). **g** *Euglena gracilis* cytochrome c-558 (Pettigrew 1973; Lin et al. 1973)

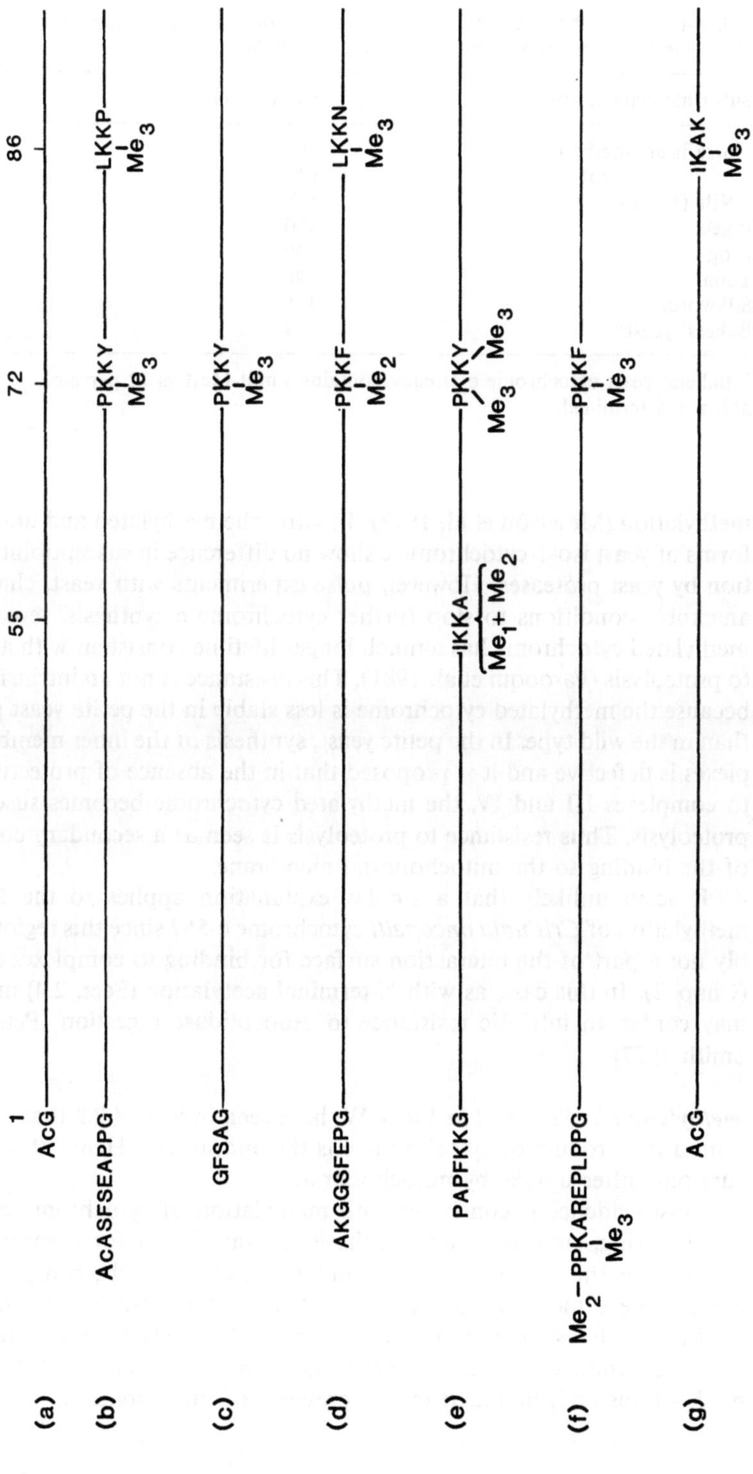
1
55
72
86
(a)
AcG
(b)
AcASFSEAPPG
PKKY
Me3
LKKP
Me3
(c)
GFSAG
PKKY
Me3
(d)
AKGGSFEPG
PKKF
Me2
LKKN
Me3
(e)
PAPFKKG
IKKA
Me1 + Me2
PKKY
Me3
Me3
(f)
Me2 – PPKAREPLPPG
Me3
PKKF
Me3
(g)
AcG
IKAK
Me3

Table 4.1. The substrate specificity of *Neurospora* S-adenosyl methionine cytochrome c lysine methyl transferase. From Durban et al. (1978)

Substrate cytochrome	% Activity	K_m
Horse heart (native)	100	0.32 mM
(apo)	nD	0.03 mM
CNBr (1 – 80)	nD	0.007 mM
Pigeon	120	
Frog	40	
Tuna	96	
Silkworm	166	
Bakers' yeast[a]	4	

[a] Bakers' yeast cytochrome c already contains 1 mol Me_3Lys 72 per mol.
nD, not determined.

methylation (Meselson et al. 1972). In vitro, the methylated and unmethylated forms of yeast iso-1-cytochrome c show no difference in susceptibility to digestion by yeast proteases. However, pulse experiments with yeast, chased under anaerobic conditions to stop further cytochrome c synthesis, reveal that the methylated cytochrome has a much longer lifetime consistent with a resistance to proteolysis (Farooqui et al. 1981). This resistance is not an intrinsic property because the methylated cytochrome is less stable in the petite yeast phenotype than in the wild type. In the petite yeast, synthesis of the inner membrane complexes is defective and it is proposed that in the absence of protective binding to complexes III and IV, the methylated cytochrome becomes susceptible to proteolysis. Thus resistance to proteolysis is seen as a secondary consequence of the binding to the mitochondrial membrane.

It seem unlikely that a similar explanation applies to the N-terminal methylation of *Crithidia oncopelti* cytochrome c-557 since this region is probably not a part of the interaction surface for binding to complexes III and IV (Chap. 2). In this case, as with N-terminal acetylation (Sect. 2.4) methylation may confer an intrinsic resistance to exopeptidase digestion (Pettigrew and Smith 1977).

Methylation in Vivo and in Vitro. We have seen in Sect. 4.2.2 that the primary translation product of cytochrome c is the apoprotein. Heme attachment occurs only after uptake by mitochondria.

Most evidence is consistent with methylation of cytochrome c occurring either during, or soon after, synthesis as was shown for myosin (Reporter 1973). Thus the protein synthesis inhibitor, cycloheximide, completely blocks cytochrome c methylation (Farooqui et al. 1980). Also SAM cytochrome c methyl transferase activity is cytoplasmic in *Neurospora* and prefers apo- to holocytochrome c as a substrate (Durban et al. 1978; Table 4.1). Furthermore methyl transfer is inhibited by the presence of mitochondria.

The study of Scott and Mitchell (1969), however, is anomalous in that the unmethylated holocytochrome of *Neurospora* gave rise to the methylated form in a cycloheximide-insensitive process. This is incompatible with cytoplasmic methylation of the apoprotein but may be a peculiarity of the "poky" phenotype which accumulates 16 times the normal level of cytochrome c.

Cytochrome c-specific protein-lysine methyl transferases have been isolated from *Neurospora, Saccharomyces cerevisiae* and wheat germ (Durban et al. 1978; DiMaria et al. 1979, 1982). All specifically transfer the methyl of S adenosyl methionine to Lys 72 of unmethylated cytochromes c (Table 4.1). Although the native horse heart cytochrome c was used as a standard substrate for assays, its K_m is much higher than that for the apocytochrome which may be the natural substrate (see above). Indeed, the unusually high pH optimum of 9 for the reaction with holocytochrome may reflect a requirement for substrate unfolding rather than enzyme activity. The specificity for Lys 72 therefore probably resides in recognition of a primary sequence rather than a conformation. This is borne out by the high activity, yet unaltered acceptor specificity, of CNBr fragments 1–80 and 66–100.

It is probable that only one enzyme is responsible for the addition of the three methyl groups to form trimethyllysine since the ratios of methylated lysines formed do not vary during the course of enzyme purification. It is also probable that successive methyl transfers occur without dissociation of the cytochrome substrate because trimethyllysine was observed even at short incubation times. Since 99.8% of the cytochrome remains unmethylated in the case of horse cytochrome c, further methylation of a dissociated molecule would be an unlikely event.

Although the enzymes show high specificity for cytochrome c as a substrate, the synthesis of the yeast enzyme and iso-1-cytochrome c are not coordinately regulated (DiMaria et al. 1979). In anaerobic yeast, cytochrome c synthesis is completely depressed but the methyl transferase activity is little altered. This may indicate that the methyl transferase also acts on substrates not yet identified.

As shown in Fig. 4.11, wheat cytochrome c contains an additional site of methylation at Lys 86. Yet the wheat methyltransferase did not methylate Lys 86 of horse cytochrome c (DiMaria et al. 1982). This cannot be due to loss of a second enzyme activity during purification because crude extracts also could not catalyse this reaction. One possibility is that there is a stringent sequence requirement for methylation at Lys 86 which is not met by horse cytochrome c. Another is that the true substrate is a nascent conformation on the ribosome.

In the case of *Euglena* cytochrome c-558, Lys 86 is the only point of methylation (Pettigrew 1973). Yet purification of cytochrome c-specific methyl transferase activity from *Euglena* surprisingly yielded not a lysine-specific enzyme but one which methylated arginine and a second which methylated methionine (Farooqui et al. 1985). Their physiological significance is not known.

The two methyls of the N-terminal dimethylproline of *Crithidia* cytochrome c-557 are also derived from methionine (Valentine and Pettigrew 1982).

A soluble extract of *Crithidia* can methylate cytochrome c-557 at both the N-terminal proline and the nearby lysine −8. Allowing for the fact that cytochrome c-557 is already almost fully methylated at these positions, the methyl transfer is much more efficient than that involving the *Neurospora* enzyme and horse cytochrome c.

4.2.7 The Folding of Cytochrome c

We have seen that it is the apocytochrome that is taken up by the mitochondrion and heme attachment probably occurs in the intermembrane space. As judged by viscosity, CD, NMR and Trp fluorescence measurements, apocytochrome c is unstructured (Stellwagen et al. 1972; Fisher et al. 1973; Cohen et al. 1974) and addition of free heme does not allow folding into a compact shape (Fisher et al. 1973) unlike the case of myoglobin (Breslow et al. 1965). However, Ehrenberg and Theorell (1955) noted that an N-terminal α-helix would bring cysteines 14 and 17 into a correct orientation for heme attachment and it may be that such a secondary structure might form on association with the heme attachment enzyme in a way analogous to that proposed for the binding of some flexible polypeptide hormones to their receptors (Blundell and Wood 1982). Once the heme is attached it probably acts as a nucleus around which the polypeptide chain folds.

It is now accepted that the folding of proteins generally is more complex than the original two-state model U (unfolded)⇌N (native). Folding occurs along pathways of intermediate stages and is influenced by proline isomerisation (Kim and Baldwin 1982).

The influence of proline isomerisation is recognised by the presence of fast and slow phases in refolding which are due to the unfolded forms U_F and U_S. The relationship between U_F, U_S and N is represented in Fig. 4.12. Thus the equilibrium between U_F and U_S can be observed during unfolding and has the features expected from model studies of proline isomerisation (Kim and

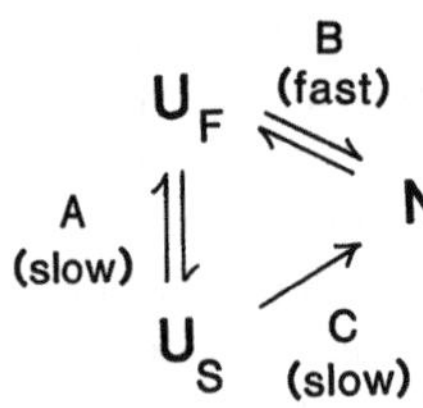

Fig. 4.12. The processes of protein folding and unfolding. U_F Unfolded species with fast folding kinetics; U_S unfolded species with slow folding kinetics; *N* native protein. None of the reactions *A*, *B* and *C* are simple single-step processes. *A* is proline isomerisation but may include isomerisations of several U_S species depending on the number of critical prolines. *B* involves a sequential series of intermediate forms which may or may not be the same in the folding and unfolding directions. *C* also involves a series of intermediate forms and includes proline isomerisation

Baldwin 1982). Not all prolines in a molecule may influence the kinetics of folding; those that do are termed critical.* The greater the number of critical prolines, the larger the magnitude of the slow kinetic phase. Notice that the unfolding of N to give U_S is along a different pathway from the folding of U_S to give N. This is clearly illustrated by ribonuclease refolding at low temperature which gives rise to an active enzyme with the wrong proline isomer (Cook et al. 1979).

In horse cytochrome c which contains four *trans* proline residues (Dickerson and Timkovich 1975) the slow phase of refolding amounts to 20% of the total process (Ridge et al. 1981) and this may reflect a single critical proline. Indeed, tuna cytochrome c with only three prolines has very similar refolding kinetics (Babul et al. 1978). The finding that porphyrin cytochrome c refolds rapidly (Henkens and Turner 1979; Brems et al. 1982; Brems and Stellwagen 1983) at first seems to rule out proline isomerisation and implicate methionine iron ligation as the structural basis for the slow phase. However, this phase is present in the refolding from urea of imidazole cytochrome c, a structure in which no methionine iron coordination can occur (Myer 1984). It may be that the porphyrin cytochrome c conformation is flexible enough to accommodate wrong proline isomers (Ridge et al. 1981).

Although the scheme of Fig. 4.12 indicates single reactions for $U_F \rightarrow N$ and $U_S \rightarrow N$ there is now evidence for many proteins that these represent sequential pathways with defined structural intermediates (Kim and Baldwin 1982). However, the problems in determining the nature of these intermediates are great because structural methods such as X-ray diffraction and NMR spectroscopy have too slow a time scale. The presence of intermediates is established by the kinetic ratio test (Kim and Baldwin 1982) in which different probes are used to monitor the kinetics of refolding and the rates obtained are compared.

For cytochrome c, useful probes have been UV spectroscopy, which detects changes in the exposure of aromatic residues to solvent; Trp fluorescence, which reflects the distance between the donor Trp59 and the quenching heme group; and appearance of the 695-nm band, which indicates methionine-iron coordination (Vol. 2, Chap. 2). During the refolding of yeast iso-2-cytochrome c, the quenching of Trp fluorescence by the heme group is complete before recovery of the 695-nm band and changes in UV absorption (Nall 1983). This is consistent with an intermediate in which the tryptophan has taken up a position close to the heme but the methionine-iron bond has not been formed and some aromatic residues are still exposed to solvent. A similar sequence of formation of Trp59-heme interaction followed by coordination of methionine to iron is found during reconstitution of cytochrome c heme peptides with apoprotein (Parr and Taniuchi 1980).

The intermediate observed in these studies may be related to the "molten globule" proposed by Ohgushi and Wada (1983) for the structure of cytochrome c at pH 2 in the presence of 0.5 M KCl. Under these conditions, cytochrome c adopts a compact structure in which Trp fluorescence is quenched (as with the native protein) and α-helical structure is present, but the NMR

* See appendix note 25

spectrum is that of a random coil. Goldberg (1985) proposed that the "molten globule" may be a common feature of protein-folding pathways in order to stabilise elements of nascent secondary structure.

The difficulties inherent in characterising the folding of cytochrome c have led several authors to explore pathways of denaturation of the native protein. As with the folding pathway, various intermediate stages are observed but these have not been structurally characterised (see Vol. 2, Chap. 4 for further details).

References

Arcangioli B, Lescure B (1985) Identification of proteins involved in the regulation of yeast iso-1 cytochrome c expression by oxygen. EMBO J 4:2627–2633

Babul J, Nakagawa A, Stellwagen E (1978) An examination of the involvement of proline peptide isomerisation in protein folding. J Mol Biol 126:117–121

Bassford P, Beckwith J (1979) *E. coli* mutants accumulating the precursor of a secreted protein in the cytoplasm. Nature (London) 277:538–541

Becam AM, Lederer F (1981) Amino acid sequence of the cytochrome c from the yeast *Hansenula anomala* – identification of three methylated positions. Eur J Biochem 118:295–302

Bitar KG, Vinogradov SN, Nolan C, Weiss LJ, Margoliash E (1972) The primary structure of cytochrome c from the rust fungus, *Ustilago sphaerogena.* Biochem J 129:561–569

Blobel G (1980) Intracellular protein topogenesis. Proc Natl Acad Sci USA 77:1496–1500

Blundell T, Wood S (1982) The conformation, flexibility and dynamics of polypeptide hormones. Annu Rev Biochem 51:123–154

Boehni PC, Daum G (1983) Processing of mitochondrial polypeptide precursors in yeast. Meth Enzymol 97:311–323

Bohner H, Böger P (1978) Reciprocal formation of a cytochrome c-553 and plastocyanin in *Scenedesmus.* FEBS Lett 85:337–340

Bohner H, Bohme H, Böger P (1980) Reciprocal formation of plastocyanin and cytochrome c-553 and the influence of cupric ions on photosynthetic electron transport. Biochim Biophys Acta 592:103–112

Borun TW, Pearson D, Paik WK (1972) Studies of histone methylation during the HeLa cell cycle. J Biol Chem 247:4288–4298

Boss JM, Gillam S, Zitomer RS, Smith M (1981) Sequence of the yeast iso-1 cytochrome c mRNA. J Biol Chem 256:12958–12961

Brems DN, Stellwagen E (1983) Manipulation of the observed kinetic phases in the refolding of denatured ferri-cytochromes c. J Biol Chem 258:3655–3660

Brems DN, Cass R, Stellwagen E (1982) Conformational transitions of frog heart ferricytochrome c. Biochemistry 21:1488–1493

Breslow E, Beychok S, Hardman KD, Gurd FRN (1965) Relative conformations of sperm whale metmyoglobin and apomyoglobin in solution. J Biol Chem 240:304–309

Brown RH, Boulter D (1973) The amino acid sequence of a cytochrome c from *Allium porrum* (leek). Biochem J 131:247–251

Brown RH, Richardson M, Scogin R, Boulter D (1973) The amino acid sequence of cytochrome c from Spinacea oleracea (spinach). Biochem J 131:253–256

Burstein Y, Schechter I (1976) Amino acid sequence of the precursors of mouse immunoglobulin λ-type and k-type light chains. Proc Natl Acad Sci USA 74:716–720

Clavilier L, Perc-Aubert G, Somlo M, Slonimski PP (1976) Réseau d'interactions entre des gênes non liés: régulation synergique ou antagoniste de la synthèse de l'iso-1-cytochrome c, de l'iso-2-cytochrome c et du cytochrome b. Biochimie 58:155–172

Cohen JS, Fisher WR, Schechter AN (1974) Spectroscopic studies on the conformation of cytochrome c and apocytochrome c. J Biol Chem 249:1113–1118

Colleran EM, Jones OTG (1973) Studies on the biosynthesis of cytochrome c. Biochem J 134:89–96

Cook KH, Schmidt FX, Baldwin RL (1979) Role of proline isomerisation in refolding of ribonuclease A at low temperatures. Proc Natl Acad Sci USA 76:6157–6161

Daldal F, Cheng S, Applebaum J, Davidson E, Prince RC (1986) Cytochrome c is not essential for photosynthetic growth of *Rhodopseudomonas capsulata.* Proc Natl Acad Sci USA 83:2012–2016

Daum G, Bohni PC, Schatz G (1982a) Cytochrome b_2 and cytochrome c peroxidase are located in the intermembrane space of yeast mitochondria. J Biol Chem 257:13028–13033

Daum G, Gasser SM, Schatz G (1982b) Energy-dependent 2 step processing of the intermembrane space enzyme – cytochrome b_2 – by isolated yeast mitochondria. J Biol Chem 257:13075–13080

DeLange RJ, Glazer AN, Smith EL (1969) Presence and location of an unusual amino acid ε-N-trimethyllysine in cytochrome c of wheat germ and *Neurospora.* J Biol Chem 244:1385–1388

DeLange RJ, Glazer AN, Smith EL (1970) Identification and location of ε-N-trimethyllysine in yeast cytochromes c. J Biol Chem 245:3325–3327

Dickerson RE, Timkovich R (1975) Cytochrome c. In: Boyer PD (ed) The enzymes, vol 11. Academic Press, London New York, pp 397–547

DiMaria P, Polastro E, DeLange RJ, Kim S, Paik WK (1979) Studies on cytochrome c methylation in yeast. J Biol Chem 254:4645–4652

DiMaria P, Kim S, Paik WK (1982) Cytochrome c specific methylase from wheat germ. Biochemistry 21:1036–1044

Donovan JJ, Simon MI, Montal M (1982) Insertion of diphtheria toxin into and across membranes – role of phosphoinositide asymmetry. Nature (London) 298:669–672

Downie JA, Stewart JW, Brockman N, Schweingruber AM, Sherman F (1977) Structural gene for yeast iso-2-cytochrome c. J Mol Biol 113:369–384

Durban E, Nochumson S, Kim S, Paik WK, Chan SK (1978) Cytochrome c-specific protein lysine methyl transferase from *Neurospora.* J Biol Chem 253:1427–1433

Ehrenberg A, Theorell H (1955) Stereochemical structure of cytochrome c. Acta Chem Scand 9:1193–1205

Ellis RJ (1981) Chloroplast proteins – synthesis, transport and assembly. Annu Rev Plant Physiol 32:111–137

Ernst JF, Stewart JW, Sherman F (1981) The cyc1–11 mutation in yeast reverts by recombination with a non-allelic gene – composite genes determining the isocytochromes c. Proc Natl Acad Sci USA 78:6334–6338

Farooqui J, Kim S, Paik WK (1980) In vivo studies on yeast cytochrome c methylation in relation to protein synthesis. J Biol Chem 255:4468–4473

Farooqui J, DiMaria P, Kim S, Paik WK (1981) Effect of methylation on the stability of cytochrome c of *Saccharomyces cerevisiae* in vivo. J Biol Chem 256:5041–5045

Farooqui JZ, Tuck M, Paik WK (1985) Purification and characterisation of enzymes from *Euglena gracilis* that methylate methionine and arginine residues of cytochrome c. J Biol Chem 260:537–545

Faye G, Leeung DW, Tatchell K, Hall BD, Smith M (1981) Deletion mapping of sequences essential for in vitro transcription of the iso-1-cytochrome c gene. Proc Natl Acad Sci USA 78:2258–2262

Fisher WR, Taniuchi H, Anfinsen CB (1973) On the role of heme in the formation of the structure of cytochrome c. J Biol Chem 248:3188–3195

Fouchere M, Verdiere J, Lederer F, Slonimski PP (1972) On the presence of a non-trimethylated iso-1-cytochrome c in a wild-type strain of *Saccharomyces cerevisiae.* Eur J Biochem 31:139–143

Garrard WT (1972) Synthesis, assembly and localisation of periplasmic cytochrome c. J Biol Chem 247:5935–5943

Gasser SM, Ohashi A, Daum G, Bohni PC, Gibson J, Reid GA, Yonetani T, Schatz G (1982) Imported mitochondrial proteins, cytochrome b_2 and c_1 are processed in two steps. Proc Natl Acad Sci USA 79:267–271

Gilbert W (1978) Why genes in pieces? Nature (London) 271:501

Godson GN, Barrell BG, Staden R, Fiddes JC (1978) Nucleotide sequence of bacteriophage T4 DNA. Nature (London) 276:236–247

Goldberg E, Sbema D, Wheat TE, Urbanski GJ, Margoliash E (1977) Cytochrome c – immunofluorescent localisation of the testis specific form. Science 196:1010–1011

Goldberg M (1979) Sequence analysis of Drosophila histone genes. PhD Thesis, Stanford Univ

Goldberg ME (1985) The second translation of the genetic message: protein folding and assembly. Trends Biochem Sci 10:388–391

Gorman DS, Levine RP (1965) Cytochrome f and plastocyanin – their sequence in the photosynthetic electron transport chain of *Chlamydomonas reinhardii.* Proc Natl Acad Sci USA 54:1665–1669

Gray JC (1980) Maternal inheritance of cytochrome f in interspecific *Nicotiana* hybrids. Eur J Biochem 112:39–46

Guarente L (1986) Regulation of the yeast CYC1 gene. 13th Int Conf Yeast Genetics and Molecular Biology. Wiley, S137

Guarente L, Mason T (1983) Heme regulates the transcription of the CYC1 gene of *Saccharomyces cerevisiae* via an upstream activator site. Cell 32:1279–1286

Guarente L, Ptashne M (1981) Fusion of *E. coli* lacZ to the cytochrome c gene of *Saccharomyces cerevisiae.* Proc Natl Acad Sci USA 78:2199–2203

Guarente L, Lalonde B, Gifford P, Alani E (1984) Distinctly regulated tandem upstream activation sites mediate catabolite repression of the CYC1 gene of *Saccharomyces cerevisiae.* Cell 36:503–511

Henkens RW, Turner SR (1979) Kinetic studies of the effect of the heme iron III on the protein folding of ferricytochrome. J Biol Chem 254:8110–8112

Hennig B (1975) Change of cytochrome c structure during development of the mouse. Eur J Biochem 55:167–183

Hennig B, Neupert W (1981) Assembly of cytochrome c – apocytochrome c is bound to specific sites on mitochondria before it is converted to holocytochrome c. Eur J Biochem 121:203–212

Hennig B, Koehler H, Neupert W (1983) Receptor sites involved in posttranslational transport of apocytochrome c into mitochondria – specificity and number of sites. Proc Natl Acad Sci USA 80:4963–4967

Hill GC, Pettigrew GW (1975) Evidence for the amino acid sequence of *Crithidia fasciculata* cytochrome c-555. Eur J Biochem 57:265–271

Hurt EC, Loon APGM van (1986) How proteins find mitochondria and intramitochondrial compartments. Trends Biochem Sci 11:204–207

Iborra F, Francingues MC, Guerineau M (1985) Localisation of the upstream regulatory sites of yeast iso-2-cytochrome c gene. Mol Gen Genet 199:117–122

Inoue S, Hiroyashi T, Matsubara H, Yamanaka T (1984) Complete amino acid sequences of two iso-cytochromes c of the housefly, *Musca domestica* L., and their developmental variation in different tissues. Biochim Biophys Acta 790:188–195

Inouye H, Barnes W, Beckwith J (1982) Signal sequence of alkaline phosphatase of *E. coli.* J Bacteriol 149:434–439

Jahn CL, Hutchison CA, Phillips SJ, Weaver S, Haigwood NL, Voliva CF, Egdell MH (1980) DNA sequence organisation of the *β*-globin complex in the BALB/c mouse. Cell 21:159–168

Kaput J, Goltz S, Blobel G (1982) Nucleotide sequence of the yeast nuclear gene for cytochrome c peroxidase precursor. J Biol Chem 257:15054–15058

Kim CS, Kueppers F, DiMaria P, Farooqui J, Kim S, Paik WK (1980) Enzymatic trimethylation of residue 72 lysine in cytochrome c. Biochim Biophys Acta 622:144–150

Kim IC (1980) Isolation and properties of somatic and testicular cytochromes c from rat tissues. Arch Biochem Biophys 203:519–528

Kim PS, Baldwin RL (1982) Specific intermediates in the folding reactions of small proteins and the mechanism of protein folding. Annu Rev Biochem 51:459–489

Komar E, Weber H, Tanner W (1979) Greatly decreased susceptibility of non-metabolising cells towards detergents. Proc Natl Acad Sci USA 76:1814–1818

Korb H, Neupert W (1978) Biogenesis of cytochrome c in *Neurospora crassa.* Eur J Biochem 91:609–620

Kozak M (1978) How do eukaryotic ribosomes select initiation regions in mRNA. Cell 15:1109–1123

Kunert KJ, Boger P (1975) The absence of plastocyanin in the alga *Bumilleriopsis filiformis* and its replacement by cytochrome c-553. Z Naturforsch 30C:190–200

Laycock MV (1972) The amino acid sequence of cytochrome c-553 from the Chrysophycean alga *Monochrysis lutheri.* Can J Biochem 50:1311–1325

Laz TM, Pietras DF, Sherman F (1984) Differential regulation of the duplicated isocytochrome c genes in yeast. Proc Natl Acad Sci USA 81:4475–4479

Lescure B, Arcangioli B (1984) Yeast RNA polymerase II initiates transcription in vitro at TATA sequences proximal to potential non-B forms of the DNA template. EMBO J 3:1067–1073

Leuders K, Leder A, Leder P, Kuff E (1982) Association between a transposed α-globin pseudogene and retrovirus-like elements in the BALBYc mouse genome. Nature (London) 295:426–428

Limbach KJ, Wu R (1983) Isolation and characterisation of two alleles of the chicken cytochrome c gene. Nucl Acids Res 11:8931–8950

Limbach KJ, Wu R (1985 a) Characterisation of mouse somatic cytochrome c gene and three cytochrome c pseudogenes. Nucl Acids Res 13:617–630

Limbach KJ, Wu R (1985 b) Characterisation of two *Drosophila melanogaster* cytochrome c genes and their transcripts. Nucl Acids Res 13:631–644

Lin DK, Niece RL, Fitch W (1973) The properties and amino acid sequence of cytochrome c from *Euglena gracilis.* Nature (London) 241:533–535

Looze Y, Polastro E, Gielens C, Leonis J (1976) Iso-cytochrome c species from Bakers yeast – analysis of their CD spectra. Biochem J 157:773–775

Matner RR, Sherman F (1982) Differential accumulation of the two apoisocytochromes c in processing mutants of yeast. J Biol Chem 257:9811–9821

Matsuura S, Arpin M, Hannum C, Margoliash E, Sabatini DD, Morimoto T (1981) In vitro synthesis and posttranslational uptake of cytochrome c into isolated mitochondria – role of a specific addressing signal in the apocytochrome. Proc Natl Acad Sci USA 78: 4368–4372

McKnight GL, Cardillo TS, Sherman F (1981) An extensive deletion causing overproduction of yeast iso-2-cytochrome c. Cell 25:409–419

Meatyard BT, Boulter D (1974) The amino acid sequence of cytochrome c from *Enteromorpha intestinalis.* Phytochemistry 13:2777–2782

Merchant S, Bogorad L (1986) Regulation by copper of the expression of plastocyanin and cytochrome c-552 in *Chlamydomonas reinhardii.* Mol Cell Biol 6:462–469

Meselson M, Yuan R, Heywood J (1972) Restriction modification of DNA. Annu Rev Biochem 41:447–464

Michaelis S, Beckwith J (1982) Mechanisms of incorporation of cell envelope proteins in *E. coli.* Annu Rev Microbiol 36:435–465

Montgomery DL, Leung DW, Smith M, Shalit P, Faye G, Hall BD (1980) Isolation and sequence of the gene for iso-2-cytochrome c in *Saccharomyces cervisiae.* Proc Natl Acad Sci USA 77:541–545

Montgomery DL, Boss JM, McAndrew SJ, Marr L, Walthall DA, Zitomer RS (1982) The molecular characterisation of three transcriptional mutations in the yeast iso-2-cytochrome c gene. J Biol Chem 257:7756–7761

Morgan WT, Hensley CP, Riehm JP (1972) Isolation and amino acid sequence of a cytochrome c from a thermophilic fungus *Humicola lanuginosa.* J Biol Chem 247: 6555–6565

Myer YP (1984) Ferricytochrome c – refolding and the Met 80-S-Fe linkage. J Biol Chem 259:6127–6133

Nall BT (1983) Structural intermediates in folding of yeast iso-2-cytochrome c. Biochemistry 22:1423–1429

Nelson N, Schatz G (1979) Energy-dependent processing of cytoplasmically made precursors to mitochondrial proteins. Proc Natl Acad Sci USA 76:4365–4369

Neupert W, Schatz G (1981) How proteins are transported into mitochondria. Trends Biochem Sci 6:1–4

Ohashi A, Gibson J, Gregor I, Schatz G (1982) The precursor of cytochrome c_1 is processed in two steps, one of them heme-dependent. J Biol Chem 257:13042–13047

Ohgushi M, Wada A (1983) Molten globule state – compact form of globular proteins with mobile side chains. FEBS Lett 164:21–24

Paik WK, Kim S (1967) ε-N-dimethyllysine in histones. Biochem Biophys Res Commun 27:479–483

Paik WK, Farooqui J, Gupta A, Smith HT, Millet F (1983) Enzymatic trimethylation of lysine 72 in cytochrome c. Eur J Biochem 135:259–262

Palmiter RD, Gagnon J, Walsh KA (1978) Ovalbumin, a secreted protein without a transient hydrophobic leader sequence. Proc Natl Acad Sci USA 75:94–98

Parr GR, Taniuchi H (1980) Kinetic intermediates in the formation of ordered complexes from cytochrome c fragments – evidence that methionine ligation is a late event in the folding process. J Biol Chem 255:8914–8918

Pettigrew GW (1973) The amino acid sequence of cytochrome c from *Euglena gracilis.* Nature (London) 241:531–533

Pettigrew GW, Smith G (1977) Novel N-terminal protein blocking group identified as dimethylproline. Nature (London) 265:661–662

Pettigrew GW, Leaver JL, Meyer TE, Ryle AP (1975) Purification, properties and amino acid sequence of atypical cytochrome c from two Protozoa, *Euglena gracilis* and *Crithidia oncopelti.* Biochem J 147:291–302

Polastro ET, Looze Y, Leonis J (1976) Thermal, acid and guanidinium hydrochloride denaturation of Bakers yeast ferricytochromes c. Biochim Biophys Acta 446:310–320

Polastro ET, Looze Y, Leonis J (1977) Biological significance of methylation of cytochromes from Ascomycetes and plants. Phytochemistry 16:639–641

Polastro ET, Deconnick MM, DeVogel MR, Mailier EL, Looze YR, Schnek AG, Leonis J (1978) Evidence that trimethylation of iso-1-cytochrome from *Saccharomyces cerevisiae* affects interaction with the mitochondrion. FEBS Lett 86:17–20

Ramshaw JAM. Boulter D (1975) Amino acid sequence of cytochrome c from Niger seed *Guizotia abyssincia.* Phytochemistry 14:1945–1949

Randall L (1983) Translocation of domains of nascent periplasmic proteins across the cytoplasmic membrane is independent of elongation. Cell 33:231–240

Reid GA, Yonetani T, Schatz G (1982) Import and maturation of the mitochondrial intermembrane space enzymes – cytochrome b_2 and cytochrome c peroxidase in intact yeast cells. J Biol Chem 257:13068–13074

Reporter M (1973) Protein synthesis in cultured muscle cells – methylation of nascent proteins. Arch Biochem Biophys 158:577–585

Ridge JA, Baldwin RL, Labhardt AM (1981) Nature of the fast and slow refolding reactions of iron III cytochrome c. Biochemistry 20:1622–1630

Rietveld A, Kruiff B (1984) Is the mitochondrial precursor protein – apocytochrome c – able to pass a lipid barrier? J Biol Chem 259:6704–6707

Rietveld A, Ponjee GAE, Schiffers P, Jordi W, VandeCollwijk PJFM, Demel RA, Marsh D, Kruiff B (1985) Investigations on the insertion of the mitochondrial precursor protein, apocytochrome c into model membranes. Biochim Biophys Acta 818:398–409

Roggenkamp P, Kustermann-Kuhn B, Hollenberg CP (1981) Expression and processing of bacterial β-lactamase in the yeast *Saccharomyces cerevisiae.* Proc Natl Acad Sci USA 78:4466–4470

Rothstein SJ, Gatenby AA, Willey DL, Gray JC (1985) Binding of pea cytochrome f to the inner membrane of *E. coli* requires the bacterial *sec A* gene product. Proc Natl Acad Sci USA 82:7955–7959

Russell PR, Hall HD (1982) Structure of the *Schizosaccharomyces pombe* cytochrome c gene. Mol Cell Biol 2:106–116

Sadler I, Suda K, Schatz G, Kaudewitz F, Haig A (1984) Sequencing of the nuclear gene for the yeast cytochrome c_1 precursor reveals an unusually complex amino terminal presequence. EMBO J 3:2137–2143

Sandman G, Boger P (1980) Copper-induced exchange of plastocyanin and cytochrome c-553 in cultures of *Anabaena variabilis* and *Plectonema boryanum.* Plant Sci Lett 17:417–424

Sano S, Granick S (1961) Mitochondrial coproporphyrinogen oxidase and protoporphyrin formation. J Biol Chem 236:1173–1180

Sano S, Tanaka K (1964) Recombination of protoporphyrinogen with cytochrome c apoprotein. J Biol Chem 239:3109–3110

Scarpulla RC (1984) Processed pseudogenes for rat cytochrome c are preferentially derived from one of three alternate mRNAs. Mol Cell Biol 4:2279–2288

Scarpulla RC (1985) Association of a truncated cytochrome c processed pseudogene with a similarly truncated member from a long interspersed repeat family of rat. Nucl Acid Res 13:763–775

Scarpulla RC, Wu R (1983) Non-allelic members of the cytochrome c multigene family of the rat may arise through different messenger mRNAs. Cell 32:473–482

Scarpulla RC, Agne KM, Wu R (1981) Isolation and structure of a rat cytochrome c gene. J Biol Chem 256:6480–6486

Scarpulla RC, Agne KM, Wu R (1982) Cytochrome c gene related sequences in mammalian genomes. Proc Natl Acad Sci USA 79:739–743

Schatz G, Mason TL (1974) The biosynthesis of mitochondrial proteins. Annu Rev Biochem 43:51–87

Schmidt GW, Mishkind ML (1986) The transport of proteins into chloroplasts. Annu Rev Biochem 55:879–912

Scott WA, Mitchell HK (1969) Secondary modification of cytochrome c by *Neurospora crassa.* Biochemistry 8:4282–4289

Sherman F, Stewart JW (1971) Genetics and biosynthesis of cytochrome c. Annu Rev Genet 5:257–296

Sherman F, Stewart JW, Margoliash E, Parker J, Campbell W (1966) The structural gene for yeast cytochrome c. Proc Natl Acad Sci USA 55:1498–1504

Sherman F, Stewart JW, Helms C, Downie JA (1978) Chromosome mapping of the CYC7 gene determining yeast iso-2-cytochrome c – structural and regulatory regions. Proc Natl Acad Sci USA 75:1437–1441

Sherman F, Stewart JW, Schweingruber AM (1980) Mutants of yeast initiating translation of iso-1-cytochrome c within a region spanning 37 nucleotides. Cell 20:215–222

Smeekens S, DeGroot M, Binsbergen JV, Weisbeek P (1985) Sequence of the precursor of the chloroplast thylakoid lumen protein plastocyanin. Nature (London) 317:456–458

Smith GM, Pettigrew GW (1980) Identification of N,N-dimethylproline as the N-terminal blocking group of *Crithidia oncopelti* cytochrome c-557. Eur J Biochem 110:123–130

Smith M, Leung DW, Gillam S, Astell CR, Montgomery DL, Hall BD (1979) Sequence of the gene for iso-1-cytochrome c in *Saccharomyces cerevisiae.* Cell 16:753–761

Steitz JA (1979) Genetic signals and nucleotide sequences in messenger RNA. In: Goldberger R (ed) Biological regulation and development. I.: Gene expression. Plenum Press, New York, pp 349–399

Stellwagen E, Rysavy R, Babul G (1972) The conformation of horse heart apocytochrome c. J Biol Chem 247:8074–8077

Stewart JW, Sherman F (1974) Yeast frameshift mutations identified by sequence changes in iso-1-cytochrome c. In: Prakash L, Sherman F, Miller MW, Lawrence CW, Taber HW (eds) Molecular and environmental aspects of mutagenesis. Thomas, Springfield, pp 102–127

Stewart JW, Sherman F, Shipman NA, Jackson H (1971) Identification and mutational relocation of the AUG codon initiating translocation of iso-1-cytochrome c in yeast. J Biol Chem 246:7429–7445

Sugeno K, Narita K, Titani K (1971) The amino acid sequence of cytochrome c from *Debaromyces kloeckeri.* J Biochem 70:659–682

Talmadge K, Stahl S, Gilbert W (1980) Eukaryotic signal sequence transports insulin antigen in *E. coli.* Proc Natl Acad Sci USA 77:3369–3373

Taniuchi H, Basile G, Taniuchi M, Veloso D (1983) Evidence for formation of two thioether bonds to link heme to apocytochrome c by partially purified cytochrome c synthase. J Biol Chem 258:10963–10966

Tosteson MT, Tosteson DC (1981) The sting – mellitin forms channels in lipid bilayers. Biophys J 36:109–116

Tsunasawa S, Sakujama F (1984) Amino-terminal acetylation of proteins – an overview. Methods Enzymol 106:165–170

Tsunasawa S, Stewart JW, Sherman F (1985) Amino terminal processing of mutant forms of yeast iso-1-cytochrome c. J Biol Chem 260:5382–5391

Valentine J, Pettigrew GW (1982) A cytochrome c methyltransferase from *Crithidia oncopelti.* Biochem J 201:329–338

Verdiere J, Creusot F, Guerineau M (1985) Regulation of the expression of iso-2-cytochrome c gene in *Saccharomyces cerevisiae* – cloning of the positive regulatory gene cyp1 and identification of the region of its target sequence on the structural gene cyp3. Mol Gen Genet 199:524–533

Visco C, Taniuchi H, Berlett BS (1985) On the specificity of cytochrome c synthetase in recognition of the amino acid sequence of apocytochrome c. J Biol Chem 260:6133–6138

Walter P, Blobel G (1982) Signal recognition particle contains a 7S RNA essential for protein translocation across the endoplasmic reticulum. Nature (London) 299:691–698

Weiner AM, Deininger PL, Efstratiadis A (1986) Non-viral retroposons, genes, pseudogenes and transposable elements generated by the reverse flow of genetic information. Annu Rev Biochem 55:631–661

Wickner WT (1980) Assemply of proteins into membranes. Science 210:861–862

Wickner WT, Lodish HF (1985) Multiple mechanisms of protein insertion into membranes. Science 230:400–407

Willey DL, Auffret AD, Gray JC (1984) Structure and topology of cytochrome f in pea chloroplast membranes. Cell 36:555–562

Wolfe PB, Wickner W, Goodman JM (1983) Sequence of the leader peptidase gene of *E. coli* and the orientation of leader peptidase in the bacterial envelope. J Biol Chem 258:12073–12080

Wood PM (1978) Interchangeable copper and iron proteins in algal photosynthesis. Eur J Biochem 87:9–19

Wood PM (1983) Why do c-type cytochromes exist? FEBS Lett 164:223–226

Yamanaka T, Inoue S, Hiroyoshi T (1980) Structural differences between larval and adult cytochromes c of the housefly *Musca domestica.* J. Biochem 88:601–604

Yaniv M (1984) Regulation of eukaryotic gene expression by trans-acting proteins and cis-acting DNA elements. Biol Cell 50:203–216

Young ET, Pilgrim D (1985) Isolation and DNA sequence of ADH3, a nuclear gene encoding the mitochondrial isozyme of alcohol dehydrogenase in *Saccharomyces cerevisiae.* Mol Cell Biol 5:3024–3034

Zimmerman R, Paluch U, Neupert W (1979) Cell-free synthesis of cytochrome c. FEBS Lett 108:141–146

Zitomer RS, Hall BD (1976) Yeast cytochrome c mRNA in vitro translation and specific immunoprecipitation of the CYC1 gene product. J Biol Chem 251:6320–6326

Zitomer RS, Nichols DL (1978) Kinetics of glucose repression of yeast cytochrome c. J Bacteriol 135:39–44

Zitomer RS, Montgomery DL, Nichols DL, Hall BD (1979) Transcriptional regulation of the yeast cytochrome c gene. Proc Natl Acad Sci USA 76:3627–3631

Appendix

Notes on Recent Papers Added in Proof

Note 1 (p. 21). The flavocytochrome c, p-cresol methylhydroxylase, from *Pseudomonas putida* contains a cytochrome c subunit of 96 amino acids that shows some sequence similarity to Class I c-type cytochromes.

McIntyre W, Singer TP, Smith AJ, Mathews FS (1986) Amino acid sequence analysis of the cytochrome and flavoprotein subunits of p-cresol methyl hydroxylase. Biochemistry 25:5975–5981

Note 2 (p. 36). Bovine and *Paracoccus* cytochromes aa_3 contain three coppers. Steffens GCM, Biewald R, Buse G (1987) Cytochrome c oxidase is a three-copper, two-heme A protein. Eur J Biochem 164:295–300

Note 3 (pp. 53, 90). A similar conclusion is reached by Sinjorgo et al. (1986) from a study of the ionic strength dependence of cytochrome c oxidase steady state kinetics at different pH values.

Sinjorgo KMC, Steinebach OM, Dekker HL, Muijsers AO (1986) The effects of pH and ionic strength on cytochrome c oxidase steady state kinetics reveal a catalytic and a non-catalytic interaction domain for cytochrome c. Biochim Biophys Acta 850:108–115

Note 4 (p. 56). ATP induces conformational changes in cytochrome oxidase that modify the cytochrome c binding site.

Bisson R, Shiavo G, Montecucco C (1987) ATP induces conformational changes in mitochondrial cytochrome c oxidase. J Biol Chem 262:5992–5998

Note 5 (p. 75). Carboxydinitrophenylation of lysines 8, 13, 27, 72, 73, 86 and 87 decreases the rate of reduction of horse cytochrome c by cytochrome b_2 of *Saccharomyces cerevisiae.* The enzyme can bind two molecules of cytochrome c.

Matsushima A, Yoshimura T, Aki K (1986) Region of cytochrome c interacting with yeast cytochrome b_2: determination with singly modified carboxydinitrophenyl cytochromes c. J Biochem 100:543–551

Note 6 (p. 80). As with cytochrome c peroxidase, a single histidine and several acidic residues are important in the interaction domain of sulfite oxidase for cytochrome c.

Ritzmann M, Bosshard HR (1986) Sulfite oxidase from chicken liver – the role of imidazole and carboxyl groups for the reaction with cytochrome c. Eur J Biochem 159:493–497

Note 7 (p. 97). It is the "low affinity reaction" which is abolished at high ionic strength: the "high affinity reaction" persists but is weakened.

Sinjorgo et al. (1986): see note 3

Note 8 (p. 122). Purified cytochrome c_1 from *Rps. sphaeroides* closely resembles its mitochondrial counterpart in many properties but does not form a stable complex with cytochrome c_2 although electron transfer is rapid.

Yu L, Dong JH, Yu CA (1986) Characterisation of purified cytochrome c_1 from *Rhodobacter sphaeroides* R26. Biochim Biophys Acta 852:203–211

Note 9 (p. 127). A reinvestigation of *Nitrosomonas* cytochrome c-554 shows it to contain four c-type hemes in a polypeptide chain of M_r 25 K.

Andersson KK, Lipscomb JD, Valentine M, Munck E, Hooper AB (1986) Tetraheme cytochrome c-554 from *Nitrosomonas europaea* – heme-heme interactions and ligand binding. J Biol Chem 261:1126–1138

Note 10 (p. 127). The midpoint potentials of the hemes of *Nitrosomonas* hydroxylamine oxidase range from 295 mV to −390 mV. The midpoint potential of P460 is −260 mV.

Prince RC, Hooper AB (1987) Resolution of the hemes of hydroxylamine oxidoreductase by redox potentiometry and electron spin resonance spectroscopy. Biochemistry 26:970–974

Note 11 (p. 130). Cytochrome c-554 but not c-552 is reactive with purified *Nitrosomonas* cytochrome aa_3.

Dispirito AA, Lipscomb JD, Hooper AB (1986) Cytochrome aa_3 from *Nitrosomonas europaea.* J Biol Chem 261:17048–17056

Note 12 (p. 131). The enzyme methylamine dehydrogenase of bacterium W3Al contains pyrroloquinoline quinone and reduces a cytochrome c-552.

Chandrasekar R, Klapper MH (1986) Methylamine dehydrogenase and cytochrome c-552 from the bacterium W3Al. J Biol Chem 261:3616–3619

Cytochromes c-551 (M_r 22 K) and c-553 (M_r 30 K) are inducible by methylamine in *Paracoccus denitrificans* and could be reduced by methylamine dehydrogenase via the blue copper protein amicyanin.

Husain M, Davidson VL (1986) Characterisation of two inducible periplasmic c-type cytochromes from *Paracoccus denitrificans.* J Biol Chem 261:8577–8580

Note 13 (p. 139). The p-cresol dehydrogenase is now known to contain two cytochrome c subunits each of M_r 8 K, and two flavoprotein subunits each of M_r 49 K.

McIntyre W et al. (1986): see note 1

Note 14 (p. 146). Cytochrome aa_3 of *Nitrosomonas europaea* contains three subunits, two heme a and two coppers and is reactive with cytochrome c-554 and horse cytochrome c but not c-552.

Dispirito et al. (1986): see note 11

Note 15 (p. 151). The gene for subunit III is present in *Paracoccus*. Purification of cytochrome aa_3 may lead to loss of this subunit and lower proton translocation efficiency.

Saraste M, Raitio M, Jalli T, Peramua A (1986) A gene in *Paracoccus* for subunit III of cytochrome oxidase. FEBS Lett 206:154–161

Note 16 (p. 152). A cytochrome o (M_r 29 K), free of c-type heme, can be prepared from *Azotobacter vinelandii* but is inactive with added cytochrome c_4.

Yang T (1986) Biochemical and biophysical properties of cytochrome o of *Azotobacter vinelandii*. Biochim Biophys Acta 848:342–351

Note 17 (p. 154). The mutants studied by Zannoni et al. (1980) lack the bc_1 complex.

Davidson E, Prince RC, Daldal F, Hauska G, Marrs BL (1987) *Rhodobacter capsulatus* MT113 a single mutation results in the absence of c-type cytochromes and in the absence of the cytochrome bc_1 complex. Biochim Biophys Acta 890:292–301

Note 18 (p. 176). However, cytochrome c′ is not present in *Rps. sphaeroides* (f. *denitrificans*) grown under denitrifying conditions, although it is present in cells grown photoheterotrophically without nitrate.

Michalski WP, Miller DJ, Nicholas DJD (1986) Changes in cytochrome composition of *Rhodopseudomonas sphaeroides* f. sp. *denitrificans* grown under denitrifying conditions. Biochim Biophys Acta 849:304–315

Note 19 (p. 181). A rearrangement of the Rhodospirillaceae is proposed. The new genus *Rhodobacter* includes the two *Rhodopseudomonads – capsulata* and *sphaeroides. Rhodopseudomonas gelatinosa* and *Rhodospirillum tenue* enter the genus *Rhodocyclus*.

Imhoff JF, Truper HG, Pfennig N (1984) Rearrangement of the species and genera of the phototrophic purple non-sulfur bacteria. Int J Syst Bacteriol 34:340–343

Note 20 (pp. 190, 191). The four hemes of the reaction centre cytochrome c of *Rps. viridis* are kinetically, thermodynamically and spectrally distinct. Kinetic analysis suggests an oxidation sequence of c-556→c-559→$P960^+$ for the two high potential hemes proximal to the photooxidised special pair.

Dracheva SM, Drachev LA, Zaberezhnaya SM, Konstantinov AA, Semenov AY, Skulachev VP (1986) Spectral, redox and kinetic characteristics of high potential cytochrome c hemes in *Rhodopseudomonas viridis* reaction centre. FEBS Lett 205:41–46

Note 21 (p. 243). The gene sequences of cytochrome c_3, azurin, reaction centre cytochrome c and *Rhodopseudomonas sphaeroides* cytochrome c_2 encode hydrophobic, positively charged presequences.

Voordouw G, Brenner S (1986) Cloning and sequencing of the gene encoding cytochrome c_3 from *Desulfovibrio vulgaris* (Hildenborough). Eur J Biochem 159:347–351.

Canters GW (1987) The azurin gene from *Pseudomonas aeruginosa* codes for a pre-protein with a signal peptide. FEBS Lett 212:168–172.

Weyer KA, Schafer W, Lottspeich F, Michel H (1987) The cytochrome subunit of the photosynthetic reaction centre from *Rhodopseudomonas viridis* is a lipoprotein. Biochemistry 26:2909–2914.

Donohue TJ, McEwan AG, Kaplan S (1986) Cloning, DNA sequence and expression of the *Rhodobacter sphaeroides* cytochrome c_2 gene. J Bact 168:962–972

Note 22 (p. 247). The heme attachment enzyme, cytochrome c heme lyase, is nuclear encoded, located in the intermembrane space and acts on cytochrome c but not cytochrome c_1. The enzyme requires hemin, NADH and an uncharacterised heat stable factor. It is distinct from the apo-c binding protein.

Nicholson DW, Kohler H, Neupert W (1987) Import of cytochrome c into mitochondria. Eur J Biochem 164:147–157.

Dumont ME, Ernst JF, Hampsey DM, Sherman F (1987) Identification and sequence of the gene encoding cytochrome c heme lyase in the yeast *Saccharomyces cerevisiae.*

A conflicting report finds that the enzyme is located at the cytoplasmic surface of the inner membrane.

Enosawa S, Onashi A (1986) Localisation of enzyme for heme attachment to apocytochrome c in yeast mitochondria. Biochem Biophys Res Commun 141:1145–1150

Note 23 (p. 249). Attachment of the N-terminal half of the cytochrome c_1 presequence to dihydrofolate reductase allows transport of this enzyme to the matrix.

Van Loon APGM, Brandli AW, Schatz G (1986) The presequences of two imported mitochondrial proteins contain information for intracellular and intramitochondrial sorting. Cell 44:801–812

Note 24 (p. 253). The N-terminal cysteine of the reaction centre cytochrome c from *Rps. viridis* is linked to a diglyceride via a thioether bond. Weyer et al. (1987): see note 21

Note 25 (p. 259). The unvaried proline 71 of mitochondrial cytochrome c is not a 'critical' residue for the slow folding process.

Ramdas L, Nall BT (1986) Folding/unfolding kinetics of mutant forms of iso-l-cytochrome c with replacement of proline 71. Biochemistry 25:6959–6964

Subject Index

MIX
Papier aus verantwortungsvollen Quellen
Paper from responsible sources
FSC® C105338

If you have any concerns about our products,
you can contact us on
ProductSafety@springernature.com

In case Publisher is established outside the EU,
the EU authorized representative is:
Springer Nature Customer Service Center GmbH
Europaplatz 3, 69115 Heidelberg, Germany

Printed by Libri Plureos GmbH
in Hamburg, Germany